Kinetic Theory of Granular Gases

Kinetic Theory of Granular Gases

Nikolai V. Brilliantov and Thorsten Pöschel
Humboldt University Berlin – Charité

UNIVERSITY PRESS

UNIVERSITY PRESS

Great Clarendon Street, Oxford, OX2 6DP,
United Kingdom

Oxford University Press is a department of the University of Oxford.
It furthers the University's objective of excellence in research, scholarship,
and education by publishing worldwide. Oxford is a registered trade mark of
Oxford University Press in the UK and in certain other countries

© Oxford University Press 2004

The moral rights of the authors have been asserted

First published 2004
First published in paperback 2010

All rights reserved. No part of this publication may be reproduced, stored in
a retrieval system, or transmitted, in any form or by any means, without the
prior permission in writing of Oxford University Press, or as expressly permitted
by law, by licence or under terms agreed with the appropriate reprographics
rights organization. Enquiries concerning reproduction outside the scope of the
above should be sent to the Rights Department, Oxford University Press, at the
address above

You must not circulate this work in any other form
and you must impose this same condition on any acquirer

Published in the United States of America by Oxford University Press
198 Madison Avenue, New York, NY 10016, United States of America

British Library Cataloguing in Publication Data
Data available

Library of Congress Cataloging in Publication Data
Data available

ISBN 978–0–19–958813–8

PREFACE

Kinetic gas theory and the transport theory of molecular gases are part of any undergraduate course on classical statistical mechanics. Among others, one of the main preconditions of these theories is the elastic interaction of colliding particles. In 1993 it was first expressed in pioneering articles by Goldhirsch and Zanetti and by McNamara that a gas of particles which collide *inelastically* behaves qualitatively different compared to a molecular gas. Inelasticity of colliding particles is based on the fact that these particles are themselves macroscopic bodies with many degrees of freedom. We call such particles grains and a gas of such particles a *granular gas*. Mainly triggered by the above-mentioned publications, a persistently increasing number of scientific publications has appeared which reflects an interest in the physics of granular gases that is still growing. By now we have known many interesting properties of granular gases, such as an inherent cluster instability, non-Maxwellian velocity distribution, spontaneous formation of vortices, anomalous diffusion and others.

The present book is intended to serve as an introduction to statistical mechanics and the kinetic theory of granular gases as it has been developed mainly during the past decade. The text is self-evident, in the sense that (for very few exceptions) the understanding of its content does not require mathematical or physical knowledge that goes beyond the knowledge needed to follow undergraduate physics courses being taught at any university. The material is adequate for a one-semester course in 'Kinetic Theory of Dissipative Gases'.

In early publications on the theory of granular gases, quite an arbitrary mechanism of energy dissipation at collision was assumed, characterized by a constant coefficient of restitution. Although this assumption simplifies the mathematics considerably, it does not agree with physical reality. Therefore, throughout the book we have put special emphasis on a physically correct microscopic collision law.

Many of the techniques described are not restricted to granular gases, but can be applied to other systems of dissipative particles or quasi-particles. The authors would be glad if scientists working in different fields feel inspired by the text.

Nikolai V. Brilliantov and Thorsten Pöschel   Berlin, April 2003

CONTENTS

1	**Introduction**	1
	1.1 Kinetic gas theory for dissipative particles	1
	1.2 Atomic level of material deformation	3
	1.3 Continuum description of particles	4
	1.4 Pairwise collision of particles	5
	1.5 Many-particle systems	7
	1.6 Hydrodynamics description	9
	1.7 Experimental investigation of granular gases	10
	Some remarks	13

I. MECHANICS OF PARTICLE COLLISIONS

2	**Particle collisions**	17
	2.1 Collision on a line	17
	2.2 Collision in space	18
	2.3 Does the coefficient of restitution suffice to describe granular gas dynamics?	19
3	**Coefficient of restitution**	21
	3.1 Forces between colliding spheres	21
	3.2 Derivation of the coefficient of restitution	23
	3.3 Padé approximation	30
	3.4 Coefficient of tangential restitution	31
4	**Application to few-particle systems**	36
	4.1 Inelastic collapse	36
	4.2 Collision cannon	40
	Summary	48

II. GRANULAR GASES — VELOCITY DISTRIBUTION

5	**Cooling granular gas — Haff's law**	51
	5.1 Homogeneous cooling state	51
	5.2 Haff's law for the evolution of the granular temperature	52
6	**Boltzmann equation**	54
	6.1 Velocity distribution function	54
	6.2 Direct and inverse collisions	55
	6.3 Collision integral and Boltzmann–Enskog equation	58
	6.4 An important property of the collision integral	60

7 Sonine polynomials expansion of the velocity distribution function — 62

8 Velocity distribution and temperature of a granular gas for the case $\varepsilon =$ const. — 67
- 8.1 Decomposition of the Boltzmann equation — 67
- 8.2 The second Sonine coefficient and the moments of the collision integral — 69
- 8.3 Linear approximation of the second Sonine coefficient — 71
- 8.4 Complete solution for the second Sonine coefficient — 78
- 8.5 Time-dependent scaled velocity distribution function — 81
- 8.6 Evolution of temperature for $\varepsilon =$ const. — 84
- 8.7 Stability analysis of the Boltzmann equation — 85
- 8.8 High-order coefficients of the Sonine polynomials expansion — 87

9 Velocity distribution function and temperature for viscoelastic particles — 90
- 9.1 Why do we expect qualitatively different distribution functions for the cases $\varepsilon =$ const. and $\varepsilon = \varepsilon(g)$? — 90
- 9.2 Collision integral for a gas of viscoelastic particles — 92
- 9.3 Moments of the collision integral for viscoelastic particles — 95
- 9.4 Equations for temperature and for the shape of the velocity distribution function — 97
- 9.5 Velocity distribution and temperature in $\mathcal{O}\left(\delta^0\right)$ — 98
- 9.6 Velocity distribution and temperature in $\mathcal{O}\left(\delta^1\right)$ — 99
- 9.7 Beyond the linear theory — 103
- 9.8 Age of granular gases — 104

10 High-energy tail of the velocity distribution function — 108
- 10.1 Overpopulation of the high-velocity tail — 108
- 10.2 High-velocity tail for $\varepsilon =$ const. — 109
- 10.3 High-velocity tail for viscoelastic particles — 112

11 Two-dimensional granular gases — 115

Summary — 119

III. SINGLE-PARTICLE TRANSPORT. SELF-DIFFUSION AND BROWNIAN MOTION

12 Diffusion and self-diffusion — 123
- 12.1 Transport in granular gases — 123
- 12.2 Diffusion coefficient and mean square displacement — 124
- 12.3 Diffusion coefficient and velocity-time correlation function — 125

13 Pseudo-Liouville and binary collision operators in dissipative gas dynamics — 127
13.1 Liouville operator in classical mechanics — 127
13.2 Derivation of the binary-collision operator — 129
13.3 Application of the pseudo-Liouville operator to standard problems — 134

14 Coefficient of self-diffusion — 137
14.1 Velocity correlation time — 137
14.2 Constant coefficient of restitution — 139
14.3 Coefficient of self-diffusion for gases of viscoelastic particles — 142
14.4 Inherent time scales — 144
14.5 Coefficient of self-diffusion beyond the adiabatic approximation — 146

15 Brownian motion in granular gases — 149
15.1 Boltzmann equation for the velocity distribution function of Brownian particles — 149
15.2 Fokker–Planck equation for Brownian particles — 151
15.3 Velocity distribution function for Brownian particles — 155
15.4 Diffusion of Brownian particles — 158

16 Two-dimensional granular gases — 162

Summary — 164

IV. TRANSPORT PROCESSES AND KINETIC COEFFICIENTS

17 Granular gas as a continuum: hydrodynamic equations — 167
17.1 Macro- and microscales of inhomogeneous granular gas — 167
17.2 Hydrodynamic fields — 168
17.3 Hydrodynamic equations for granular gases — 169

18 Chapman–Enskog approach for non-uniform granular gases — 175
18.1 Basic idea of the Chapman–Enskog scheme — 175
18.2 Equations of zeroth order of the Chapman–Enskog expansion — 178
18.3 First-order equations of the Chapman–Enskog expansion — 179
18.4 Solution of the first-order equation — 181
18.5 Kinetic coefficients expressed by the velocity distribution function — 183

19 Kinetic coefficients and velocity distribution for gases of elastic particles — 186
19.1 First-order Chapman–Enskog equations — 186
19.2 Coefficient of viscosity — 186
19.3 Coefficient of thermal conductivity — 191
19.4 Velocity distribution function of an inhomogeneous gas of elastic particles — 194

20 Kinetic coefficients for granular gases of simplified particles ($\varepsilon =$const) — 195
20.1 Viscosity coefficient — 195
20.2 Thermal conductivity coefficient κ and transport coefficient μ — 198
20.3 Velocity distribution function — 201

21 Kinetic coefficients for granular gases of viscoelastic particles — 202
21.1 Chapman–Enskog approach for gases of viscoelastic particles — 202
21.2 Viscosity coefficient — 205
21.3 Coefficients κ and μ — 207

22 Chapman–Enskog method for the self-diffusion coefficient — 211
22.1 Constant coefficient of restitution — 211
22.2 Viscoelastic particles — 215

23 Two-dimensional granular gases — 218
23.1 Constant coefficient of restitution — 218
23.2 Granular gases of viscoelastic particles — 219

Summary — 221

V. STRUCTURE FORMATION

24 Instability of the homogeneous cooling state — 225
24.1 Arguments for the instability of the homogeneous cooling state — 225
24.2 Linearized hydrodynamic equations — 227

25 Structure formation for $\varepsilon =$const. — 230
25.1 Linearized hydrodynamic equations for $\varepsilon =$ const. — 230
25.2 Hydrodynamic modes — 232
25.3 Vortex formation due to the instability of the transverse modes — 233
25.4 Cluster formation due the instability of the other hydrodynamic modes — 236

26 Structure formation in granular gases of viscoelastic particles — 239
26.1 Linearized equations for the hydrodynamic modes — 239
26.2 Stability analysis of the hydrodynamic modes and structure formation — 241
26.3 Structure formation as a transient process — 243

27 Nonlinear mechanisms of structure formation — 246

28 Two-dimensional granular gases — 249

Summary — 251

APPENDIX

A	**Functions of the collision integral**	252
	A.1 Kinetic integrals and basic integrals	252
	A.2 Evaluation of the basic integral	253
	A.3 Computational formula manipulation of kinetic integrals	259
B	**Molecular dynamics of granular gases**	269
	B.1 Event-driven molecular dynamics	269
	B.2 A simple event-driven algorithm	270
	B.3 A simple program for event driven simulations	271
	B.4 Efficient algorithms	280
C	**Solutions to problems**	283
References		317
Symbol index		325
Index		327

1
INTRODUCTION

Classical kinetic gas theory considers elastic particle collisions. To account for dissipative collisions its concepts have to be generalized. We introduce the basic concepts of the kinetic theory of granular gases and mention some of the exciting phenomena found in granular gas dynamics. In this introduction we concentrate more on the general understanding rather than on scientific strictness.

1.1 Kinetic gas theory for dissipative particles

Statistical mechanics and kinetic theory of molecular gases belong to every course on theoretical physics. The gas particles are modelled as *elastically* colliding spheres, which implies conservation of mechanical energy. In this sense molecular gases are Hamiltonian systems. Once initialized it takes a molecular gas not more than few collisions per particle to relax to its equilibrium state, characterized by a Maxwellian velocity distribution function and (in the absence of external forces) a certain homogeneous density. The kinetic theory of dilute gases dates back to Maxwell and Boltzmann and has been basically completed by Chapman and Enskog. In many applications it has been proven to be a very accurate and reliable theory.

In this book we consider low-density gases of dissipatively colliding particles in the absence of external forces. We will show that slightly altering the collision law of particles, that is, making them (even slightly!) dissipative, causes drastic changes in the behaviour of the gas. The ability of the particles to dissipate energy implies that the particles are macroscopic bodies themselves which have internal degrees of freedom, as discussed below. In other words, the particles must be able to transform kinetic energy into heat. Such particles are called *grains* and the respective gases *granular gases*. To describe the kinetic theory of granular gases, the concepts of the classical kinetic gas theory have to be generalized.

Due to dissipative collisions, granular gases persistently lose mechanical energy of the grains which may be expressed by a decay of its *granular temperature* (for a careful definition see Chapter 5). Therefore, in the sense of thermodynamics, granular gases are *open systems*. Like many open systems, granular gases also reveal self-organized spatio-temporal structures. The most prominent example is the self-organized formation of clusters (Fig. 1.1).

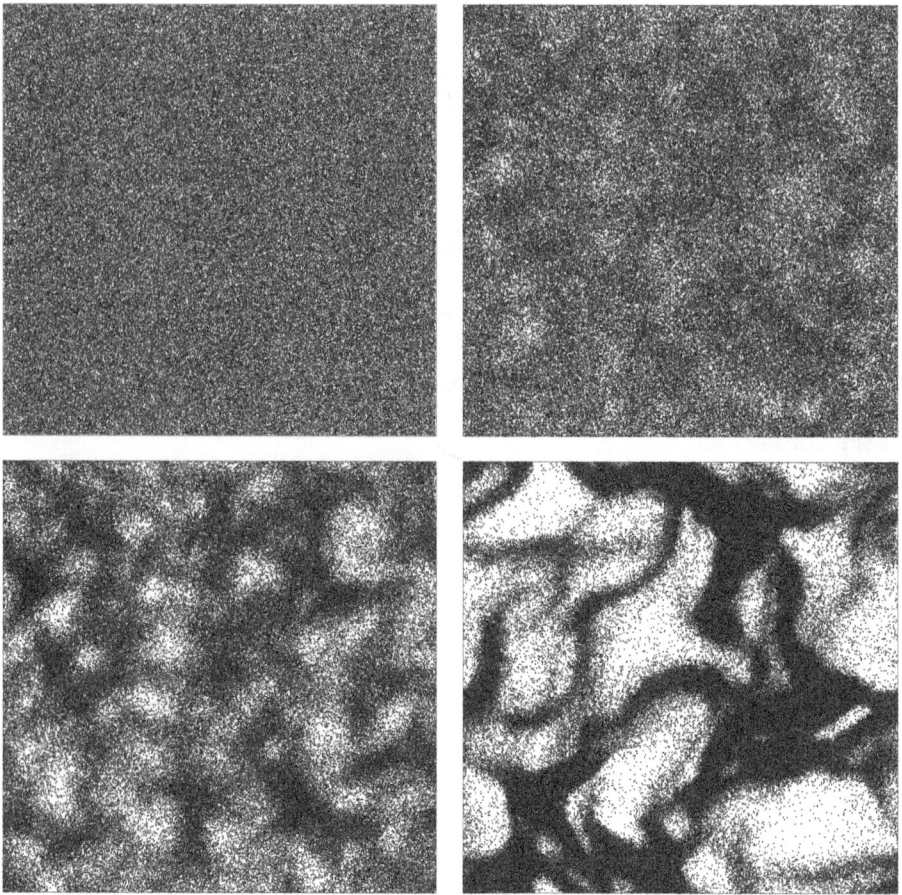

FIG. 1.1. Starting from a uniform density in the absence of any external influence granular gases develop complicated filament structures.

Besides the explanation of these structures based on the dissipative nature of particle collisions, there are many more challenges in the kinetic theory of granular gases, such as its non-Maxwellian velocity distribution function, anomalous diffusion, self-organized formation of shock-waves, violation of molecular chaos, violation of the equipartition theorem, the transition from subsonic behaviour to supersonic and back to subsonic without external influences, etc.

Similar to molecular gases, granular gases may be described by time-dependent fields of macroscopic variables such as pressure $p(\vec{r},t)$, temperature $T(\vec{r},t)$, density $n(\vec{r},t)$ and others. These fields are determined by the properties of the system, for example, by the total volume V, the number of particles in the system N and the properties of the boundaries, and by the properties of the gas particles, that is, by the details of their interaction. The field description of granular gases represents the highest level. For its derivation we have to analyse carefully the mechanics of pairwise particle collisions and the statistical properties of ensembles of such particles.

We wish to give a survey of the levels of granular gas description and to motivate the necessary model assumptions. Let us start with the most basic question: where does the dissipated mechanical energy go to?

1.2 Atomic level of material deformation

To grasp the physical idea we consider a cartoon: the bulk of particle material is a regular lattice composed of mass points linked by elastic springs. On this level the material may be described as a Hamiltonian system, that is, any deformation occurs elastically. A deformation of the material due to a collision causes compression or elongation of springs located in the deformed area of the particles. These springs gain a certain amount of energy at the expense of the centre of mass motion of the grains. This energy is then rapidly distributed amongst the other springs such that shortly after the beginning of the impact all springs get an additional energy. This transfer of energy continues until the centres of the grains approach their minimal distance.

In the language of statistical physics, a local excitation of few degrees of freedom (few springs) is equally partitioned amongst all available degrees of freedom of the colliding particles, that is, the non-uniform distribution of energy relaxes to an equidistribution at higher total energy. The mechanical energy, which was initially localized in few deformed springs, is converted in this way to thermal energy, which may be expressed by an increase of temperature of the particle material. Hence, part of the mechanical energy of the relative motion of grains is converted to heat, that is, into the internal energy stored in the springs. The larger the impact velocity the further the system is driven out of equilibrium and, thus, the more mechanical energy is converted to heat.

It is instructive to estimate the temperature increase during the evolution of a granular gas: assume that the gas consists of steel spheres of diameter 1 cm and of heat capacity $0.45 \, \text{kJ}/(\text{kg K})$ (at temperature $\mathcal{T} = 300 \, \text{K}$).[1] The typical particle velocity is assumed to be $1 \, \text{m/s}$. If the total mechanical energy was converted into heat, the temperature of the material would increase by $\Delta \mathcal{T} \approx 10^{-3} \, \text{K}$, that is, $\Delta \mathcal{T}/\mathcal{T}$ is negligibly small. Hence, the dissipative material properties will not change noticeably due to the transformation of mechanical energy into heat as long as the initial temperature of the particles is much larger than 10^{-3} K.

Based on the microscopic lattice model whose simplified version has been discussed above, Morgado and Oppenheim (1997) have derived the dissipative interaction force between colliding spheres as functions of the deformation and the deformation rate. Their results agree with earlier results, which have been obtained using a continuum description of the particle material. This description corresponds to the next level.

---

[1] The material temperature $\mathcal{T}$ is exclusively used in this section. Its meaning is different from the granular temperature T, which will be introduced later.

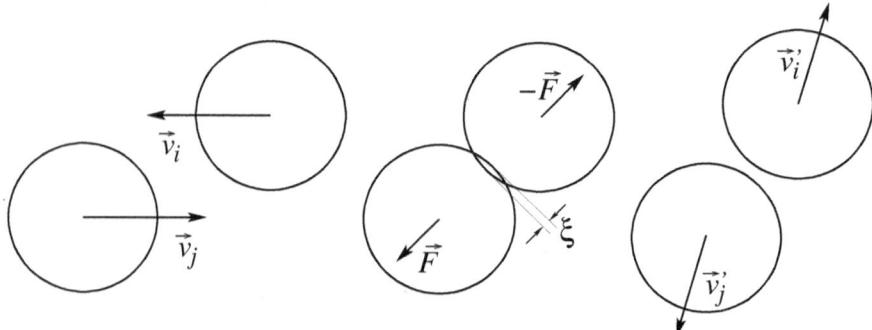

FIG. 1.2. Viscoelastic spheres before, during and after a collision.

1.3 Continuum description of particles

Particles of granular gases are macroscopic bodies which allows for a continuum description of their interaction. Any deformation of a solid body is described by a stress–strain relation. The simplest of them is Hooke's law which describes pure elastic deformation. In this law the stress depends linearly on the deformation but not on the deformation rate. The simplest stress–strain relation which incorporates dissipation is described by *viscoelastic* deformations. Here the stress is a sum of the elastic stress and the dissipative stress where the elastic stress depends linearly on the deformation and and the dissipative stress depends linearly on the deformation rate.

Elastic properties of a viscoelastic material are characterized by the Young modulus Y and the Poisson ratio ν. The former describes the resistance of the material against compression and the latter quantifies its (secondary) deformation in the direction perpendicular to the compression, see for example, Landau and Lifshitz (1965). For the description of the dissipative material properties on top of Y and ν the viscosities $\eta_{1/2}$ are needed. Therefore, the interaction forces between colliding spheres can be expressed by the above material parameters and by the particle properties, that is, their radii $R_{1/2}$, their relative position $\vec{r}_1 - \vec{r}_2$ and their relative velocity. The collision is sketched in Fig. 1.2. The elastic interaction force was derived by Heinrich Hertz (1882):

$$F_{\rm el} = \frac{2Y\sqrt{R^{\rm eff}}}{3(1-\nu^2)}\xi^{3/2} \equiv \rho\xi^{3/2}, \qquad R^{\rm eff} \equiv \frac{R_1 R_2}{R_1 + R_2}, \qquad (1.1)$$

with the compression $\xi \equiv R_1 + R_2 - |\vec{r}_1 - \vec{r}_2|$. For the dissipative force we need the additional assumption that the impact velocity is significantly smaller than the speed of sound in the material. Then it may be written in terms of the compression and the compression rate (Brilliantov *et al.*, 1996):

$$F_{\rm diss} = \frac{3}{2}A\rho\sqrt{\xi}\dot{\xi}, \qquad (1.2)$$

where $A = A\left(Y, \mu, \eta_{1/2}\right)$ is a known function of the elastic and dissipative material parameters (Brilliantov et al., 1996). The functional form of (1.2) was already suggested by Kuwabara and Kono (1987) based on physical reasoning. The mentioned careful analysis on the atomic lattice level (Morgado and Oppenheim, 1997) confirms (1.2).

The motion of particle i is determined by the total force exerted by all other particles located at positions $\vec{r}_j$ and moving with velocities $\vec{v}_j$. Its trajectory is subjected to Newton's equation of motion

$$m_i \frac{d^2 \vec{r}_i}{dt^2} = \vec{F}_i\left(\vec{r}_1, \vec{v}_1, \vec{r}_2, \vec{v}_2, \ldots, \vec{r}_N, \vec{v}_N\right). \tag{1.3}$$

The time-dependent solution of these equations for all particles i determines the evolution of the granular gas. They can be integrated, for example, by a numerical scheme, to determine the trajectories. This is the idea of molecular dynamics as widely used in the physics of many-particle systems (see, e.g. Allen and Tildesley, 1987).

Consider the collision of two isolated particles which move at initial velocities $\vec{v}_1$ and $\vec{v}_2$. If these particles undergo an elastic collision, the final velocities $\vec{v}_1'$ and $\vec{v}_2'$ may be obtained from the conservation of momentum and energy:

$$\vec{v}_1' = \vec{v}_1 - \left(\vec{v}_{12} \cdot \vec{e}\right) \vec{e}, \qquad \vec{v}_2' = \vec{v}_2 + \left(\vec{v}_{12} \cdot \vec{e}\right) \vec{e}, \tag{1.4}$$

where $\vec{v}_{12} \equiv \vec{v}_1 - \vec{v}_2$ and with the unit vector $\vec{e} \equiv |\vec{r}_1 - \vec{r}_2|/(\vec{r}_1 - \vec{r}_2)$. For dissipatively colliding identical particles these equations may be expressed as

$$\vec{v}_1' = \vec{v}_1 - \frac{1}{2}(1+\varepsilon)\left(\vec{v}_{12} \cdot \vec{e}\right)\vec{e}, \qquad \vec{v}_2' = \vec{v}_2 + \frac{1}{2}(1+\varepsilon)\left(\vec{v}_{12} \cdot \vec{e}\right)\vec{e}. \tag{1.5}$$

The *coefficient of restitution* ε characterizes the dissipative properties of the collision. It may be obtained from the elastic and dissipative interaction forces by integrating Newton's equation of motion (1.3). Hence, the details of the particle interaction are mapped onto the coefficient of restitution. Frequently, it is assumed that ε is a material constant. A detailed analysis of the mechanics of particle collisions (see Chapters 2 and 3) reveals, however, that the coefficient of restitution is a function of the impact velocity:

$$\varepsilon = 1 - D_1 |\vec{v}_{12} \cdot \vec{e}|^{1/5} + D_2 |\vec{v}_{12} \cdot \vec{e}|^{2/5} \mp, \tag{1.6}$$

where D_1 and D_2 are material constants.

1.4 Pairwise collision of particles

Before we move to the next level of description — the many-particle system — we have to discuss a very important assumption, which is perhaps, the most important precondition of the kinetic theory of granular gases:

> *We assume that the statistical properties of a granular gas are determined by pairwise and instantaneous particle collisions.*

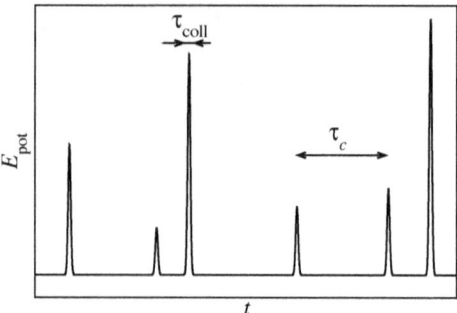

FIG. 1.3. Potential energy of particle i to illustrate the assumption of binary collisions.

Let us discuss this important item in detail. In mechanical systems the RHS of (1.3) reads

$$\vec{F}_i = \sum_{j \neq i} \vec{F}_{ij}\left(\vec{r}_{ij}, \vec{v}_{ij}\right). \qquad (1.7)$$

The forces in granular gases are of finite range, that is, colliding particles exert forces on each other only if they are in mechanical contact. If any of the particles is at any time in contact with no more than one other particle, (1.7) simplifies further:

$$\vec{F}_i = \begin{cases} 0 & \text{if } i \text{ does not contact any other particle} \\ \vec{F}_i\left(\vec{r}_{ij}, \vec{v}_{ij}, t\right) & \text{if } i \text{ is in contact with particle } j \,. \end{cases} \qquad (1.8)$$

Note that in (1.7) the force depends on time only implicitly via the time-dependent relative positions $\vec{r}_{ij}(t)$ and velocities $\vec{v}_{ij}(t)$. In (1.8) there appears an additional time-dependence since the index j varies with time. The form of (1.8) represents the requirement that particle i collides with another particle not before the preceding collision has been accomplished. It is said that all collisions occur pairwise. Figure 1.3 sketches the potential energy of particle i as a function of time where we introduce the typical duration of a collision τ_{coll} and mean free flight time τ_c. Later we will define these quantities more precisely.

Obviously, the more dilute the granular gas and the shorter the duration of the collisions, the better the assumption of pairwise (or binary) collisions is justified. The duration of the collisions is mainly determined by the interaction potential: the collision of soft particles lasts longer than the collision of hard ones. Therefore, the assumption of binary collisions is also called *hard-particle approximation*.

For finite duration of the collisions (determined by the finite elastic constant Y) there is a finite probability that during the time of the contact one of the collision partners comes into contact with a third particle. The latter probability is given by the ratio of the collision frequencies of binary and three-particle (or ternary) collisions. For identical elastic particles of number density n we obtain

MANY-PARTICLE SYSTEMS

$$\frac{\text{number of triple collisions}}{\text{number of binary collisions}} \sim nR^2 \xi_{\max}, \tag{1.9}$$

where $\xi_{\max}$ is the maximal compression of the colliding particles which is vanishingly small for hard spheres. This result will be reconsidered later after introducing some necessary prerequisites (see Exercise 5.1). From (1.9) we see that in strict mathematical sense multi-body collisions can be excluded only for either *infinitely steep* interaction potential or *vanishing* density. Both conditions are not physically reasonable. Hence, we explicitely admit that there occur multi-particle collisions and the number of them may be large.

Nevertheless, throughout this book it is assumed that there are *exclusively* pairwise collisions. The precondition for this assumption is that three-particle collisions, four-particle collisions, etc. are so rare that they do not affect the *statistical* properties of the system. Those collisions may be neglected, provided there are many more binary collisions than multi-particle collisions, as specified by the ratio in (1.9).

1.5 Many-particle systems

With the assumption on binary collisions the description of the dynamics of a many-body system simplifies considerably. Instead of integrating Newton's equation of motion (1.3) for the many-body system we can treat each collision separately. Therefore, the dynamics of the many-particle system is exhaustively described by a sequence of binary collisions which are characterised by the collision rule (1.5) together with the coefficient of restitution.

The dynamics of a granular gas of N particles may be obtained by a simple algorithm:

1. Consider the system at the present time t_k. Compute the time t_{k+1} of the next collision in the system if all particles move with their present velocities and determine the indices of the collision partners i and j.
2. Propagate all particles due to $\vec{r}_l := \vec{r}_l + (t_{k+1} - t_k) \vec{v}_l$, $(l = 1, \ldots, N)$.
3. Modify the velocities of the particles i and j due to the collision law.
4. Proceed with step 1.

This type of algorithm, where the integration of Newton's equations is replaced by the application of a collision rule for pairwise collisions, is called an *event-driven algorithm*. With the assumption of pairwise collisions it is identical with numerically integrating Newton's equation. Let us apply an event-driven algorithm to a granular gas of uniform density. Figure 1.4 shows snapshots of a granular gas of $N = 10^4$ particles with periodic boundary conditions. The simulation starts at homogeneous density. During the first stage of its evolution the granular gas stays homogeneous while the kinetic energy decays. This stage of the evolution is called *homogeneous cooling*. After about 1.2×10^7 collisions spatial inhomogeneities appear which grow to pronounced clusters. For $\varepsilon = $ const. these clusters persist while for more realistic collision models the system eventually returns to the homogeneous cooling state. At several places we will show

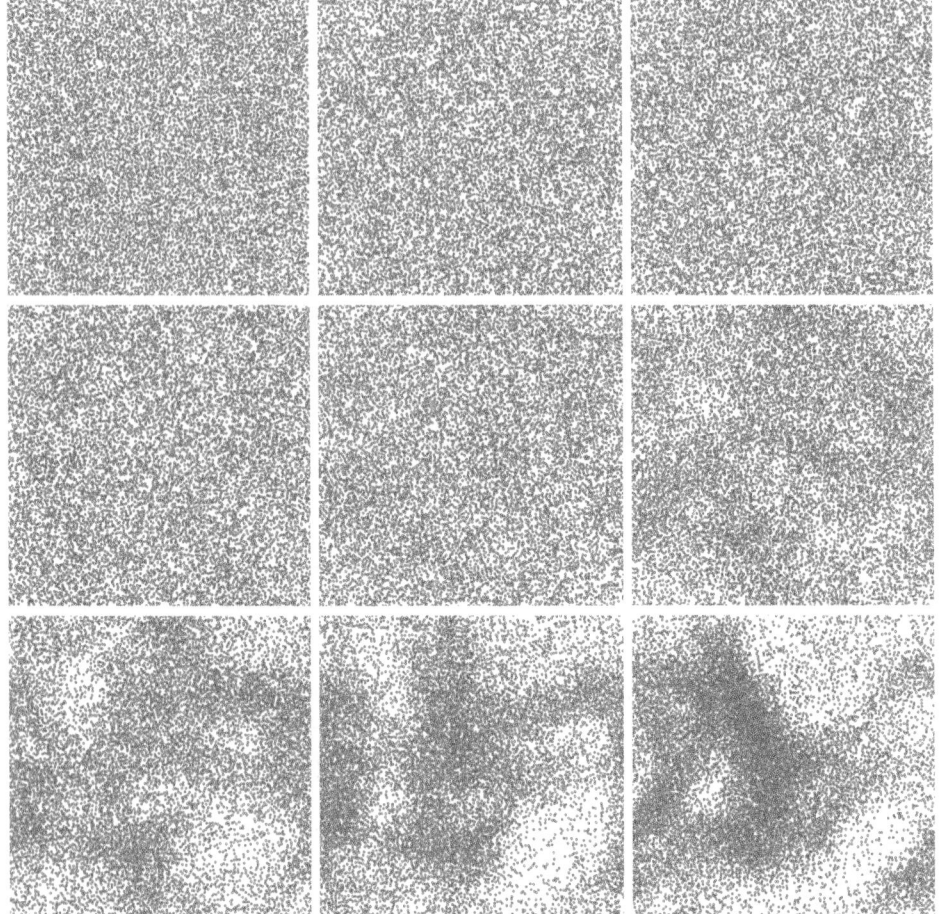

FIG. 1.4. Evolution of a granular gas of $N = 10^4$ particles of initially uniform density in the absence of external forces. For the coefficient of restitution $\varepsilon =$ const. was assumed. The snapshots have been taken after 0, 3×10^6, 6×10^6, 12×10^6, 15×10^6, 18×10^6, 21×10^6 and 24×10^6 collisions (top left to bottom right).

that the approximation $\varepsilon =$ const. may be too crude and sometimes it leads even to qualitatively incorrect system behaviour (see Chapters 4, 14 and 26).

In the homogeneous cooling state the system is completely described by the velocity distribution function of the particles. If the particles were elastic we would expect a Maxwell distribution. For granular gases, however, there are characteristic time-dependent deviations from the Maxwell distribution. Another consequence of dissipative particle collision is anomalous diffusion (Chapters 12 – 14).

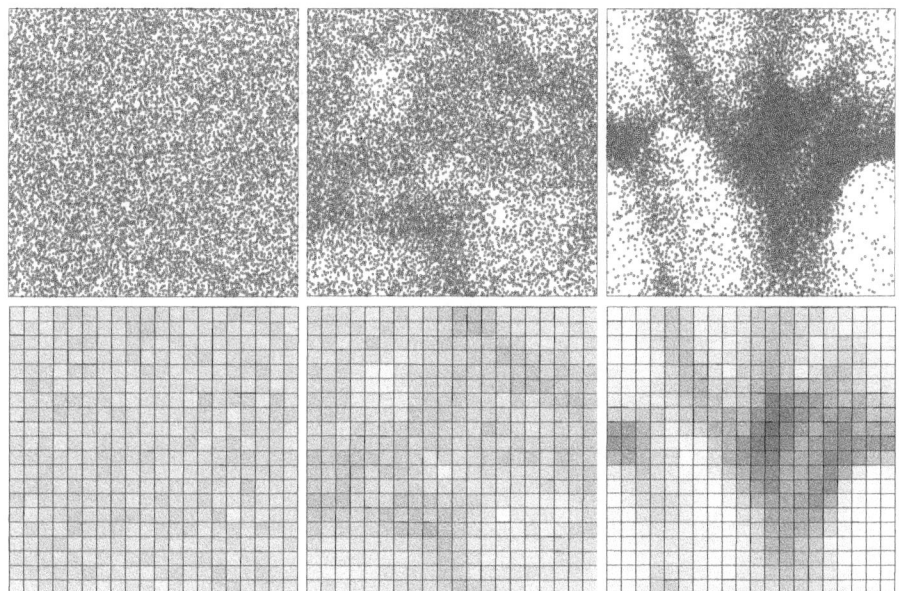

FIG. 1.5. Particle and field representation of a granular gas. Top row: particle representation, bottom row: the field of particle number density $n(\vec{r},t)$. Left column: the homogeneous cooling state, middle column: weak inhomogeneities appear, right column: the gas is characterized by pronounced clusters.

1.6 Hydrodynamics description

So far we have described a granular gas by the positions and velocities of the particles. Alternatively, a granular gas can be described by the coarse grained time-dependent fields of particle number density $n(\vec{r},t)$, flow velocity $\vec{u}(\vec{r},t)$ and temperature $T(\vec{r},t)$, as illustrated in Fig. 1.5. As for molecular gases, the gradients of the fields $n(\vec{r},t)$, $\vec{u}(\vec{r},t)$ and $T(\vec{r},t)$ induce macroscopic flows of mass, momentum and energy. The transition from the particle representation to the field representartion allows for the application of the powerful tools of hydrodynamics. The hydrodynamic equations which govern the evolution of the fields and the transport coefficients for granular gases have to be derived from the laws of microscopic particle collisions (1.5).

The hydrodynamic level of description allows for the explanation of self-organized structures which have been mentioned in the very beginning of this introduction. Analysing the hydrodynamic equations it is shown that the homogeneous cooling state is unstable with respect to cluster and vortex formation. At very late times granular gases of viscoelastic particles will eventually return to the homogeneous cooling state (see Chapter 26).

1.7 Experimental investigation of granular gases

The comparison of theoretical results with experiments is always highly desirable. In this book we deal with systems of dissipatively colliding particles under the following main preconditions:

1. The loss of energy due to particle collisions is small enough such that the preconditions for the application of kinetic theory are fulfilled. This implies that the coefficient of restitution ε is close to one.
2. The system is dilute, that is, the mean free path exceeds the particle diameter significantly.
3. The dynamics of the particles is determined by exclusively pairwise collisions. In particular, there are no long-range interactions such as electromagnetic or gravitational forces.
4. There are no external forces such as gravity.
5. The system is large enough to neglect the effects of boundary conditions.

Whereas the first two conditions can be easily satisfied experimentally, the latter three conditions are problematic. The third condition allows for the assumption of exclusively binary collisions which is crucial for the presented theory. Systems of charged particles as well as gravitating particles are beyond the scope of the book.

In Earth-bound experiments there is always gravity which leads to qualitatively different behaviour of a granular gas as compared with force-free gases. This statement applies even for a one-particle system: a particle which bounces on a table comes to rest after *finite* time (see Exercises 2.3 and 3.1), whereas a gas of non-gravitating particles preserves for any finite time a certain amount of kinetic energy (see Chapter 5).

Boundary conditions may affect the granular gas dynamics significantly (e.g. Goldhirsch, 2001), nevertheless, there is still no closed kinetic theory which incorporates boundary conditions. In this book we have developed the kinetic theory of granular gases without considering boundary conditions, that is, the theory is addressed to large systems.

There are three experimental setups which may be suitable to check the theoretical predictions:

- Natural granular gases in the absence of gravity, that is, clouds of cosmic dust.
- Experiments under conditions of micro gravity.
- Earth-bound experiments where the effect of gravity is balanced by an external excitation of the system.

Let us discuss these experimental setups in detail.

1.7.1 *Natural granular gases*

Observations of the evolution of interstellar dust clouds would be an ideal test ground for granular gas theory, however, at present (and in near future) data on the gas properties, such as the velocity distribution function, are not accessible.

The large planets of our Solar system, Jupiter, Saturn and Neptune are surrounded by rings of particles that range from micrometre to metre size (Greenberg and Brahic, 1984). The particles of which planetary rings consist, orbit around the central planet with a period of some hours to few days. The gravitational force in such systems is balanced by the centrifugal force. Since in the central field of the planet the inner particles move slower than the outer ones the gas is heated by permanent shear. The physics of planetary rings is subject of intensive research (Greenberg and Brahic, 1984). Statistical data on dust particles became available from the use of dust detectors which are carried by space probes such as Galileo and Ulysses, orbiting Jupiter. These dust detectors allow for measuring a certain combination of single particle mass and velocity. There exists a large amount of data, however, the dynamics of particles is affected by the intensive magnetic field of Jupiter which mainly determines the dynamics of the (charged) particles. Moreover, the dust detector allows for a resolution of the particle velocities with a large tolerance only. Therefore, a direct comparison with theoretical predictions, such as deviations from the Maxwell distribution (see Chapter 8) and others is presently not possible. Much more precise data is expected from the *Cosmic Dust Analyzer* of the space probe Cassini which was launched in 1997 and which will reach Saturn in 2004.

Planetary rings offer a variety of challenging problems for the application of granular gas theory, such as structure formation on many scales, ranging from metres to the 10^8 m scale on which the system spanning arcs are formed. A collection of open problems can be found in Borderies *et al.* (1984) and Brahic (2001). Nevertheless, planetary rings are hardly a suitable test ground for kinetic granular gas theory due to the low fidelity of the present measurement technology. Moreover the influence of shear (tidal forces) and electromagnetic forces is unclear.

Consequently, at present astrophysical observations do not allow for checking any prediction of granular gas theory.

1.7.2 *Experiments under conditions of micro-gravity*

Granular systems have been investigated under conditions of micro-gravity (i.e. weightlessness), either aboard spacecrafts (e.g. Jenkins and Louge, 1997; Sture *et al.*, 1998), in rockets on a parabolic trajectory (Évesque *et al.*, 2001), in fall experiments (Falcon *et al.*, 1999) or in parabolic-flight experiments (Pöschel, 1999). These experiments concern, however, dense granular systems. Micro-gravity experiments of granular gases have not been performed so far.

1.7.3 *Earth-bound experiments of excited granular systems*

In Earth-bound experiments the gaseous state may be sustained by external supply of energy due to vertical or horizontal vibration of a container or by shear of rough walls. There exists a variety of experimental results where the statistical properties of particle motion in shaken two- and three-dimensional containers have been investigated (e.g. Huntley, 1998; Falcon *et al.*, 1999, 2001;

Losert et al., 1999; Wildman et al., 1999, 2001; Kudrolli and Henry, 2000; Blair and Kudrolli, 2001, 2003; Feitosa and Menon, 2002). The dynamics of particles in such systems is determined not only by the inherent dynamics, due to particle collisions, but also by the properties of the external excitation. Thus, neither the temperature decay law nor the inherent velocity distribution law for these systems is adequate for a comparison with theoretical results for the force-free gas.

So far, the theoretical understanding of the phenomena observed in excited granular gases is still lacking, moreover, some experimental results seem to contradict each other. As an example let us mention the velocity distribution in shaken granular systems: Olafson and Urbach (1998, 1999, 2001) have determined statistical properties of granular gases in two dimensions. In their experiments a horizontal sub-monolayer of grains was vertically shaken such that the loss of kinetic energy due to collisions (in horizontal direction) was balanced by excitation in vertical direction. Although in these experiments an overpopulation of the high-velocity tail of the distribution function was found as predicted by the kinetic theory of *force-free* gases, a direct comparison with kinetic theory is questionable since the details of energy transformation from the motion in vertical direction into horizontal motion as well as the influence of rolling of the particles remain unclear. An overpopulation of the high-velocity tail was also reported by Losert et al. (1999) and by Rouyer and Menon (2000) for experiments on a driven two-dimensional gas in a vertical plane. The observed overpopulation of the high-velocity tails is in agreement with the theory of gases driven by a *white noise* (van Noije and Ernst, 1998). Again, there is no theory which relates the properties of the noise with the characteristics of the excitation. Contrarily, in experiments of a vertically shaken very shallow granular system of only two particle layers, Baxter and Olafsen (2003) observed an overpopulation of the high velocity tail for particle velocities of the lower layer, whereas for the upper layer a Maxwell distribution was found.

Consequently, so far none of the available experimental data are suitable to verify or falsify the results of the kinetic theory of granular gases presented in this book. We hope that the theoretical predictions may inspire experimentalists to probe the limits of the theory.

The theoretical results can be, however, checked by means of molecular dynamics. In many places in this book we will refer to such simulations. The basic idea of these computer experiments are outlined in Appendix B.

Some remarks

The kinetic theory of granular gases is by far not complete yet and we do not pretend to give a comprehensive review. The present book is an introduction to the theory in its present state. A criterion for selecting the material was to keep the level of the book accessible for undergraduate and graduate students and to keep its volume adequate for a one-semester course. With these restrictions we had to exclude several important topics which are certainly relevant but which would require more space and more complicated mathematics. We believe that the presented material will enable beginners to understand the modern literature on granular gases and to start their own research.

We wish to mention some of the topics which are definitely important but could not be included. We consider only spherical particles and neglect their rotational degrees of freedom. The coupling of the translational and rotational degrees of freedom leads to interesting effects (Cafiero et al., 2002). However, practically nothing is known about the tangential coefficient of restitution as a function of material parameters and impact velocities, which is of primary importance for studying these phenomena.

We consider granular gases exclusively in the absence of external forces, that is, we consider neither driven gases, nor gases in external fields. From experiments it is known that driven granular gases reveal very rich physics (e.g. Clement and Labous, 2001; Wildman et al., 2001), while the influence of the external forces is important for astrophysical applications of the granular gas theory (e.g. Greenberg and Brahic, 1984; Salo et al., 1988; Salo, 1992, 2001; Spahn et al., 2001). So far, for driven granular gases there exist plenty of experimental (or observational) data and numerical results, but the complete kinetic theory is still lacking.

We also do not discuss long-range interaction between particles. Long-range forces (self-gravity, electrostatic interaction) affects the behaviour of astrophysical systems (Borderies et al., 1984; Greenberg and Brahic, 1984; Weidenschilling, 1995) where the particles interact via gravitational forces and possibly electrostatic forces. Electrostatically interacting assemblies of granular particles are sometimes called dusty plasma (e.g. Piel and Melzer, 2002; Scheffler and Wolf, 2002; Vranjes et al., 2002). In addition, further forces may be of importance, such as van der Waals interaction, etc.

Deriving the hydrodynamic equations from the Boltzmann equation we do not discuss the problem of boundary conditions. For granular gases this is a rather complicated issue, that requires a profound mathematical analysis (see e.g. Goldhirsch, 2001) which is beyond the scope of this book.

The physics of granular gases is a rapidly developing field and a lot of new findings have been reported during the very last time. To incorporate all of them is impossible — it would require a permanent updating of the book. We believe, however, that the selected material provides a solid basis for a beginner to start a new field, which is rich of intriguing phenomena and which will definitely provide many surprises.

About the notation

There are certain symbols such as $\propto$, $\simeq$, $\sim$, $\approx$ all of which have similar meaning. Since there is no generally acknowledged standard of their usage we wish to specify them by means of an example. Given the equation

$$z = a\pi^2 x y \qquad \text{with} \qquad a = 1.499$$

the table describes the meaning of the notation:

$z \propto x$	z is proportional to x (but not to x^2)
$z \simeq \pi^2 x y$	only numerical factors which are close to 1 are omitted
$z \sim x y$	any numerical factors are omitted
$a \approx 1.5$	this symbol is used only for numerical constants.

For the convenience of the reader there is an index of frequently used variables at the end of the book.

I. Mechanics of Particle Collisions

The dynamics of a granular gas is completely determined by pure mechanical collisions of its particles. Therefore, any theoretical description of a granular gas requires detailed understanding of the collision of two isolated grains.

The only characteristics of pairwise particle collisions are the particles' masses and radii and their mechanism of dissipation of mechanical energy during collisions, expressed by a single number – the coefficient of restitution.

In this part we discuss in detail the coefficient of restitution as a function of material properties, particle size and impact velocity. We analyse the conservative and dissipative parts of the interaction force of colliding viscoelastic spheres and derive the coefficient of restitution by integrating Newton's equation of motion.

To illustrate the importance of the correct description of particle collisions for the overall behaviour of granular systems, we analyse the dynamics of two simple few-particle systems. It is shown that the simplifying assumption of a constant coefficient of restitution may lead to an inadequate description of granular systems.

2

PARTICLE COLLISIONS

We introduce the notation to describe the collision of particles by using the coefficient of restitution and raise the question whether the coefficient of restitution is adequate enough to describe the collisions of particles in a granular gas. We look at the motion of two colliding spheres of mass m_1 and m_2 whose velocities are termed $\vec{v}_1$ and $\vec{v}_2$ before the collision and $\vec{v}_1'$ and $\vec{v}_2'$ after the collision.

2.1 Collision on a line

First, we look at the one-dimensional problem, that is, two particles suffering a collision when moving along a line. Before the collision their relative velocity is $v_{12} \equiv v_1 - v_2 < 0$. Due to conservation of the total momentum, the collision cannot change the centre of the mass motion, but can alter the relative velocity. Elastic particles will suffer a rebound with $v_{12}' = -v_{12}$,[2] conserving the kinetic energy of the relative motion. At a dissipative collision, however, part of the energy of the relative motion is lost. Therefore, after the contact the relative velocity will only partly be restored. The coefficient of restitution quantifies this phenomenon:

$$\varepsilon \equiv -v_{12}'/v_{12}. \qquad (2.1)$$

According to the dissipative nature of particle collisions the kinetic energy of the relative motion decreases and, hence, $0 \le \varepsilon \le 1$. For $\varepsilon = 1$ the total kinetic energy is conserved, that is, the collision occurs elastically. If $\varepsilon = 0$, the complete energy of the relative motion is lost, that is, after a collision the particles are bound together.

Conservation of the total momentum and (2.1) yield the closed set

$$\begin{aligned} m_1 v_1' + m_2 v_2' &= m_1 v_1 + m_2 v_2, \\ v_1' - v_2' &= -\varepsilon (v_1 - v_2), \end{aligned} \qquad (2.2)$$

which determines the velocities of the particles after the collision:

$$v_1' = v_1 - \frac{m^{\text{eff}}}{m_1}(1+\varepsilon) v_{12}, \qquad v_2' = v_2 + \frac{m^{\text{eff}}}{m_2}(1+\varepsilon) v_{12} \qquad (2.3)$$

---

[2]Throughout this book we have marked particle properties such as velocities, angular velocities, etc. after a collision with a prime.

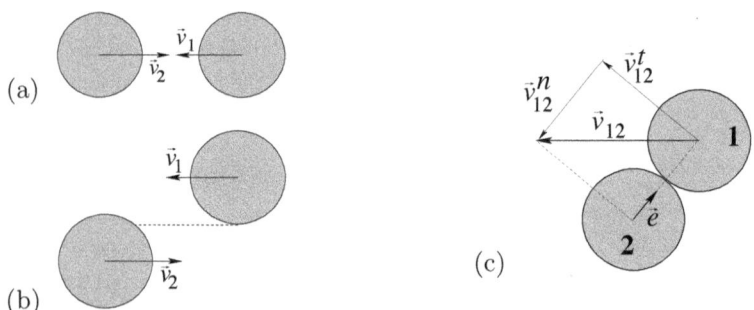

FIG. 2.1. The particles approach each other with relative velocity $\vec{v}_{12} = \vec{v}_1 - \vec{v}_2$. The geometry of the collision may range from a head-on collision (a) to a glancing one (b). It is characterized by the unit vector $\vec{e}$ (c).

with the effective mass

$$m^{\text{eff}} \equiv \frac{m_1 m_2}{m_1 + m_2}. \tag{2.4}$$

For elastic particles ($\varepsilon = 1$) and identical masses ($m^{\text{eff}}/m_1 = m^{\text{eff}}/m_2 = 1/2$) the collision causes exchange of the velocities, $v_1' = v_2$, $v_2' = v_1$.

We have focused on the simple one-dimensional motion to illustrate the basic concept. The geometry of the impact is trivial in this case. We now turn to the more relevant three-dimensional case.

2.2 Collision in space

Similarly to the one-dimensional case, the collective motion of the particles does not influence the collision, that is, we can restrict to consider the relative velocity $\vec{v}_{12} = \vec{v}_1 - \vec{v}_2$. The geometry of the collision, however, may be very different from the one-dimensional case. It may range from a head-on collision (Fig. 2.1(a)) to a glancing collision (Fig. 2.1(b)). We specify the geometry of the collision by the unit vector

$$\vec{e} \equiv \frac{\vec{r}_1 - \vec{r}_2}{|\vec{r}_1 - \vec{r}_2|} = \frac{\vec{r}_{12}}{|\vec{r}_{12}|}, \tag{2.5}$$

where $\vec{r}_{12} = \vec{r}_1 - \vec{r}_2$ is the vector which joins the centres of the spheres at the very moment when they come into contact. The relative velocity of the particles can be decomposed into the normal part, $\vec{g} \equiv (\vec{v}_{12} \cdot \vec{e})\,\vec{e} = \vec{g}_{12}^n$ and the tangential part, $\vec{g}_{12}^t \equiv \vec{v}_{12} - \vec{g}^n$. In this section we only look at the normal part and assume that the tangential velocity is not affected by the collision, that is, $(\vec{g}_{12}^t)' = \vec{g}_{12}^t$. More general analysis of the tangential motion at the impact is given in Section 3.4. In normal direction, the three-dimensional collision may be considered as an effective 1d collision which defines the (normal) coefficient of restitution

$$\vec{g}\,' = -\varepsilon\,\vec{g}. \tag{2.6}$$

This, together with the velocity of the centre of mass $m^{\text{eff}} (\vec{v}_1/m_2 + \vec{v}_2/m_1)$, which remains constant, constitute the velocities of the particles after the collision

$$\begin{aligned} \vec{v}_1' &= \vec{v}_1 - \frac{m^{\text{eff}}}{m_1}(1+\varepsilon)(\vec{v}_{12}\cdot\vec{e})\vec{e} \\ \vec{v}_2' &= \vec{v}_2 + \frac{m^{\text{eff}}}{m_2}(1+\varepsilon)(\vec{v}_{12}\cdot\vec{e})\vec{e}. \end{aligned} \qquad (2.7)$$

Exercise 2.1 *Derive the collision law (2.7)!*

Exercise 2.2 *Two spheres collide with an angle of their paths α. Express the angle of their traces after the collision as function of the coefficient of restitution!*

2.3 Does the coefficient of restitution suffice to describe granular gas dynamics?

The coefficient of restitution, albeit defined for a collision of an isolated pair of colliding particles, is the central characteristics of a granular gas which reflects its dissipative properties. All effects which give granular gases a special position with regard to molecular gases are consequences of dissipative particle collisions, expressed by the coefficient $\varepsilon < 1$.

Granular gases are complex many-particle systems. Hence, the question arises whether such a simple concept as the coefficient of restitution is adequate enough to characterise the dissipative gas behaviour. In general, the dynamics of a many particle system is governed by Newton's equations of motion

$$m_i \ddot{\vec{r}}_i = \vec{F}(\vec{r}_1,\ldots,\vec{r}_N,\vec{v}_1,\ldots,v_N)\,, \qquad i = 1,\ldots,N\,, \qquad (2.8)$$

which is a system of N coupled differential equations.

Granular particles exert forces only if they get in contact, that is, there are no long-range forces. Moreover, in dilute granular gases the typical duration of collisions is negligible as compared to the mean duration of free flight in between successive collisions. Hence, in the absence of external fields such as gravity the particles only feel forces during short periods of contact. In between the collisions they move with constant velocity along straight lines. Consequently, for dilute granular gases the dominating number of contacts occurs between pairs of particles. Contacts between three or more particles occur extremely rarely and may be neglected (see also the discussion in Section 1.4 and Exercise 5.1). Instead of the full N-particle problem (2.8) the dynamics may be described by successively occurring two-particle collisions. The mechanical interaction of two macroscopic bodies i and j is described by

$$m^{\text{eff}}\ddot{\vec{r}}_{ij}(t) = \vec{F}\left[\vec{r}_{ij}(t),\vec{v}_{ij}(t)\right] \qquad (2.9)$$

with $\vec{r}_{ij} \equiv \vec{r}_i - \vec{r}_j$ and $\vec{v}_{ij} \equiv \vec{v}_i - \vec{v}_j$. Equation (2.9) is in fact a one-body problem of motion of an effective particle of mass m^{eff} subjected to the force $\vec{F}\left[\vec{r}_{ij}(t),\vec{v}_{ij}(t)\right]$.

Only the full solution of (2.9) together with appropriate initial conditions describes the collision strictly. Nevertheless the concept of the coefficient of restitution can be motivated to characterize particle interaction in granular gases: The assumption that the number of three-particle collisions is negligible corresponds to the assumption that particle collisions are instantaneous events. Any particle i collides with particle j not before the preceding collision of particle i with another particle k has been accomplished. Therefore, the detailed dynamics of a single collision due to (2.9) is irrelevant for the dynamics of a granular gas. Instead, only the result of a collision matters, that is, the transformation of the velocities $\vec{v}_i$ and $\vec{v}_j$ immediately before the collision into the after-collisional velocities $\vec{v}'_i$ and $\vec{v}'_j$. It is this information that is precisely provided by the coefficient of restitution and the corresponding collision rule (2.7).

Since for an isolated pair of colliding particles the descriptions (2.7) and (2.9) must agree, there is a direct relation between the coefficient of restitution and the interaction force in (2.9). Therefore, the coefficient of restitution can be derived from Newton's equation of motion which is focused upon in the next section. Obviously, since the motion of colliding bodies depends on material properties, sizes of the particles, initial conditions, etc., the coefficient of restitution which describes the collision consequently depends on these values as well. Thus, the coefficient of restitution of realistic particles *does not qualify as a constant*. The fact that ε depends on the impact velocity is experimentally known for a long time (e.g. Hodkinson, 1835). In Section 3.2.1 we prove that the assumption of a constant coefficient of restitution does not agree with the mechanics of materials.

The assumption of a constant coefficient of restitution $\varepsilon = $ const. is very helpful when performing calculations since it simplifies the mathematics significantly. We will show, however, that a physically justifiable coefficient, which is a function of the impact velocity $\varepsilon = \varepsilon(g)$, leads to many exciting effects.

We wish to mention that situations arise when the assumption of a constant coefficient of restitution is well justified, provided there is a narrow velocity distribution which is, moreover, constant in time. This applies, for example, to dilute shear flow in narrow containers or to small containers with heated[3] walls.

A granular gas, as considered in this book, is an ensemble of purely mechanically interacting particles. This restriction might be decisive for the application of the theory. For Earth-bound systems it is known that grains may be charged (i.e. interact not only by contacts) and for astrophysical systems it is plausible that particles are not electrically neutral either. In the subsequent sections we will see that even with this restriction the kinetic theory of granular gases cannot be qualified as trivial. If it is wished to deal with charged particles, that is, with long-range interactions, large part of the theory has to be revised.

Exercise 2.3 *Assume $\varepsilon = $ const. A particle falls from height H and rebounces recurrently from the floor. At which time t_∞ does it come to rest? Can t_∞ be finite?*

---

[3]The term *heated* is meant with respect to the *granular temperature*, see Sec. 5.1.

3

COEFFICIENT OF RESTITUTION

For colliding spheres we derive the coefficient of restitution from the interaction forces. It turns out that the coefficient is not a material constant but rather depends on the impact velocity. An analytic expression for the coefficient of restitution of viscoelastic spheres is derived.

3.1 Forces between colliding spheres

When two elastic spheres of radii R_1 and R_2 collide they deform each other and, hence, a repulsive interaction force is developed there. The deformation is described by

$$\xi \equiv R_1 + R_2 - |\vec{r}_1 - \vec{r}_2|, \qquad (3.1)$$

(see Fig. 3.1). The spheres are entirely characterized by their radii and elastic material properties, that is, the Young modulus Y and the Poisson ratio ν. Hence, the elastic force as derived by Hertz (1882) is a function of R_1, R_2, Y and ν too. It reads

$$F_{\text{el}} = \frac{2Y\sqrt{R^{\text{eff}}}}{3(1-\nu^2)} \xi^{3/2} \equiv \rho \xi^{3/2}, \qquad (3.2)$$

where we introduce ρ – a short-hand notation for the elastic constant of the material. The effective radius R^{eff} is defined analogously to the effective mass

$$R^{\text{eff}} = \frac{R_1 R_2}{R_1 + R_2}. \qquad (3.3)$$

For dissipative interactions between colliding spheres appears a force which depends on the compression ξ and on the velocity of the relative motion $g = \dot{\xi}$. There are several possibilities to incorporate dissipative forces. We assume viscoelastic interaction for two reasons:

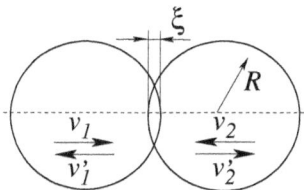

FIG. 3.1. Two colliding spheres. The figure illustrates the compression $\xi \equiv 2R - |\vec{r}_1 - \vec{r}_2|$ and the compression rate $\dot{\xi} = \vec{v}_1 - \vec{v}_2$. For simplicity, the head-on collision of identical spheres is shown.

1. Viscoelastic interaction means that the dissipative stress tensor expresses proportionally the deformation rate tensor (Landau and Lifshitz, 1965). Obviously this is the simplest form of relation between the dissipative stress and the deformation rate that one can think of. If the deformation is slow enough, that is, for small collision velocity, the viscoelastic law always applies to granular particles. Correspondingly, slow collisions of smooth (in mathematical sense) bodies, are governed by viscoelastic forces provided other forces, for example, due to surface adhesion, fluid covering, etc., can be neglected.
2. In literature it is sometimes assumed that dissipation may occur due to plastic deformation of the material (e.g. Thornton and Ning, 1998; Thornton et al., 2001). However, the simultaneous assumption of spherical shape of the particles and of plastic deformation is not consistent. If we assumed plastic deformations, the particles would not stay spheres but may change their shape considerably. It is not even clear whether the spherical shape persists *on average* (Busse and Starr, 1960). Viscoelastic interactions preserve the shape of the colliding bodies, i.e. their form is completely recovered after the impact.

    In the following sections we develop the kinetic theory of granular gases where the assumption of the spherical shape of the particles is essential. Most of the theory would become extremely difficult (if not impossible) if we abandoned this assumption.

Equation (3.2) has been extended to the case of viscoelastic interaction (Brilliantov et al., 1996) resulting in the dissipative force

$$F_{\text{diss}} = \frac{3}{2} A \rho \sqrt{\xi} \dot{\xi}, \qquad (3.4)$$

with

$$A \equiv \frac{1}{3} \frac{(3\eta_2 - \eta_1)^2}{(3\eta_2 + 2\eta_1)} \left[\frac{(1 - \nu^2)(1 - 2\nu)}{Y \nu^2} \right]. \qquad (3.5)$$

The constant A stands for the dissipative material parameter, which itself is a function of the viscous constants η_1, η_2 which relate the dissipative stress tensor to the deformation rate tensor. The details of the derivation of (3.4) are described by Brilliantov et al. (1996).

The functional form of (3.4) was guessed (but not derived) by Kuwabara and Kono (1987) and has been derived independently by Brilliantov et al. (1996) and then by Morgado and Oppenheim (1997) who used very different approaches. However, only the strict analysis of the viscoelastic deformation (Brilliantov et al., 1996) yields the pre-factors A and ρ as functions of the material parameters. The knowledge about these pre-factors is necessary to be able to obtain the coefficient of restitution as a function of the parameters of the particle material. We want to remark that equation (3.4) is a particular case of a more general relation between the elastic and dissipative forces

$$F_{\text{diss}} = A\dot{\xi}\frac{\partial F_{\text{el}}(\xi)}{\partial \xi}, \qquad (3.6)$$

which applies to contacts of arbitrarily shaped particles, provided their surface is smooth in the mathematical sense.

Combining (3.2) and (3.4) we obtain the total force which acts between colliding viscoelastic spheres. Since the deformation ξ characterizes the distance, (see (3.1)), we come to the equation of relative motion

$$\ddot{\xi} + \frac{\rho}{m^{\text{eff}}}\left(\xi^{3/2} + \frac{3}{2}A\sqrt{\xi}\dot{\xi}\right) = 0, \qquad (3.7)$$

with initial conditions (applying at the beginning of a collision at time $t = 0$)

$$\xi(0) = 0, \qquad \dot{\xi}(0) = g \qquad (3.8)$$

with $g = |\vec{g}|$ being the normal component of the impact velocity. To be precise, this is the value of g at the very instant when the particles are just starting to touch.

Equation (3.4) for the dissipative force has been deducted with the quasistatic approximation (Brilliantov et al., 1996), which requires that the characteristic velocity (the impact rate g) is much smaller than the speed of sound c in the material and that the viscous relaxation time τ_{vis} is much smaller than the duration of the collisions τ_c. The range of validity of our approach is, therefore, limited from both sides: the collisions should not be too rapid to assure $g \ll c$, $\tau_{\text{vis}} \ll \tau_c$, and not too slow to avoid influences of surface effects such as adhesion.

3.2 Derivation of the coefficient of restitution

3.2.1 Dimensionless equation of motion and dimension analysis

The coefficient of restitution as introduced above embodies the ratio of the relative velocities (in normal direction) before and after the collision, that is,

$$\varepsilon = \dot{\xi}(t_c)\Big/\dot{\xi}(0) = \dot{\xi}(t_c)/g, \qquad (3.9)$$

where t_c is the duration of the collision. It can be found numerically by integrating the equation of motion (3.7).

To derive an expression for ε in a closed form is not that simple. It can be carried out straightforwardly, but requires considerable efforts (Schwager and Pöschel, 1998). Here we present a derivation which is based on a dimension analysis of the equation of motion (3.7) (Ramírez et al., 1999). The method of dimension analysis was first used by Tanaka et al. (1991) to prove that a constant coefficient of restitution is not consistent with physical reality.

Although we already know the interaction forces between colliding spheres, (3.2, 3.4), we want to perform a general analysis and, thus, we assume a more general form of the forces

$$F_{\rm el} = m^{\rm eff} D_1 \xi^\alpha,$$
$$F_{\rm diss} = m^{\rm eff} D_2 \xi^\gamma \dot{\xi}^\beta, \qquad (3.10)$$

with D_1 and D_2 being material parameters. With these notations the equation of motion for colliding particles reads

$$\ddot{\xi} + D_1 \xi^\alpha + D_2 \xi^\gamma \dot{\xi}^\beta = 0, \qquad \xi(0) = 0, \qquad \dot{\xi}(0) = g. \qquad (3.11)$$

As the characteristic length ξ_0 of the problem we have chosen the maximal compression for the elastic case. It may be found by equating the initial kinetic energy of the relative motion $m^{\rm eff} g^2/2$ and the elastic energy at the instant of maximal compression

$$\int_0^{\xi_0} F_{\rm el}(\xi)\, d\xi = m^{\rm eff} D_1 \frac{\xi_0^{\alpha+1}}{\alpha+1}. \qquad (3.12)$$

From (3.12) one obtains

$$\xi_0 \equiv \left(\frac{\alpha+1}{2 D_1}\right)^{1/(1+\alpha)} g^{2/(1+\alpha)}. \qquad (3.13)$$

For the characteristic time we have chosen the time the particles need to cover the distance ξ_0 when they move with relative velocity g

$$\tau_0 \equiv \frac{\xi_0}{g}. \qquad (3.14)$$

Having defined the characteristic values we can now construct dimensionless variables

$$\hat{\xi} \equiv \xi/\xi_0, \qquad \hat{\dot{\xi}} \equiv \dot{\xi}/g, \qquad \hat{\ddot{\xi}} = \frac{\xi_0}{g^2}\ddot{\xi} \qquad (3.15)$$

and recast the equation of motion (3.7) and its initial conditions (3.8) into dimensionless form:

$$\hat{\ddot{\xi}} + \varkappa \hat{\xi}^\gamma \hat{\dot{\xi}}^\beta + \frac{1+\alpha}{2} \hat{\xi}^\alpha = 0, \qquad \hat{\xi}(0) = 0, \qquad \hat{\dot{\xi}}(0) = 1 \qquad (3.16)$$

with

$$\varkappa = \varkappa(g) = D_2 \left(\frac{1+\alpha}{2 D_1}\right)^{(1+\gamma)/(1+\alpha)} g^{2(\gamma-\alpha)/(1+\alpha)+\beta}. \qquad (3.17)$$

The terms in (3.16) depend neither on material properties nor on impact velocity except for $\varkappa$. Therefore, if the motion of the particles depends on material properties and on impact velocity, it may depend only via $\varkappa$, that is, in the combination of the parameters as given by (3.17). Hence, any dependence on the impact velocity $f(g)$ must be $f[\varkappa(g)]$.

Contrarily, if the equation of motion does not depend on the impact velocity, $\varkappa$ must be independent of g, that is, the condition

$$2(\gamma - \alpha) + \beta(1 + \alpha) \stackrel{!}{=} 0 \quad (3.18)$$

does apply. The same reasoning may be applied to the coefficient of restitution: If the equation of motion does not depend on g, its solution, which yields ε (see (3.9)) is independent of g as well. Thus, we conclude that (3.18) is the condition for the coefficient of restitution to be a constant.

Now we assume a linear dependence of the dissipative force on the impact velocity (i.e. $\beta = 1$), which seems to be the most realistic case for small $\dot{\xi}$ and then analyse how the dissipative force should depend on ξ to fulfil the condition (3.18). We focus on two important cases: The case of linear elastic interaction, $F_{\text{el}} \propto \xi$, valid for one-dimensional problems such as for cubes which collide with a flat surface, and the case of colliding spheres $F_{\text{el}} \propto \xi^{3/2}$, according to (3.2).

1. For the elastic force $F_{\text{el}} \propto \xi$, (i.e. $\alpha = 1$), the condition (3.18) is fulfilled for a linear dissipative force $F_{\text{diss}} \propto \dot{\xi}$, that is, for $\gamma = 0$. This is the linear one-dimensional dashpot force model.

2. Assuming the Hertz law for three-dimensional spheres (3.2), (i.e. $\alpha = 3/2$), the condition (3.18) requires $\gamma = \frac{1}{4}$, that is, $F_{\text{diss}} \propto \dot{\xi}\xi^{1/4}$. As far as we can see there is no physical argument by which this functional form of the dissipative force could be supported.

Therefore, we conclude that a constant coefficient of restitution is in agreement with physical mechanics only in case of (quasi-) one-dimensional systems. For three-dimensional spheres the assumption $\varepsilon = $ const. disagrees with basic mechanical laws.

We now ask the question: What kind of $\varepsilon(g)$ dependence corresponds to the collision of viscoelastic spheres? As it follows from (3.2, 3.4), for this case $\alpha = 3/2$, $\beta = 1$ and $\gamma = 1/2$. From (3.17) the following can be obtained

$$\varkappa = \frac{3}{2}\left(\frac{5}{4}\right)^{3/5} A \left(\frac{\rho}{m^{\text{eff}}}\right)^{2/5} g^{1/5} \quad (3.19)$$

and, therefore,

$$\varepsilon = \varepsilon\left[A\left(\frac{\rho}{m^{\text{eff}}}\right)^{2/5} g^{1/5}\right]. \quad (3.20)$$

If we assume that the function $\varepsilon(g)$ is a sufficiently smooth function which can be expanded into a Taylor series, and if we realize that $\varepsilon(0) = 1$ and $\varepsilon(g)$ decreases with g,[4] we can already anticipate the functional form of the coefficient of restitution as a function of the impact velocity:

$$\varepsilon = 1 + k_1 A \left(\frac{\rho}{m^{\text{eff}}}\right)^{2/5} g^{1/5} + k_2 A^2 \left(\frac{\rho}{m^{\text{eff}}}\right)^{2/5} g^{2/5} + \cdots \quad (3.21)$$

Expression (3.21) is one of the central foundations for many calculations in the following sections. We will use it in a slightly different representation:

$$\varepsilon = 1 - C_1 A \kappa^{2/5} g^{1/5} + C_2 A^2 \kappa^{4/5} g^{2/5} \mp \cdots \quad (3.22)$$

with

$$\kappa \equiv \left(\frac{3}{2}\right)^{5/2} \left(\frac{\rho}{m^{\text{eff}}}\right) = \left(\frac{3}{2}\right)^{3/2} \frac{Y \sqrt{R^{\text{eff}}}}{m^{\text{eff}} (1 - \nu^2)}. \quad (3.23)$$

The coefficients $C_1, C_2, \ldots$ are pure numbers. The first of them read

$$C_1 \approx 1.15344, \quad C_2 \approx 0.79826, \quad C_3 \approx -0.483582, \quad C_4 \approx 0.285279. \quad (3.24)$$

Analytic expressions for these coefficients are given below in (3.41–3.44). They were found by the rigorous integration of the equation of motion (3.7) (Schwager and Pöschel, 1998). In the following section we suggest a simpler method to obtain the coefficients C_i (Ramírez et al., 1999).

Exercise 3.1 *Assume ε depends on the impact velocity as $\varepsilon(v) = 1 - Cv^{1/5}$ (This formula is an approximation of (3.22)). A particle is again dropped from height H to a rigid floor and rebounces recurrently. After which time t_∞^v this sphere will come to rest? Compare the bouncing times for the cases $\varepsilon = \text{const.}$ and $\varepsilon = \varepsilon(v)$! Assume that at the first contact with the floor the velocity dependent coefficient of restitution and the constant one have the same value!*

3.2.2 The coefficients C_i

For quantitative calculations of the granular gas dynamics we need the numerical values of the coefficients C_i in the expansion (3.22). With the identity

$$\frac{d}{dt} = \dot{\hat{\xi}} \frac{d}{d\hat{\xi}} \quad (3.25)$$

the equation of motion (3.16) reads

$$\frac{d}{d\hat{\xi}}\left(\frac{1}{2}\dot{\hat{\xi}}^2 + \frac{1}{2}\hat{\xi}^{5/2}\right) = -\varkappa \dot{\hat{\xi}} \sqrt{\hat{\xi}} = \frac{dE(\hat{\xi})}{d\hat{\xi}}, \quad \hat{\xi}(0) = 0, \quad \dot{\hat{\xi}}(0) = 1, \quad (3.26)$$

---

[4]This may be seen from (3.16): For $\varkappa \sim g^{1/5} \to 0$ the second term in the LHS of (3.16) vanishes and it converts in this limit to the equation for elastic colliders.

DERIVATION OF THE COEFFICIENT OF RESTITUTION

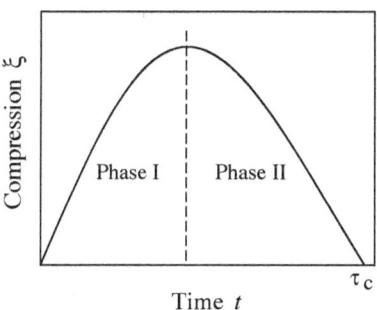

FIG. 3.2. The collision consists of the compression phase (I) and the expansion phase (II).

where we introduce the mechanical energy

$$E \equiv \frac{1}{2}\dot{\hat{\xi}}^2 + \frac{1}{2}\hat{\xi}^{5/2}. \tag{3.27}$$

We subdivide the collision of particles into two phases (see Fig. 3.2): (I) the compression phase and (II) the expansion phase. The compression phase starts at the instant of first contact of the spheres $t=0$ and ends when the turning point of maximal compression $\hat{\xi}_0$ has been reached. The expansion phase covers the rest of the collision.

The loss of mechanical energy during the first phase is according to (3.26)

$$\int_0^{\hat{\xi}_0} \frac{dE}{d\hat{\xi}} d\hat{\xi} = -\varkappa \int_0^{\hat{\xi}_0} \dot{\hat{\xi}} \sqrt{\hat{\xi}} \, d\hat{\xi}. \tag{3.28}$$

To be able to evaluate the RHS of (3.28) one needs to know the dependence of the compression rate on the compression $\dot{\hat{\xi}} = \dot{\hat{\xi}}(\hat{\xi})$.

For the case of elastic collisions, the maximal compression is $\hat{\xi}_0 = 1$, according to the definition of our dimensionless variables. Hence, the desired function follows from the conservation of energy:

$$\dot{\hat{\xi}}(\hat{\xi}) = \sqrt{1 - \hat{\xi}^{5/2}}. \tag{3.29}$$

The velocity $\dot{\hat{\xi}}$ vanishes at the turning point $\hat{\xi} = 1$. For inelastic collisions the maximal compression $\hat{\xi}_0$ is smaller than 1, therefore, an approximative relation for the inelastic case is

$$\dot{\hat{\xi}}(\hat{\xi}) \approx \sqrt{1 - (\hat{\xi}/\hat{\xi}_0)^{5/2}} \tag{3.30}$$

which also yields vanishing velocity $\dot{\hat{\xi}}$ at the turning point $\hat{\xi}_0$. Using (3.30) the integration in (3.28) may be performed yielding

$$\frac{1}{2}\hat{\xi}_0^{5/2} - \frac{1}{2} = -\varkappa b\, \hat{\xi}_0^{3/2} \qquad (3.31)$$

where we take into account

$$E(\hat{\xi}_0) = \frac{1}{2}\hat{\xi}_0^{5/2}$$
$$E(0) = \frac{1}{2}\dot{\xi}^2(0) = \frac{1}{2} \qquad (3.32)$$

and introduce a constant

$$b \equiv \int_0^1 \sqrt{x}\sqrt{1-x^{5/2}}\, dx = \frac{\sqrt{\pi}\,\Gamma(3/5)}{5\,\Gamma(21/10)}. \qquad (3.33)$$

We now define the *inverse collision* which is a collision that starts with velocity $\varepsilon\, g$ and ends with velocity g. During the inverse collision the system gains energy, that is, it is characterized by negative damping. The maximal compression $\hat{\xi}_0$ is, of course, the same in both collisions, the direct and the inverse since the inverse collision equals the direct collision except for the fact that time runs backwards.

Analogous to (3.26) one obtains for the inverse collision

$$\frac{dE(\hat{\xi})}{d\hat{\xi}} = +\varkappa \dot{\hat{\xi}}\sqrt{\hat{\xi}}, \qquad \hat{\xi}(0) = 0, \qquad \dot{\hat{\xi}}(0) = \varepsilon. \qquad (3.34)$$

This suggests an approximative relation for the inverse collision,

$$\dot{\hat{\xi}}(\hat{\xi}) \approx \varepsilon \sqrt{1-(\hat{\xi}/\hat{\xi}_0)^{5/2}}, \qquad (3.35)$$

with the additional pre-factor ε, which is the initial velocity for the inverse collision.

Integration of the energy *gain* for the first phase of the inverse collision (which equals up to its sign the energy loss in the second phase of the direct collision (Schwager and Pöschel, 1998)) may be performed just in the same way as was done for the direct collision, leading to the result

$$\frac{1}{2}\hat{\xi}_0^{5/2} - \frac{\varepsilon^2}{2} = +\varepsilon\, \varkappa b\, \hat{\xi}_0^{3/2}, \qquad (3.36)$$

Exercise 3.2 *Derive (3.36)!*

From (3.31, 3.36) we obtain a simple approximative relation between the coefficient of restitution and the (dimensionless) maximal compression:

$$\varepsilon = \hat{\xi}_0^{5/2}. \qquad (3.37)$$

Substituting this into (3.31) we arrive at an equation for the coefficient of restitution

$$\varepsilon + 2\varkappa b\, \varepsilon^{3/5} = 1. \qquad (3.38)$$

The formal solution to this equation may be written as a continuous fraction (which does not diverge in the limit $g \to \infty$):

DERIVATION OF THE COEFFICIENT OF RESTITUTION

$$\varepsilon^{-1} = 1 + \frac{2\varkappa b}{\left(1 + \frac{2\varkappa b}{\left(1 + \frac{2\varkappa b}{(1 + \cdots)^{2/5}}\right)^{2/5}}\right)^{2/5}} \quad (3.39)$$

Another and, perhaps more appropriate way of representation of the coefficient of restitution ε is a series expansion in terms of $\varkappa$. For practical applications it is convenient to return to dimensional units. We define the characteristic velocity g^* such that

$$\varkappa \equiv \frac{1}{2b}\left(\frac{g}{g^*}\right)^{1/5}, \quad (3.40)$$

with b being defined in (3.33). Using the definition (3.17) together with (3.2, 3.4), which provide the values of D_1 and D_2 and (3.33) for b, we find for the characteristic velocity

$$(g^*)^{-1/5} = \frac{\sqrt{\pi}}{2^{1/5} 5^{2/5}} \frac{\Gamma(3/5)}{\Gamma(21/10)} \left(\frac{3}{2}A\right) \left(\frac{\rho}{m^{\text{eff}}}\right)^{2/5}. \quad (3.41)$$

With this new notation the coefficient of restitution reads

$$\varepsilon = 1 - a_1 \left(\frac{g}{g^*}\right)^{1/5} + a_2 \left(\frac{g}{g^*}\right)^{2/5} - a_3 \left(\frac{g}{g^*}\right)^{3/5} + a_4 \left(\frac{g}{g^*}\right)^{4/5} \mp \cdots, \quad (3.42)$$

with

$$a_1 = 1, \quad a_2 = 3/5, \quad a_3 \approx 0.315, \quad a_4 \approx 0.161. \quad (3.43)$$

We have to remark that the last two coefficients in a_3 and a_4 in (3.43) stem from the straightforward brute-force theory (Schwager and Pöschel, 1998; Ramírez et al., 1999). The simple theory presented here yields the correct values for a_1 and a_2 but not for higher coefficients. The coefficients C_i of the expansion (3.22) can be obtained via

$$C_i = a_i C_1^i = a_i (g^*)^{-i/5}. \quad (3.44)$$

In particular,

$$C_1 = \frac{\sqrt{\pi}}{2^{1/5} 5^{2/5}} \frac{\Gamma(3/5)}{\Gamma(21/10)}$$

$$C_2 = \frac{3}{5} C_1^2. \quad (3.45)$$

The numerical values are given in (3.24). The convergence of the series is rather slow, hence, accurate results can be expected only for small enough g/g^*.

3.3 Padé approximation

The expansions given in (3.22, 3.42) together with the coefficients (3.43, 3.44) represent the full solution for the coefficient of restitution as a function of the impact velocity. However, a problem arises when this series is used in numerical simulations: whenever the series is cut after a certain order n, the function $\varepsilon(g)$ diverges as $\lim_{g \to \infty} \varepsilon(g) \to \pm\infty$, depending whether the order n is even or odd.

If one deals with finite velocities only, one can choose the order n so that the series gives sufficiently correct values for all velocities of interest. Frequently, however, the maximum collision velocity cannot be controlled as, for instance, in a granular gas with a certain velocity distribution which always contains very fast particles with small but finite probability.

Hence, for practical applications such as molecular dynamics simulations, one is interested in a representation of $\varepsilon(g)$ which does not diverge for large impact velocity. Such a representation can be found as a Padé approximation. A Padé approximation represents a fracture of two polynomials of order m and n whose coefficients are chosen in a way that the limits of the approximated functions are correctly represented and that in between the limits the function is represented as well as possible.

The function $\varepsilon(g)$ has the limits $\varepsilon(0) = 1$ and $\varepsilon(\infty) = 0$. Since the dependence $\varepsilon(g)$ is expected to be a smooth, monotonically decreasing function, therefore, the order of the numerator must be smaller than the order of the denominator. We choose a 1–4 Padé approximation

$$\varepsilon = \frac{1 + d_1 \left(g/g^*\right)^{1/5}}{1 + d_2 \left(g/g^*\right)^{1/5} + d_3 \left(g/g^*\right)^{2/5} + d_4 \left(g/g^*\right)^{3/5} + d_5 \left(g/g^*\right)^{4/5}}. \quad (3.46)$$

Standard analysis (e.g. Baker, 1970) yields the coefficients d_k in terms of the coefficients a_k

$$d_0 = a_4 - 2a_3 - a_2^2 + 3a_2 - 1 \quad (3.47)$$
$$d_1 = [1 - a_2 + a_3 - 2a_4 + (a_2 - 1)(3a_2 - 2a_3)]/d_0 \quad \approx 2.583$$
$$d_2 = [(a_3 - a_2)(1 - 2a_2) - a_4]/d_0 \quad \approx 3.583$$
$$d_3 = [a_3 + a_2^2(a_2 - 1) - a_4(a_2 + 1)]/d_0 \quad \approx 2.983$$
$$d_4 = [a_4(a_3 - 1) + (a_3 - a_2)(a_2^2 - 2a_3)]/d_0 \quad \approx 1.148$$
$$d_5 = \left[2(a_3 - a_2)(a_4 - a_2 a_3) - (a_4 - a_2^2)^2 - a_3(a_3 - a_2^2)\right]/d_0 \quad \approx 0.326$$

We want to compare the theoretical result for the velocity dependence of the coefficient of restitution with experimental values taken from Bridges et al. (1984). Using the characteristic velocity $g^* = 0.32$ cm/s for ice at very low temperature as a fitting parameter, the theoretical curve $\varepsilon(g)$ agrees fairly well with the impact velocity in the whole range of the impact velocity g (Fig. 3.3). The discrepancy with the experimental data at small g is based on the fact that the extrapolation expression, $\varepsilon = 0.32/g^{0.234}$ used by Bridges et al. (1984) to fit the experimental

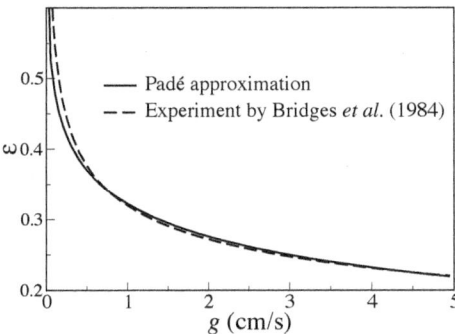

FIG. 3.3. Dependence of the coefficient of normal restitution on the impact velocity for ice particles. The dashed line — experimental (Bridges et al., 1984), solid line — the Padé approximation (3.46) with the constants given by (3.47) and with the characteristic velocity for ice $g^* = 0.32$ cm/s.

data has an unphysical divergence at $g \to 0$ and does not imply the failing of the theory for this region. The scattering of the experimental data presented by Bridges et al. (1984) is large for small impact velocity according to experimental complications, therefore the fit formula by Bridges et al. (1984, 2001) cannot be expected to be accurate enough for velocities that are too small. Moreover, in the region of very small velocity other than viscoelastic interactions possibly influence the collision behaviour as, for example, adhesion. Similarly, for very high velocities effects such as brittle failure, fracture and others may contribute to dissipation.

3.4 Coefficient of tangential restitution

In the previous section we defined the normal coefficient of restitution for colliding spheres, that is, only the normal component of the relative velocity was taken into account. To describe the change of the tangential component of the relative velocity, the coefficient of tangential restitution has to be introduced. Rotation of the particles contributes to the tangential component of their relative motion at contact. Hence, the collision is completely described by two coefficients of restitution, $\varepsilon = \varepsilon^n$, for the normal motion, and ε^t for the tangential motion:

$$\begin{aligned} (\vec{g}^n)' &= -\varepsilon^n \vec{g}^n \quad &\text{with} \quad &(0 \leq \varepsilon^n \leq 1) \\ (\vec{g}^t)' &= \varepsilon^t \vec{g}^t \quad &\text{with} \quad &(-1 \leq \varepsilon^t \leq 1), \end{aligned} \quad (3.48)$$

where $\vec{g} = \vec{g}^n$ and $\vec{g}^t$ are the relative velocities of the particle surfaces at the point of contact in normal and tangential direction.[5] These read for identical

---

[5] The limits of the coefficients ε^n and ε^t as given in (3.48) are deduced from the natural assumption that both components of the relative velocity diminish after a dissipative impact. It has been recently shown, however, that the normal coefficient of restitution can be larger than unity under conditions of strong coupling between the rotational and translational motion (Louge and Adams, 2002).

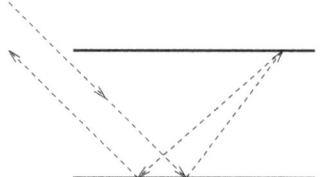

FIG. 3.4. The surprising motion of a superball.

particles of radius R:

$$\vec{g}^n = (\vec{v}_{12} \cdot \vec{e})\,\vec{e} = \vec{g}$$
$$\vec{g}^t = \vec{v}_{12} - \vec{g}^n + R\,[\vec{e} \times (\vec{\omega}_1 + \vec{\omega}_2)]\,, \qquad (3.49)$$

where $\vec{\omega}_{1/2}$ are the rotational velocities of the particles. For $\varepsilon^t = 1$, the tangential motion is not altered by a collision. Sometimes it is said that this behaviour corresponds to ideally smooth spheres, which is wrong as discussed below. In the opposite case, $\varepsilon^t = -1$, a complete reversion of the tangential velocity can be observed. If $\vec{g}^t$ is attributed mostly to the rotation velocity, the condition $\varepsilon^t \to -1$ implies that the spin of the particle almost reverts. One can convince oneself of this surprising fact by throwing a very elastic rubber ball with a sticky surface (superball) into a channel as sketched in Fig. 3.4. Interesting discussions on the motion of superballs can be found, for example, in Garwin (1969) and Stroink (1983). If rotational motion is taken into account, the propagation rule (2.7) modifies. To derive a new collision rule we use Newton's equation for the rotation motion of the colliding particles,

$$\frac{dI\vec{\omega}_1}{dt} = \vec{r} \times \vec{F}_{12}\,. \qquad (3.50)$$

Here I represents the (scalar) moment of inertia of a sphere. $\vec{F}_{12}$ is the interparticle force, and $\vec{r} = -R\vec{e}$ gives the point (with respect to the centre of the first particle), where force $\vec{F}_{12}$ is applied. A similar equation may be written for the rotational motion of the second particle. Let us integrate (3.50) over the collision. If the compression is small we can neglect the change of $\vec{r}$ during the collision, then the integration of the force will give the change of linear momentum at the collision. Thus we arrive at the following set of equations:

$$I\,(\vec{\omega}_1' - \vec{\omega}_1) = -R\,[\vec{e} \times m\,(\vec{v}_1' - \vec{v}_1)]$$
$$I\,(\vec{\omega}_2' - \vec{\omega}_2) = R\,[\vec{e} \times m\,(\vec{v}_2' - \vec{v}_2)] \qquad (3.51)$$
$$\vec{v}_1' + \vec{v}_2' = \vec{v}_1 + \vec{v}_2\,,$$

where the last equation in (3.51) stands for the linear momentum conservation. Equations (3.51) together with (3.49) completely determine the dissipative collision including the rotational degrees of freedom. Solving this set of equations yields the velocities and angular velocities after the collision:

$$\vec{v}_1' = \vec{v}_1 - \frac{1+\varepsilon^n}{2}\vec{g}^n - \frac{1-\varepsilon^t}{2(1+q^{-1})}\vec{g}^t$$
$$\vec{v}_2' = \vec{v}_2 + \frac{1+\varepsilon^n}{2}\vec{g}^n + \frac{1-\varepsilon^t}{2(1+q^{-1})}\vec{g}^t$$
$$\vec{\omega}_1' = \vec{\omega}_1 + \frac{1-\varepsilon^t}{1+q}\frac{1}{2R}[\vec{e}\times\vec{g}^t]$$
$$\vec{\omega}_2' = \vec{\omega}_2 + \frac{1-\varepsilon^t}{1+q}\frac{1}{2R}[\vec{e}\times\vec{g}^t]\,,$$
(3.52)

where $q \equiv I/(mR^2)$ denotes the reduced moment of inertia. For $\varepsilon^t = 1$ these equations reduce to (2.7) which were used previously.

Exercise 3.3 *Derive the propagation rule (3.52) by means of (3.51, 3.49)!*

The tangential coefficient of restitution is a complicated parameter since the tangential restitution is not independent of the normal restitution: according to the Coulomb friction law the tangential force F^t which is exerted on a body from a surface is related to the normal force F^n via

$$|F^t| \leq \mu |F^n|\,,$$
(3.53)

where μ is the Coulomb friction coefficient. This means that for small normal force the friction force is also small and cannot prevent the contacting surfaces from sliding. Contrary, for a large normal force the tangential force is strong enough to suppress relative motion of the surfaces. Such a contact implies rolling of particles, just as gear wheels. During a collision the normal force is not a constant, according to (3.2, 3.4). It increases from zero at the beginning of the collision to the moment of maximal compression and decreases during the rest of the collision. Therefore, the type of contact, sliding or rolling, may change during the impact. The complicated dynamics of such collisions is discussed, for example, by Stronge (1990).

Hence, in general, the coefficient of tangential restitution is a function of both components of the relative velocity of the particles, $\varepsilon^t = \varepsilon^t(g^n, g^t)$. While the normal coefficient of restitution, ε^n, is determined by the bulk dissipation of the deformed material, it is evident that ε^t is determined by both, the bulk dissipation and the surface properties of the particle. This experimentally known fact can be easily understood by touching a superball: its surface is adhesive and sticky, which mainly gives rise to its entertaining motion.

To derive a formula for $\varepsilon^t = \varepsilon^t(g^n, g^t)$ similar to $\varepsilon^n = \varepsilon^n(g^n)$, a model of the surface properties of colliding particles has been proposed by Brilliantov et al. (1996). It was assumed that the rough and sticky surface can be mimicked by irregular asperities of a certain average size ζ_0 which cover the surfaces of the spheres. These asperities give rise to a tangential force since they inhibit tangential relative motion of the surfaces. Each of the asperities can be overcome by applying a certain critical shear stress. The details of this model can be found

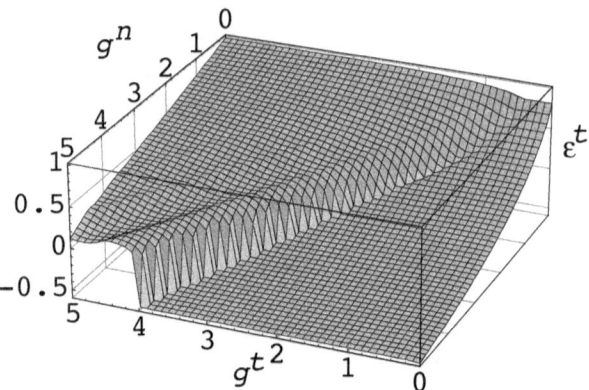

FIG. 3.5. Coefficient of tangential restitution as function of the normal and tangential component of the impact velocity.

in Brilliantov et al. (1996); here we only want to briefly discuss the velocity dependence of the tangential coefficient.

From Fig. 3.5 one notices that ε^t depends on both components of the velocity. As expected, for large normal component g^n and small tangential component g^t the coefficient has a negative value $\varepsilon^t \approx -0.5$. This means that during the contact the increasing tangential force suppresses sliding and the surfaces of the grains stick together. The tangential relative motion of the two particles, nevertheless, continues and gives rise to the shear stress in their bulk, which accumulates the energy of the relative tangential motion. Contrary to the case of sliding, a part of this energy may be returned, causing the rebound in tangential direction, that is negative ε^t. The kinematics of the impact may be very complicated, for example, particles might roll on each other, when sliding is suppressed. In the opposite case of large g^t and small g^n, one observes that $\varepsilon^t \to 1$, that is, the small normal force may cause only small tangential force, according to the condition described in (3.53). Naturally, the small tangential force can not alter the tangential velocity noticeably, that is the reason why $\varepsilon^t \to 1$ in this case. Let us fix the tangential component for a moment at say $g^t = 3$. For small g^n we have $\varepsilon^t \to 1$, that is, the particles slide during the entire collision. As g^n grows, ε^t decreases. This is based on the fact that during some part of the contact time the particles can roll (at high compression ξ) and the other time they slide. At $g^n \approx 4$ we observe a steep decay of ε^t, which corresponds to the transition between stick-slip motion and regular sliding.

We want to make a few comments on the coefficient of tangential restitution:

1. The coefficient depends on the impact velocity and is definitely not a constant. From our discussion follows directly that $\varepsilon^t = $ const. does not comply with Coulomb's friction law (3.53).

2. Even for the case of *ideally smooth* spheres $\varepsilon^t \neq 1$, except for perfectly elastic particles ($\varepsilon^n = 1$) for the following reason: the normal coefficient as derived in Section 3.2 is a direct consequence of the deformation rate $\dot{\xi}$ in the particle's material. Hence if the particle rotate in compressed state during the collision, an additional deformation rate arises due to the rotation. This causes a viscous stress which counteracts the rotation (Brilliantov and Pöschel, 1998), that is, a torque arises which decelerates the rotation and, thus, slows down the relative tangential motion. According to the definition of ε^t, the slowing down of the tangential motion at the collision implies $\varepsilon^t \neq 1$. Therefore, two spheres which meet by a glancing collision with $g^t \neq 0$ always exert a torque, even if they are perfectly smooth, that is, the collision always changes their rotation velocities. The simultaneous assumptions $\varepsilon^n \neq 1$ and $\varepsilon^t = 1$ are, thus, inconsistent.
3. In the following chapters where we discuss the kinetics of granular gases it will become obvious that it requires considerable effort to incorporate the physically correct velocity dependence of the coefficient of normal restitution ε^n due to (3.22). The coefficient ε^t depends in a much more complicated way on both components of the velocity $\varepsilon^t(g^n, g^t)$ as shown in Fig. 3.5. Therefore, we believe that for technical purposes presently it is not feasible to consider the dynamics of a granular gas with a physically correct expression for ε^t. Hence, all results presented in this book have been obtained by neglecting particle rotation, that is, with the assumption $(\vec{v}_{12}^t)' = \vec{v}_{12}^t$, which leads to the collision rule (2.7). We are aware of the fact that, according to the previous comment, this simplification is inconsistent.

Exercise 3.4 *The reduced moment of inertia $q \equiv I/(mR^2)$ characterizes the distribution of particle material inside the grain. How does this quantity affect the coupling between the rotational and translational motion? Look at two opposite cases: (i) all mass is distributed in a very thin shell of radius R, and (ii) the mass is concentrated in a very small volume around the centre of the particle. What is the value of q for the grains of a uniform density? Does q depend on mass, density, or radius in the latter case?*

4
APPLICATION TO FEW-PARTICLE SYSTEMS

The concept of the coefficient of restitution is illustrated by two examples of few-particle systems. It is shown that the detailed mechanics of particle collisions is of major importance to their overall behaviour: even small systems consisting of only few viscoelastic particles behave qualitatively different from equivalent systems when a constant coefficient of restitution is assumed. Therefore, it can be expected that the details of pairwise particle collisions determine the properties of a granular many-body system, that is, a granular gas, qualitatively too.

4.1 Inelastic collapse

Look at three identical inelastically colliding particles that move in one dimension. At initialization time, the particles move with velocities v_0, v_1 and v_2. Are there conditions for the initial velocities and for the coefficient of restitution $\varepsilon = $ const. for which the energy of the relative motion is entirely consumed by dissipative collisions? In this case the particles would finally move as a cluster with common velocity. The answer to the problem is 'yes', and the corresponding scenario is called *inelastic collapse*. Before performing the mathematical analysis where we follow the arguments by Shida and Kawai (1989), we would like to give some further explanation of the problem.

The relative motion of three particles along a line is completely described by two relative velocities, $g_1 \equiv v_1 - v_0$ and $g_2 \equiv v_2 - v_1$. Hence, the problem may be

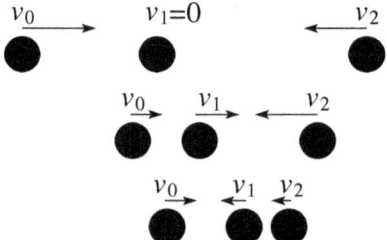

FIG. 4.1. The inelastic collapse problem: assume $\varepsilon = $ const. Are there initial conditions g_1, g_2 for which the energy of the relative motion is completely dissipated? The top row shows the system at the time of initialization, the second and third rows show the system after the first collision (0,1) and after the second one (1,2), respectively.

mapped to an equivalent problem where at initial time the particle in the middle rests, $v_1(0) = 0$. The problem is sketched in Fig. 4.1.

Assume that the first collision in the system occurs between particles 0 and 1. Then according to the collision rule (2.3) their velocities become

$$v'_0 = \frac{1-\varepsilon}{2} v_0, \qquad v'_1 = \frac{1+\varepsilon}{2} v_0. \tag{4.1}$$

In contrast to elastic particles (where we would obtain $v'_0 = 0$, $v'_1 = v_0$), both particles now move in the same direction and the relative velocity is now $g_1 > 0$. The next collision must occur between particles 1 and 2, hence, the condition $g_2 < 0$ is to be fulfilled. Clearly, we will never find two consecutive collisions that occur between the same particles. The collision sequence is always $\cdots \to (0,1) \to (1,2) \to (0,1) \to \cdots$. To obtain an infinite sequence of collisions at each time instant either $g_1 < 0 < g_2$ or $g_1 > 0 > g_2$ must be fulfilled. As shown below, this necessary condition is fulfilled only for a certain interval of the coefficient of restitution ε and for appropriately chosen initial conditions g_1 and g_2.

For the mathematical analysis we scale the initial velocities with a constant factor to obtain $g_1 = -1$. This scaling does not reduce generality; it only corresponds to scaling of the time unit. For the second initial relative velocity we have chosen $g_2 = 0$. This condition, which allows for a simplified notation, does not affect the final results but it reduces the generality of the calculation. Since the simplified case with $g_2 = 0$ contains the physical spirit of the problem and may be easily extended to $g_2 \neq 0$ for the sake of clearness we do not wish to discuss the general case here but refer to the literature (Shida and Kawai, 1989; Constantin et al., 1995).

To describe the relative velocities after a certain number i of collisions has occurred, we add an upper index. Then in vector notation the initial condition reads $\vec{g}^{(0)} = (-1, 0)$. Due to the collision rules a collision between particles 0 and 1 changes the relative velocities

$$g_1^{(i)} = -\varepsilon g_1^{(i-1)}, \qquad g_2^{(i)} = g_2^{(i-1)} + \frac{1+\varepsilon}{2} g_1^{(i-1)}, \tag{4.2}$$

whereas collisions between particles 1 and 2 cause

$$g_1^{(i)} = g_1^{(i-1)} + \frac{1+\varepsilon}{2} g_2^{(i-1)}, \qquad g_2^{(i)} = -\varepsilon g_2^{(i-1)}. \tag{4.3}$$

The propagation rules (4.2, 4.3) can be written in a compact matrix form:

$$\vec{g}^{(i)} = \begin{pmatrix} -\varepsilon & 0 \\ (1+\varepsilon)/2 & 1 \end{pmatrix} \vec{g}^{(i-1)} \equiv \hat{A} \vec{g}^{(i-1)} \quad \text{for 0-1 collisions}$$

$$\vec{g}^{(i)} = \begin{pmatrix} 1 & (1+\varepsilon)/2 \\ 0 & -\varepsilon \end{pmatrix} \vec{g}^{(i-1)} \equiv \hat{B} \vec{g}^{(i-1)} \quad \text{for 1-2 collisions}. \tag{4.4}$$

Obviously, to achieve asymptotically vanishing energy, after each collision either g_1 or g_2 must be negative. For our initial conditions this implies that g_1 must be negative after any even number of collisions, while $g_2 < 0$ after any odd number of collisions. Moreover, to form a cluster the relative velocities must asymptotically vanish as the number of collisions i tends to infinity. Thus, the conditions for a collapse read

$$g_1^{(2i)} < 0; \qquad g_2^{(2i+1)} < 0$$
$$g_1^{(2i)} \to 0; \qquad g_2^{(2i+1)} \to 0 \qquad \text{for} \quad i \to \infty. \tag{4.5}$$

Since the collisions 0–1 and 1–2 occur alternatively,

$$\vec{g}^{(2i)} = \left(\hat{B}\hat{A}\right)^i \vec{g}^{(0)} = \begin{pmatrix} (1-\varepsilon)^2/4 & (1+\varepsilon)/2 \\ -\varepsilon(1+\varepsilon)/2 & -\varepsilon \end{pmatrix}^i \vec{g}^{(0)}. \tag{4.6}$$

Similarly, $\vec{g}^{(2i+1)} = \hat{A}\left(\hat{B}\hat{A}\right)^i \vec{g}^{(0)}$. To express the velocities $\vec{g}^{(2i)}$ and $\vec{g}^{(2i+1)}$ in terms of the eigenvalues of $\hat{B}\hat{A}$ we use the common rules of matrix algebra[6] and obtain

$$g_1^{(2i)} = c_1 \lambda_1^i + c_2 \lambda_2^i \tag{4.7}$$
$$g_2^{(2i+1)} = c_1' \lambda_1^i + c_2' \lambda_2^i, \tag{4.8}$$

provided the eigenvalues are real. The coefficients $c_{1/2}$ and $c'_{1/2}$ have to be determined from the initial velocities $\vec{g}^{(0)}$. The eigenvalues of $\hat{B}\hat{A}$ are obtained by the characteristic equation of the matrix given in (4.6):

$$\left(\frac{(1-\varepsilon)^2}{4} - \lambda\right)(-\varepsilon - \lambda) + \varepsilon\frac{(1+\varepsilon)^2}{4} = 0 \tag{4.9}$$

with the solutions

$$\lambda_{1/2} = \frac{1 - 6\varepsilon + \varepsilon^2}{8} \pm \sqrt{\frac{(1 - 6\varepsilon + \varepsilon^2)^2}{64} - \varepsilon^2}. \tag{4.10}$$

The eigenvalues are real if the expression under the square root is not negative, that is, if

$$\varepsilon \leq 7 - 4\sqrt{3} \approx 0.0718\ldots. \tag{4.11}$$

As stated above, the necessary condition for an infinite series of alternating collisions is an alternative series of collisions 0–1 and 1–2 which implies that the

---

[6] Given a matrix $\hat{M}$ with the complete set of eigenvalues λ_1 and λ_2 and eigenvectors $\vec{m}_1$ and $\vec{m}_2$. Then, if c_1 and c_2 are real constants and k is an integer constant, $\hat{M}^k(c_1\vec{m}_1 + c_2\vec{m}_2) = c_1\lambda_1^k\vec{m}_1 + c_2\lambda_2^k\vec{m}_2$. This rule is to be applied to the vector $\vec{g}_0$, decomposed into the eigenvectors of $\hat{B}\hat{A}$.

relative velocities $g_1^{(2i)}$ and $g_2^{(2i+1)}$ given by (4.7, 4.8) are all negative. The coefficients $c_{1/2}$ and $c'_{1/2}$ are determined by the first terms of the series by comparing the RHS of (4.7, 4.8, 4.2, 4.3) with $g_1^{(0)} = -1$ and $g_2^{(0)} = 0$:

$$g_1^{(0)} = c_1 + c_2 \qquad = -1 \qquad (4.12)$$

$$g_1^{(2)} = c_1 \lambda_1 + c_2 \lambda_2 \qquad = -(1-\varepsilon)^2/4 \qquad (4.13)$$

$$g_2^{(1)} = c'_1 + c'_2 \qquad = -(1+\varepsilon)/2 \qquad (4.14)$$

$$g_2^{(3)} = c'_1 \lambda_1 + c'_2 \lambda_2 \qquad = -(1+\varepsilon)\left(\varepsilon^2 - 6\varepsilon + 1\right)/8. \qquad (4.15)$$

We only need the solutions for c_1 and c'_1:

$$\begin{aligned} c_1 &= \frac{1}{\lambda_1 - \lambda_2}\left(\lambda_2 - \frac{(1-\varepsilon)^2}{4}\right) \\ c'_1 &= \frac{1}{2}\frac{1+\varepsilon}{\lambda_1 - \lambda_2}\left(\lambda_2 - \frac{(1-\varepsilon)^2}{4} + \varepsilon\right). \end{aligned} \qquad (4.16)$$

With $\lambda_1 - \lambda_2$ obtained by (4.10), one finds

$$c_1 = \frac{1}{2}\left(\lambda_2 - \frac{(1-\varepsilon)^2}{4}\right) \Big/ \sqrt{\left(\frac{1 - 6\varepsilon + \varepsilon^2}{8}\right)^2 - \varepsilon^2} \qquad (4.17)$$

$$c'_1 = \frac{1+\varepsilon}{4}\left(\lambda_2 - \frac{(1-\varepsilon)^2}{4} + \varepsilon\right) \Big/ \sqrt{\left(\frac{1 - 6\varepsilon + \varepsilon^2}{8}\right)^2 - \varepsilon^2}. \qquad (4.18)$$

From (4.10) we see that $\lambda_1 \geq \lambda_2 > 0$, that is, the values $c_1 + c_2 (\lambda_2/\lambda_1)^i$ change with increasing generation i monotonously from $c_1 + c_2$ to c_1. If we rewrite (4.7) in the form

$$\frac{g_1^{(2i)}}{\lambda_1^i} = c_1 + c_2 \left(\frac{\lambda_2}{\lambda_1}\right)^i, \qquad (4.19)$$

from (4.5) we obtain a criterion for the collapse

$$c_1 + c_2 < 0, \qquad c_1 < 0. \qquad (4.20)$$

Analogous arguments apply for the primed constants, yielding the second criterion

$$c'_1 + c'_2 < 0, \qquad c'_1 < 0. \qquad (4.21)$$

Due to (4.12, 4.14), the first conditions in (4.20, 4.21) always apply. From (4.17, 4.18) we see that the remaining two conditions are equivalent to

$$\lambda_2 - (1-\varepsilon)^2/4 < 0, \qquad \lambda_2 - (1-\varepsilon)^2/4 + \varepsilon < 0. \qquad (4.22)$$

The latter of both conditions implies the former. Inequality (4.22) applies since from (4.10) it follows that

$$\lambda_1 + \lambda_2 = -\varepsilon + \frac{(1-\varepsilon)^2}{4}, \quad \text{that is,} \quad \lambda_2 - \frac{(1-\varepsilon)^2}{4} + \varepsilon = -\lambda_1 < 0. \quad (4.23)$$

Therefore, it turns out that for all values of the coefficient of restitution which comply with condition (4.11), there exist initial velocities which cause an infinite series of collisions. During these collisions the total energy of the relative motion is dissipated, that is, the particles form a cluster. Moreover, it can be shown that the infinite number of collisions occur in finite time. This surprising effect is called *inelastic collapse* and has attracted much interest recently. Due to our understanding so far it has not become clear whether the inelastic collapse does exist in higher dimension.

In the calculation we have assumed that the coefficient of restitution is a constant. If we reconsider the final result (4.11) for particles with realistic interaction such as viscoelastic interaction, we notice that the inelastic collapse will never occur. Equating (3.42) and (4.11) we find a critical value for the relative velocity g under which the condition (4.11) is not provided anymore. Since the relative velocities decrease monotonously with time this critical value corresponds to a certain time after which the cluster dissolves. Hence, the inelastic collapse cannot exist for viscoelastic particles (see also Goldman *et al.*, 1998).

4.2 Collision cannon

We define the *collision cannon problem* as follows (Pöschel and Brilliantov, 2001): Given a linear chain of n inelastically colliding particles of masses m_i and radii R_i ($i = 0, \ldots, n$) with initial velocities $v_0 = v > 0$ and $v_i = 0$ ($i = 1, \ldots, n$) at initial positions $x_{i+1} > x_i$ and with $x_{i+1} - x_i > R_{i+1} + R_i$ (Fig. 4.2). The masses of the first and last particles m_0 and m_n are given and we ask two questions:

1. How should the masses in between be chosen to maximize the energy transfer from the first particle of the chain to the last one?
2. If the number of particles n is variable, how should n be chosen to maximize the after-collisional velocity v'_n of the last particle?

We want to restrict ourselves to the case when the energy is transmitted from the first to the last particle only by successive from-left-to-right collisions, that is, the particles are assumed to be far enough, so that multiple collisions between neighbouring particles do not occur.

We will discuss three different cases: elastic particles, particles which interact via a constant coefficient of restitution $\varepsilon = $ const. and viscoelastic particles for which the coefficient of restitution depends on the impact velocity, $\varepsilon = \varepsilon(g)$.

4.2.1 *Elastic particles*

After the collision of particle 0 with the resting particle 1 the velocity of particle 1 is

COLLISION CANNON

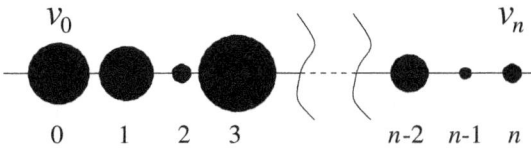

FIG. 4.2. Sketch of a collision cannon

$$v'_1 = \frac{2}{1 + (m_1/m_0)} v_0. \qquad (4.24)$$

For a chain of $n+1$ particles of masses $m_0, m_1, \ldots, m_n$ one has, analogously,

$$v'_n = 2^n \prod_{k=0}^{n-1} \left(1 + \frac{m_{k+1}}{m_k}\right)^{-1} v_0. \qquad (4.25)$$

For this system one easily finds that the choice

$$m_i = \sqrt{m_{i-1} m_{i+1}}, \qquad i = 2, \ldots, n-1 \qquad (4.26)$$

maximizes v'_n. If we fix m_0 and m_n, obviously, the mass distribution

$$m_k = \left(\frac{m_n}{m_0}\right)^{k/n} m_0 \qquad (4.27)$$

maximizes v'_n:

$$v'_n = \left(\frac{2}{1 + (m_n/m_0)^{1/n}}\right)^n v_0. \qquad (4.28)$$

The ratio v'_n/v_0 always increases with n and has the limit

$$\left(\frac{v'_n}{v_0}\right)_{n \to \infty} = \sqrt{\frac{m_0}{m_n}}, \qquad (4.29)$$

which implies that for $n \to \infty$

$$\frac{1}{2} m_n (v'_n)^2 = \frac{1}{2} m_0 v_0^2. \qquad (4.30)$$

Therefore, if the masses of the particles are chosen according to (4.27) the kinetic energy of the first particle is completely transferred to the last particle by a chain of infinite length.

As far as dissipative collisions are concerned an infinite chain cannot be optimal since in each collision energy is dissipated. Hence, we expect an optimum for the chain length for which the velocity of the last particle reaches its maximum.

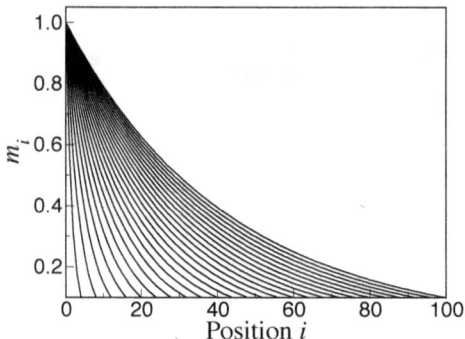

FIG. 4.3. Optimal mass distribution m_i, $i = 1,\ldots,n$ for the case $\varepsilon = $ const. Each of the lines shows the mass m_i over the index i for a specified chain length n. The masses of the first and last particles are fixed at $m_0 = 1$ and $m_n = 0.1$, respectively.

4.2.2 Particles with a constant coefficient of restitution

With the definition of the coefficient of restitution

$$\varepsilon \equiv \left|\frac{v_1' - v_0'}{v_1 - v_0}\right| \tag{4.31}$$

and from the conservation of momentum the equation analogous to (4.24) is obtained:

$$v_1' = \frac{1 + \varepsilon}{1 + (m_1/m_0)} v_0, \tag{4.32}$$

where we once again assume that the particle with velocity v_0 and mass m_0 hits a particle of mass m_1 at rest ($v_1 = 0$), which starts to move with the velocity v_1'. From the similarity of (4.24) and (4.32) it can be concluded that the optimal mass distribution for $\varepsilon = $ const. is identical to that for the elastic case (4.27). This means that the optimal mass distribution does not depend on the value of the coefficient of restitution. Each material irrespectively of its inelasticity requires the same optimal mass distribution. The velocity of the last particle in the chain reads

$$v_n' = \left(\frac{1 + \varepsilon}{1 + (m_n/m_0)^{1/n}}\right)^n v_0. \tag{4.33}$$

Figure 4.3 shows the optimal mass distributions for different chain lengths n. The mass of the first particle is $m_0 = 1$ and of the last particle $m_n = 0.1$.

In the next section we will focus on particles which interact via a velocity dependent coefficient of restitution. Since the velocities of the particles vary along the chain we do not characterise the dissipation of the colliding spheres by the

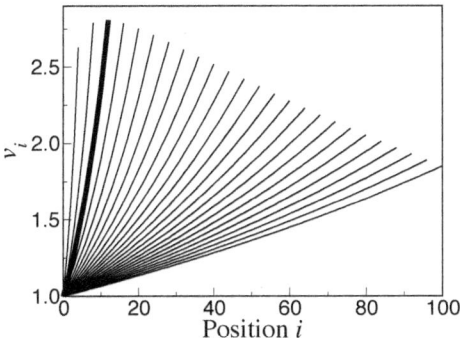

FIG. 4.4. Velocity distribution of the particles in chains with the optimal mass distribution (shown in Fig. 4.3) according to (4.27). Each of the lines shows the velocity v_i over the index i for a specified chain length n. The dissipative constant is $b = 0.032$. The last particle reaches its maximal velocity for chain length $n^* = 12$ (in bold). The velocity of the first particle of the chain is $v_0 = 1$.

coefficient of restitution itself but instead we define a dissipative constant b (see (4.37) below). For the case of $\varepsilon = $ const. it is defined as $b \equiv (1 - \varepsilon)$.

Contrary to the mass distribution, the corresponding velocity distributions do depend on the value of the coefficient of restitution ε. Figure 4.4 shows the velocity distribution for $b = 0.032$.

For the case of dissipative collisions, the ratio v'_n/v_0 does not monotonously increase with n, but rather it has an extremum which shifts to smaller chain lengths with increasing dissipative parameter b. The optimal value of n, which maximises v'_n/v_0 reads

$$n^* = \frac{\log(m_n/m_0)}{\log(x_0)} \quad (4.34)$$

where x_0 is the solution of the equation

$$1 + x_0 = (1 + \varepsilon) x_0^{x_0/(1+x_0)} \quad (4.35)$$

Correspondingly, the extremal velocity ratio reads

$$\left(\frac{v'_n}{v_0}\right)_{\text{extr}} = \left(\frac{1+\varepsilon}{1+x_0}\right)^{n^*} = \left(\frac{m_0}{m_n}\right)^{x_0/(1+x_0)} \quad (4.36)$$

Figure 4.5 shows the dependence of the extremal n^* on the coefficient of restitution. Due to (4.35) $1 > x_0 > 0$ (for $m_0 > m_n$) does not depend on the masses, but only on the coefficient of restitution. Therefore, (4.36) predicts unlimited increase of the final velocity with increasing mass ratio m_0/m_n.

Exercise 4.1 *Derive (4.34) and (4.36)!*

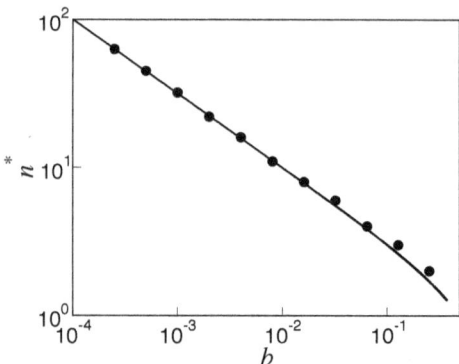

FIG. 4.5. Optimal chain length n^*, which provides the maximal transmission of energy along the chain with the fixed first and last masses, as a function of the dissipative parameter $b = 1 - \varepsilon$. The line shows the prediction of (4.34), with x_0 found numerically. Points refer to the results of a direct numerical optimization of the masses in the chain.

4.2.3 Viscoelastic particles

To simplify the understanding we assume that all particles are of the same radius R, but have different masses. The coefficient of restitution depends on the impact velocity as given by (3.22) where we only keep the first velocity-dependent term. We abbreviate

$$\varepsilon = 1 - b\frac{v_{ij}^{1/5}}{\left(m_{ij}^{\text{eff}}\right)^{2/5}} \quad \text{with} \quad b \equiv C_1 \frac{3A}{2}\left(\frac{2}{3}\frac{Y\sqrt{R/2}}{1-\nu^2}\right)^{2/5}. \tag{4.37}$$

Note that a collision with $\varepsilon =$ const. and a certain dissipative constant $b = 1 - \varepsilon$ corresponds to a viscoelastic collision with the same b and with $m_{ij}^{\text{eff}} = 1$, $v_{ij} = 1$, that is, for this choice both collisions have equal value of ε.

The equation analogous to (4.24, 4.32) reads, for viscoelastic particles,

$$v'_{k+1} = \frac{2 - bv_k^{1/5}\left(m_{k+1,k}^{\text{eff}}\right)^{-2/5}}{1 + m_{k+1}/m_k} v_k, \tag{4.38}$$

where $m_{k+1,k}^{\text{eff}} \equiv m_{k+1}m_k/(m_{k+1} + m_k)$. The masses m_k, $k = 1,\ldots, n-1$ which maximize v'_n can be determined numerically, see Fig. 4.6. For small chain length, the optimal mass distribution is very close to that for the elastic chain shown in Fig. 4.3. Again we find a monotonously decaying function. For larger chain length n or larger dissipation b, however, the mass distribution is a non-monotonous function. Note that the mass distribution and velocity distribution are related by (4.38).

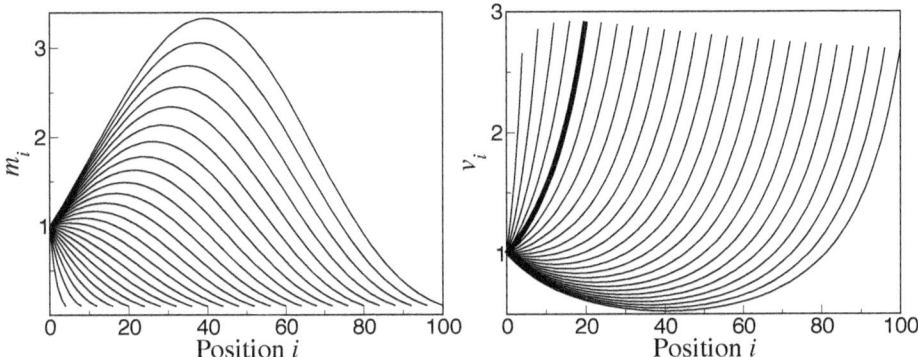

FIG. 4.6. Left: Optimal mass distribution m_i, $i = 1, \ldots, n$, for the case of viscoelastic particles with the coefficient of restitution given by (4.37) with $b = 2 \times 10^{-3}$. Each of the lines shows the mass m_i over the index i for a specified chain length n. The masses of the first and last particles are $m_0 = 1$ and $m_n = 0.1$. Right: Velocity distribution in chains with optimal mass distribution. The last particle reaches its maximal velocity for the chain length $n^* = 20$ (in bold). The velocity of the first particle of the chain is $v_0 = 1$.

A simple physical explanation can be given for the appearance of a maximum in the mass distribution (and correspondingly minimum in the velocity distribution): there are two different mechanisms for loss of energy during the transfer from the first to the last particle, that is (a) *viscous loss* due to the dissipative nature of the collisions, and (b) *inertial loss* due to mismatch of subsequent masses, which causes incomplete transfer of momentum even for elastic collision when the masses differ from each other. The inertial loss of the collision between particles $i+1$ and i is given by the energy which remains in the ith particle after the collision with the $(i+1)$th particle:

$$\Delta E_{in}^{(i)} = \frac{m_i}{2} (v_i')^2 = \frac{m_i}{2} \left(\frac{m_{i+1} - m_i}{m_{i+1} + m_i} \right)^2 v_i^2 . \qquad (4.39)$$

As (4.37) shows, the coefficient of restitution increases with decreasing impact velocity and increasing masses of colliding particles. Thus, slowing down particles by increasing their masses in the inner part of the chain leads to decrease of the viscous losses of the energy transfer. The larger the masses in the middle and the smaller their velocities, the less energy is lost due to dissipation. On the other hand, since masses m_0 and m_n are fixed, very large masses in the middle of the chain will cause large mass mismatch of the subsequent masses and thus large inertial losses. The optimal mass distribution, minimizing the *total* loss compromises (dictated by b) between these two opposite tendencies.

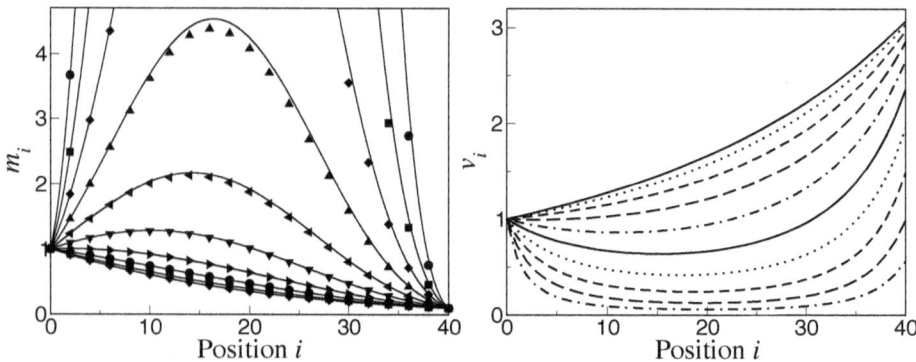

FIG. 4.7. Left: Optimal mass distribution in chains of viscoelastic particles of length $n = 40$ for different values of the dissipative parameter b. Lines: analytical results (Pöschel and Brilliantov, 2001), points: numerical optimization (from top to bottom: •, $b = 0.128$; ■, $b = 0.064$; ♦, $b = 0.032$; ▲, $b = 0.016$; ◀, $b = 0.008$; ▼, $b = 0.004$; ▶, $b = 0.002$, etc.). Right: the corresponding velocity distribution. Lines from top to bottom: $b = 2.5 \times 10^{-4}$, 5×10^{-4}, 0.001, 0.002, 0.004, 0.008, 0.016, 0.032, 0.064, 0.128. The velocity distribution is obtained from the mass distribution according to (4.38).

For long chain with small mass gradients an analytical theory based on the continuum approximation for the mass distribution (Pöschel and Brilliantov, 2001) can be developed. The derivation is, however, lengthy and the solution is not straightforward; the interested reader will find the detail in (Pöschel and Brilliantov, 2001). In Fig. 4.7 we express the results of the numerical optimization for the mass distribution. This result is illustrated together with predictions of the analytical theory for a chain of length 40 with different damping. For optimal mass distributions again we see the pronounced maximum in the middle of the chain.

In contrast, for the case $\varepsilon = $ const. the viscous losses do not depend on the impact velocity. This means that only minimization of the inertial losses, caused by the mass gradient, may play a role in the optimization of the mass distribution. Thus, only a monotonous mass distribution with minimal mass gradients along the chain may be optimal for the case $\varepsilon = $ const.

Our main intention of presenting the results in this section was to demonstrate that the detailed consideration of the two-particle interaction affects the dynamical behaviour of the collision cannon which is a simple one-dimensional few-body system. Although we have not presented here the full analytical theory, which can be found in Pöschel and Brilliantov (2001), the numerical results already show that the case $\varepsilon = $ const. is qualitatively different from the case $\varepsilon = \varepsilon(g)$ for viscoelastic particles: for $\varepsilon = $ const. the mass of each particle in a chain with optimal energy transmission is given by the geometric average of its neighbours, that is, the distribution of the masses of the spheres is a monotonous, exponentially decreasing function. This function is independent of ε where the limiting case of elastically colliding particles ($\varepsilon = 1$) is included.

The situation changes qualitatively if we assume that the chain consists of viscoelastic spheres for which the coefficient of restitution depends on the impact velocity. Here, the optimal mass distribution which leads to maximum energy transfer is not necessarily a monotonous function. Depending on the chain length n and on the material parameters of the spheres it may reveal a pronounced maximum.

In the following chapters we will focus on granular gases as thermodynamically large systems. We will find that the impact-velocity dependence of the coefficient of restitution, as it is given for viscoelastic particles, may lead to qualitatively different behaviour as compared to systems with a constant coefficient of restitution. The system investigated here clearly demonstrates that the impact-velocity dependence of the coefficient of restitution may play a crucial role even for relatively small and simple systems.

Therefore, in general, the assumption of a constant coefficient of restitution is an approximation whose justification cannot be assumed *a priori* but has to be checked for each particular application.

SUMMARY

In this chapter we have investigated in detail the pairwise particle interaction as a proposition for the description of granular gas dynamics. The essential results are:

1. The dissipative collision of particles is governed by the coefficient of (normal) restitution ε which relates the normal component of the relative particle velocities
$$\vec{g} \equiv v_{12} \cdot \vec{e} \equiv (\vec{v}_1 - \vec{v}_2) \cdot \vec{e}$$
before and after a collision:
$$\varepsilon \equiv |\vec{g}'|/|\vec{g}| \quad \text{with} \quad 0 \leq \varepsilon \leq 1.$$

2. From conservation of momentum, the collision rules
$$\vec{v}_1' = \vec{v}_1 - \frac{m^{\text{eff}}}{m_1}(1+\varepsilon)(\vec{v}_{12} \cdot \vec{e})\vec{e}$$
$$\vec{v}_2' = \vec{v}_2 + \frac{m^{\text{eff}}}{m_2}(1+\varepsilon)(\vec{v}_{12} \cdot \vec{e})\vec{e}.$$
are obtained.

3. In literature (beginning from secondary school textbooks) it is frequently assumed that $\varepsilon = $ const. This assumption significantly simplifies the analysis of granular gases, however, it is neither in agreement with the experiment nor with basic mechanics of collisions. Instead, ε is a function of the impact velocity g.

4. For the case of viscoelastic spherical particles the coefficient of restitution reads
$$\varepsilon = \varepsilon(g) = 1 - c_1 g^{1/5} + c_2 g^{2/5} \mp \cdots$$
with c_i being (known) material constants. This result agrees well with experiments.

5. Consideration of simple few-particle systems, such as three particles on a line or the collision cannon, yields qualitatively different results for the cases $\varepsilon = $ const. (idealized particles) and $\varepsilon = \varepsilon(g)$ (realistic particles). Therefore, it can be expected that for granular gases the results would be qualitatively different too.

II. Granular Gases – Velocity Distribution Function

A granular gas of almost elastic particles ($\varepsilon \lesssim 1$) that starts its evolution from a flux-free spatially uniform state remains homogeneous during the first stage of its evolution, which is called *the homogeneous cooling state*. In this state, a granular gas resembles an ordinary molecular gas of elastic particles, except that the average kinetic energy decays steadily due to inelastic collisions. For a given granular gas, the homogeneous cooling state is entirely described by the velocity distribution function. The main part of the distribution is close to the Maxwell distribution. Its deviation from the Maxwell distribution is characterized by the coefficients of the Sonine polynomials expansion. For the simplifying assumption $\varepsilon = \text{const.}$ the deviation from the Maxwell distribution is time-independent, whereas for granular gases of viscoelastic particles the deviation depends explicitly on time. This property allows to define the age of a granular gas. The high-velocity tail of the velocity distribution function differs drastically from the Maxwell distribution: instead of $\sim \exp\left(-Av^2\right)$ it obeys the law $\sim \exp\left(-Av\right)$.

5

COOLING GRANULAR GAS — HAFF'S LAW

During the first stage of its evolution a force-free, initially uniform granular gas remains homogeneous, while the mean velocity of its particles decreases continuously. This stage is termed homogeneous cooling state. In this regime the granular temperature decays according to Haff's law.

5.1 Homogeneous cooling state

We consider a homogeneous and isotropic granular gas of infinite extension in the absence of external forces. A volume V of the gas contains N particles, that is, the number density is $n \equiv N/V$. We assume that N is a large number, which allows for the application of the methods of statistical mechanics. Just as in the case of molecular gases the grains move irregularly with randomly distributed velocities. If there are no macroscopic flows, their average velocity is zero, $\langle \vec{v} \rangle = 0$. Analogously to molecular gases the intensity of this irregular motion may be characterized by the average square of the particle velocities. Thus, it is natural to introduce the *granular temperature*[7] of the gas as the average kinetic energy of the grains, just as the temperature in the theory of molecular gases:

$$\frac{3}{2} T = \left\langle \frac{1}{2} m \vec{v}^2 \right\rangle = \frac{1}{N} \sum_{i=1}^{N} \frac{1}{2} m \vec{v}_i^2 \,, \tag{5.1}$$

where m is the mass of a grain and $\vec{v}_i$ is the velocity of particle i. Over a period of time, dissipative particle collisions cause an enduring decay of the temperature of the granular gas. The state of the force-free granular gas when its temperature decays continuously but when its spatial homogeneity is preserved is called *homogeneous cooling state*. Except for Chapters 24–27, throughout this book we deal with granular gases in the regime of homogeneous cooling.

The qualitative description of the temperature decay may be obtained by simple arguments which do not require the knowledge of the velocity distribution function. The rigorous derivation of the evolution of temperature and of the velocity distribution function, which may noticeably deviate from the Maxwell distribution, is given in Chapters 8 and 9.

---

[7]In the following we call the granular temperature as just temperature. In the context of granular gases always the value given by (5.1) is meant but never the temperature of the grain material as it could be measured by a thermometer.

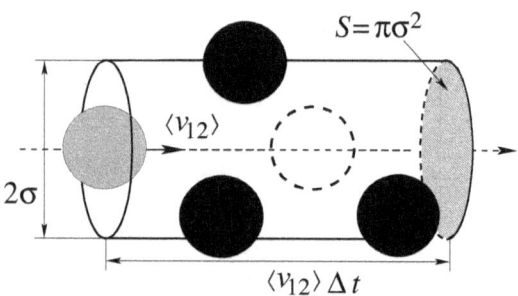

FIG. 5.1. The collision cylinder of the grey particle. Only particles whose centres are located inside the collision cylinder of radius σ and length $\langle v_{12} \rangle \Delta t$ collide with the grey particle.

5.2 Haff's law for the evolution of the granular temperature

The decay of temperature ΔT during a certain time span Δt corresponds, obviously, to the according loss of mechanical energy due to dissipative collisions. It may be calculated as the product of the average loss of kinetic energy in one collision and the average number of collisions which occur during Δt. The coefficient of restitution relates the relative particle velocities before and after a collision, $v'_{12} = -\varepsilon v_{12}$. Hence, on average a collision causes the energy loss

$$\frac{1}{2} m^{\text{eff}} \langle v'^2_{12} - v^2_{12} \rangle = -\frac{1}{2} m^{\text{eff}} \langle v^2_{12} \rangle \left(1 - \varepsilon^2\right) \propto -\left(1 - \varepsilon^2\right) T . \qquad (5.2)$$

The product of the squared average relative velocity and the effective mass $\langle v^2_{12} \rangle m^{\text{eff}}/2$ characterizes the kinetic energy of the *relative* motion. On average it is of the same order as temperature.

The average number of collisions $\nu(\Delta t)$ which occur during Δt may be estimated as follows: consider a particle moving with the average relative velocity $\langle v_{12} \rangle$, while all other particles are assumed to be fixed scatterers (see Fig. 5.1). During the time interval Δt this particle collides with all particles whose centres are located inside the collision cylinder of length $\langle v_{12} \rangle \Delta t$. Its cross-section equals $\pi \sigma^2$, where $\sigma = 2R$ is the particle diameter. The average number of scatterers in the collision cylinder equals the number of collisions during Δt:

$$\nu(\Delta t) = \left[\pi \sigma^2 \langle v_{12} \rangle \Delta t\right] n \propto n \sigma^2 \sqrt{T} \Delta t , \qquad (5.3)$$

where the factor in square brackets is the volume of the collision cylinder. From (5.2, 5.3) we obtain the temperature decay rate

$$\frac{dT}{dt} \simeq \frac{\Delta T}{\Delta t} \propto \frac{1}{\Delta t} \frac{m^{\text{eff}}}{2} \left(\langle v'^2_{12} \rangle - \langle v^2_{12} \rangle\right) \nu(\Delta t) \propto -n \sigma^2 \left(1 - \varepsilon^2\right) T^{3/2} . \qquad (5.4)$$

The solution of (5.4) yields Haff's law for the evolution of temperature of a granular gas of initial temperature T_0 (Haff, 1983)

$$T(t) = \frac{T_0}{(1+t/\tau_0)^2} \quad \text{where} \quad \tau_0^{-1} \propto n\sigma^2 \left(1-\varepsilon^2\right) \sqrt{T_0}. \tag{5.5}$$

This result was obtained for the simplified collision model where $\varepsilon = \text{const.}$ is assumed. For viscoelastic particles the coefficient of restitution depends on the impact velocity, $g \sim \langle v_{12} \rangle$, which implies that ε depends on temperature:

$$\varepsilon \simeq 1 - c_1 g^{1/5} \longrightarrow 1 - \varepsilon^2 \propto g^{1/5} \propto T^{1/10}. \tag{5.6}$$

Substituting (5.6) into (5.4) we obtain the rate equation

$$\frac{dT}{dt} \propto -n\sigma^2 T^{8/5}, \tag{5.7}$$

which yields for the evolution of temperature

$$T(t) = \frac{T_0}{(1+t/\tau_0')^{5/3}}. \tag{5.8}$$

This result, obtained by Schwager and Pöschel (1998), means that a gas of viscoelastic particles cools down noticeably slower than predicted by Haff's law for simplified particles with $\varepsilon = \text{const.}$

These simple estimates describe the main physics of granular gas cooling. Implicitly we have assumed that the kinetic state of the ensemble of grains is exhaustively described by the average of the squared velocities, that is, by the second moment of the velocity distribution function. In general, however, the kinetics of granular gases depend not only on the second moment but on the full distribution function of the particle velocities, that is, on all even (non-vanishing) moments,

$$\frac{dT}{dt} = F\left(T, \langle v^4 \rangle, \langle v^6 \rangle, \ldots\right). \tag{5.9}$$

We will show below that for the simplified model $\varepsilon = \text{const.}$ the decay rate of temperature depends only on temperature itself, that is, on the second moment of the velocity distribution function, while in general it depends on higher moments too.

Exercise 5.1 *Estimate the ratio of triple and binary collisions for a gas of soft particles of radius R and mass m which interact with a repulsive potential $\Phi(r) = A\xi^\alpha$, where ξ is the compression of particles (see Chapter 3)! The gas has temperature T and number density n.*

Exercise 5.2 *Using the results of the Exercise 5.1 estimate the temperature dependence of the ratio of triple and binary collisions for a gas of viscoelastic particles!*

6

BOLTZMANN EQUATION

The evolution of the velocity distribution function is governed by the Boltzmann equation. We derive the Boltzmann equation for the homogeneous cooling granular gas and discuss the properties of the collision integral.

6.1 Velocity distribution function

The fundamental theoretical tool of the kinetic theory of gases is the Boltzmann equation. It was derived by Ludwig Boltzmann more than 100 years ago (Boltzmann, 1896). Later, this equation was modified by Enskog to account for finite-volume effects in dense gases. The Boltzmann–Enskog equation relates the velocity distribution function of the gas particles to the microscopic properties of particle collisions. The velocity distribution function $f(\vec{r}, \vec{v}, t)$ is defined such that

$$f(\vec{r}, \vec{v}, t) \, d\vec{r} \, d\vec{v} \tag{6.1}$$

gives the number of particles in an infinitesimal volume $d\vec{r}$ located at the point $\vec{r}$ whose velocities belong to an infinitesimal volume in the velocity space $d\vec{v}$ centred around $\vec{v}$. From the definition of this function follows its normalization

$$\int_{-\infty}^{\infty} dv_x \int_{-\infty}^{\infty} dv_y \int_{-\infty}^{\infty} dv_z \iiint_{x,y,z \in V} dx \, dy \, dz \, f(\vec{r}, \vec{v}, t) = N. \tag{6.2}$$

Each of the particles has a certain velocity and is located at a certain position. Hence, if we integrate over the full phase space, we obtain the total number of particles N. For simplicity, we prefer the compact notation of this equation

$$\int f(\vec{r}, \vec{v}, t) \, d\vec{r} \, d\vec{v} = N. \tag{6.3}$$

Whenever an integration sign without explicit limits occurs, integration over the entire phase space indicated by the corresponding differentials (here $d\vec{r}$ and $d\vec{v}$) is meant.

For a homogeneous flow free system $f(\vec{r}, \vec{v}, t)$ is independent of the spatial variable $\vec{r}$. The velocity distribution function has the properties

$$\int f(\vec{v},t) \, d\vec{v} = n$$

$$\int \vec{v} f(\vec{v},t) \, d\vec{v} = n \langle \vec{v} \rangle = 0 \qquad (6.4)$$

$$\int \frac{1}{2} m v^2 f(\vec{v},t) \, d\vec{v} = n \left\langle \frac{1}{2} m v^2 \right\rangle = \frac{3}{2} n T(t).$$

Due to (6.4) the moments of f are related with certain macroscopic properties of the gas, namely with its number density n, its flow velocity (which is zero here), and its temperature T.

6.2 Direct and inverse collisions

First, we consider a very dilute homogeneous granular gas where effects due to the finite volume of the particles may be neglected. This corresponds to small packing fraction $\eta \equiv \frac{\pi}{6} n \sigma^3$ when the fraction of the total volume which is occupied by particle material is negligible compared to the gas volume V. Later, the generalization to finite packing fraction and non-homogeneous systems will be given. There exist many different ways to derive the Boltzmann equation (e.g. Resibois and de Leener, 1977; Schram, 1991), we give here a simple intuitive derivation, which, however, elicits the basic physics of this equation.

During a certain time interval Δt the number of particles $f(\vec{v}_1, t) \, d\vec{r}_1 \, d\vec{v}_1$ in a small volume of the phase space $(d\vec{r}_1 \, d\vec{v}_1)$ located at $(\vec{r}_1, \vec{v}_1)$ changes due to particle collisions. *Direct* collisions are collisions in which particles from this velocity interval are involved. After such collisions these particles, in general, will move with different velocities, that is, they will leave the velocity interval. Hence, direct collisions reduce the number of particles in the small phase-space volume $(d\vec{r}_1 \, d\vec{v}_1)$. Contrary, *inverse* (or *restituting*) collisions are collisions in which particles are involved that initially do not belong to the velocity interval $(\vec{v}_1, \vec{v}_1 + d\vec{v}_1)$. After the collision, however, they enter this interval. Hence, inverse collisions increase the number of particles in the considered phase-space volume. To obtain an equation which characterises the evolution of the velocity distribution function we have to quantify the effects of direct and inverse collisions.

Direct collisions occur for pairs of particles of velocities $\vec{v}_1$ and $\vec{v}_2$. The unit vector $\vec{e}$ specifies the geometry of the impact, see Fig. 6.1. According to (2.7) the collision law for identical particles reads

$$\boxed{\begin{aligned} \vec{v}_1' &= \vec{v}_1 - \frac{1+\varepsilon}{2} (\vec{v}_{12} \cdot \vec{e}) \vec{e} \\ \vec{v}_2' &= \vec{v}_2 + \frac{1+\varepsilon}{2} (\vec{v}_{12} \cdot \vec{e}) \vec{e} \end{aligned}} \qquad (6.5)$$

The initial velocities $\vec{v}_1$, $\vec{v}_2$ transform into $\vec{v}_1'$, $\vec{v}_2'$, that is, they reduce the number of particles in the considered phase-space volume $(d\vec{r}_1 \, d\vec{v}_1)$. To obtain the rate

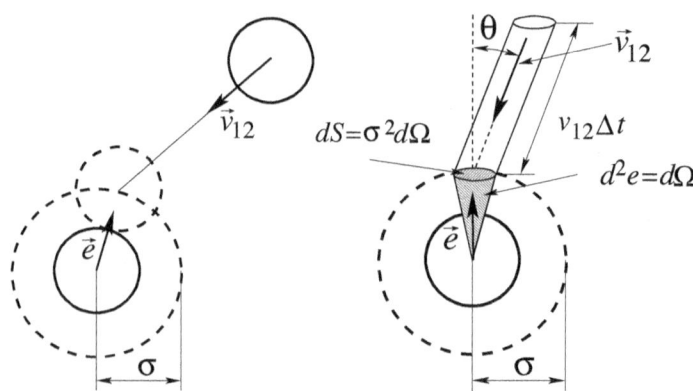

FIG. 6.1. The collision cylinder for the scattering of particles with velocity $\vec{v}_1$ at particles with velocity $\vec{v}_2$. The unit vector $\vec{e} \equiv \vec{r}_{12}/r_{12}$ ($r_{12} = \sigma$ at the instant of the collision) gives the normal to the scattering surface. The collision cylinder accounts for all possible trajectories of the incoming particle that hit the scattering surface in a small spot around $\vec{e}$. This spot is characterized by the infinitesimal angle $d\Omega \equiv d\vec{e}$.

of direct collisions, we consider the scattering of particles with velocities $\vec{v}_1$ at particles with velocity $\vec{v}_2$. These particles meet at relative velocity $\vec{v}_{12} \equiv \vec{v}_1 - \vec{v}_2$. The number of scatterers in the small volume $d\vec{r}_1$ is $f(\vec{v}_2, t)\, d\vec{v}_2\, d\vec{r}_1$, and in the dilute gas scattering at all the scatterers occurs independently. For each of the scatterers there exists an individual collision cylinder. This is shown in Fig. 6.1 and accounts for all possible collisions that may occur within the infinitesimal solid angle $d\Omega \equiv d\vec{e}$ around $\vec{e}$. The cross-section of this cylinder is $\sigma^2\, d\Omega = \sigma^2\, d\vec{e}$, its length is $v_{12}\Delta t$ and its volume is

$$dV_{cc} = \sigma^2\, d\vec{e}\, v_{12} \Delta t \cos\theta = \sigma^2\, |\vec{v}_{12} \cdot \vec{e}|\, \Delta t\, d\vec{e}. \qquad (6.6)$$

According to the definition of the distribution function the number of particles of velocity $\vec{v}_1$ in the collision cylinder is $f(\vec{v}_1, t)\, d\vec{v}_1 dV_{cc}$. Thus, we obtain the number of direct collisions ν^- that occur during Δt and whose geometry is specified by the unit vector $\vec{e}$:

$$\nu^-(\vec{v}_1, \vec{v}_2, \vec{e}, \Delta t) = f(\vec{v}_1, t)\, d\vec{v}_1\, d\vec{r}_1\, f(\vec{v}_2, t)\, d\vec{v}_2\, \sigma^2\, |\vec{v}_{12} \cdot \vec{e}|\, \Delta t\, d\vec{e}. \qquad (6.7)$$

Inverse collisions are defined as collisions of particles with $\vec{v}_1''$, $\vec{v}_2''$ and with the geometry of the collision specified by $\vec{e}$, which yield after the impact the velocities $\vec{v}_1$ and $\vec{v}_2$. Obviously, such collisions supply particles to the phase-space volume of interest. The collision rule yields

$$\vec{v}_1 = \vec{v}_1'' - \frac{1+\varepsilon}{2}(\vec{v}_{12}'' \cdot \vec{e})\,\vec{e}, \qquad \vec{v}_2 = \vec{v}_2'' + \frac{1+\varepsilon}{2}(\vec{v}_{12}'' \cdot \vec{e})\,\vec{e}, \qquad (6.8)$$

which implies the relation between the normal components of the relative velocity

DIRECT AND INVERSE COLLISIONS

$$\vec{v}_{12} \cdot \vec{e} = -\varepsilon \vec{v}''_{12} \cdot \vec{e} \quad \text{or} \quad \vec{g} = -\varepsilon \vec{g}'''. \tag{6.9}$$

Equation (6.9) defines the normal component of the impact velocity $\vec{g} \equiv (\vec{v}_{12} \cdot \vec{e})\vec{e}$. With this relation (6.8) reads

$$\boxed{\begin{aligned} \vec{v}''_1 &= \vec{v}_1 - \frac{1+\varepsilon}{2\varepsilon}(\vec{v}_{12} \cdot \vec{e})\vec{e} \\ \vec{v}''_2 &= \vec{v}_2 + \frac{1+\varepsilon}{2\varepsilon}(\vec{v}_{12} \cdot \vec{e})\vec{e} \end{aligned}} \tag{6.10}$$

Equations (6.10) may also be considered as a transformation, $\vec{v}''_1, \vec{v}''_2 \to \vec{v}_1, \vec{v}_2$, whose Jacobian will be needed below. Choosing $\vec{e}$ along the z-axis, that is, $\vec{e} = (0, 0, 1)$ we write

$$\begin{aligned} v''_{1,x} &= v_{1,x}; & v''_{2,x} &= v_{2,x}; \\ v''_{1,y} &= v_{1,y}; & v''_{2,y} &= v_{2,y}; \\ v''_{1,z} &= v_{1,z} - \frac{1+\varepsilon}{2\varepsilon}(v_{1,z} - v_{2,z}); & v''_{2,z} &= v_{2,z} + \frac{1+\varepsilon}{2\varepsilon}(v_{1,z} - v_{2,z}). \end{aligned} \tag{6.11}$$

For the case $\varepsilon = \text{const.}$ the Jacobian reads

$$\frac{D(\vec{v}''_1, \vec{v}''_2)}{D(\vec{v}_1, \vec{v}_2)} = \text{abs} \begin{vmatrix} 1 & 0 & 0 & 0 & 0 & 0 \\ 0 & 1 & 0 & 0 & 0 & 0 \\ 0 & 0 & (\varepsilon-1)/2\varepsilon & 0 & 0 & (\varepsilon+1)/2\varepsilon \\ 0 & 0 & 0 & 1 & 0 & 0 \\ 0 & 0 & 0 & 0 & 1 & 0 \\ 0 & 0 & (\varepsilon+1)/2\varepsilon & 0 & 0 & (\varepsilon-1)/2\varepsilon \end{vmatrix} = \frac{1}{\varepsilon}. \tag{6.12}$$

The number of inverse collisions, which increase the amount of particles in the phase-space volume of interest, may be derived using the same reasoning as above:

$$\nu^+(\vec{v}''_1, \vec{v}''_2, \vec{e}, \Delta t) = f(\vec{v}''_1, t) d\vec{v}''_1 d\vec{r}_1 f(\vec{v}''_2, t) d\vec{v}''_2 \sigma^2 |\vec{v}''_{12} \cdot \vec{e}| \Delta t d\vec{e}. \tag{6.13}$$

It is convenient to consider $\vec{v}''_1, \vec{v}''_2$ in (6.13) as functions of $\vec{v}_1, \vec{v}_2$ according to the transformation (6.10). Then

$$d\vec{v}''_1 d\vec{v}''_2 = \frac{D(\vec{v}''_1, \vec{v}''_2)}{D(\vec{v}_1, \vec{v}_2)} d\vec{v}_1 d\vec{v}_2. \tag{6.14}$$

With (6.14, 6.9) we recast (6.13) into the form

$$\nu^+(\vec{v}''_1, \vec{v}''_2, \vec{e}, \Delta t) = \chi f(\vec{v}''_1, t) f(\vec{v}''_2, t) \sigma^2 |\vec{v}_{12} \cdot \vec{e}| d\vec{v}_1 d\vec{v}_2 d\vec{r}_1 \Delta t d\vec{e}, \tag{6.15}$$

where

$$\chi \equiv \frac{|\vec{g}'''|}{|\vec{g}|} \frac{D(\vec{v}''_1, \vec{v}''_2)}{D(\vec{v}_1, \vec{v}_2)} = \frac{1}{\varepsilon(|\vec{g}'''|)} \frac{D(\vec{v}''_1, \vec{v}''_2)}{D(\vec{v}_1, \vec{v}_2)}. \tag{6.16}$$

The factor $|\vec{g}'''|/|\vec{g}|$ corresponds to the ratio of the lengths of the collision cylinders for the inverse and the direct collisions. With (6.10, 6.16) $\nu^+(\vec{v}''_1, \vec{v}''_2, \vec{e}, \Delta t)$

may be expressed as a function of the velocities of the direct collision $\vec{v}_1$ and $\vec{v}_2$. According to (6.12) for the case $\varepsilon = $ const. the factor χ reads

$$\chi = \frac{1}{\varepsilon^2}. \tag{6.17}$$

In general, χ is a complicated time-dependent function. For the case of viscoelastic particles it will be derived in Section 9.2.

6.3 Collision integral and Boltzmann–Enskog equation

The rates of direct and inverse collisions for particles with velocities $\vec{v}_1$ and $\vec{v}_2$ with their collision geometry characterized by the unit vector $\vec{e}$ are given by (6.7, 6.15). Integration over all possible velocities $\vec{v}_2$ and all directions of the unit vector $\vec{e}$ that correspond to an impact yields the total amount of collisions, which alter the number of particles in the small phase-space volume $(d\vec{r}_1, d\vec{v}_1)$. For the integration over the directions of the unit vector $\vec{e}$ we note that only directions with $\vec{v}_{12} \cdot \vec{e} < 0$ for the direct collision and $\vec{v}''_{12} \cdot \vec{e} < 0$ for the inverse collision lead to an impact. Thus, we can write for the increment of the number of particles in the considered phase-space volume during the time interval Δt

$$\Delta\left[f\left(\vec{v}_1, t\right)\right] d\vec{v}_1 d\vec{r}_1 = \int d\vec{v}_2 d\vec{e}\, \Theta\left(-\vec{v}''_{12} \cdot \vec{e}\right) \nu^+ \left(\vec{v}''_1, \vec{v}''_2, \vec{e}, \Delta t\right) \\ - \int d\vec{v}_2 d\vec{e}\, \Theta\left(-\vec{v}_{12} \cdot \vec{e}\right) \nu^- \left(\vec{v}_1, \vec{v}_2, \vec{e}, \Delta t\right) \tag{6.18}$$

where $\Theta(x)$ is the Heaviside step-function

$$\Theta(x) \equiv \begin{cases} 1 & \text{for} \quad x \geq 0 \\ 0 & \text{for} \quad x < 0. \end{cases} \tag{6.19}$$

According to (6.9)

$$\Theta\left(\vec{v}''_{12} \cdot \vec{e}\right) = \Theta\left(-\vec{v}_{12} \cdot \vec{e}\right), \tag{6.20}$$

where the property $\Theta(kx) = \Theta(x)$ for any $k > 0$ was used.

Changing the variable $\vec{e} \to -\vec{e}$ in the first integral in the RHS of (6.18) does not alter the transformation $\vec{v}''_1, \vec{v}''_2 \to \vec{v}_1, \vec{v}_2$, but makes the two step functions in (6.18) identical. Now we substitute ν^- and ν^+ as given by (6.7, 6.15) into (6.18), divide both sides of the equation by $d\vec{v}_1 d\vec{r}_1 \Delta t$ and take the limit $\Delta t \to 0$. We arrive at the Boltzmann equation

$$\frac{\partial}{\partial t} f\left(\vec{v}_1, t\right) = \sigma^2 \int d\vec{v}_2 \int d\vec{e}\, \Theta\left(-\vec{v}_{12} \cdot \vec{e}\right) |\vec{v}_{12} \cdot \vec{e}| \times \\ \times \left[\chi f\left(\vec{v}''_1, t\right) f\left(\vec{v}''_2, t\right) - f\left(\vec{v}_1, t\right) f\left(\vec{v}_2, t\right)\right] \\ \equiv I(f, f) \tag{6.21}$$

which defines the collision integral $I(f, f)$. While the derivation of the Boltzmann equation presented here is intuitive, there are some problems which we did not consider so far. In particular, we did not explicitly discuss the physical approximations which were used. The most severe approximation has been made deriving the rates of direct and inverse collisions. We counted independently the number of scatterers and scattered particles by simply taking the product of two distribution functions $f(\vec{v}_1, t)$ and $f(\vec{v}_2, t)$ or $f(\vec{v}_1'', t)$ and $f(\vec{v}_2'', t)$ for the inverse collisions, respectively. Rigorously, however, the two-particle distribution function $f_2(\vec{v}_1, \vec{v}_2, \vec{r}_{12}, t)$ for colliding pairs should be considered. This function does not necessarily decompose into a product of two one-particle distribution functions due to particle correlations. These correlations may be important in some cases, depending on the coefficient of restitution and on the packing fraction (see, e.g. Pagonabarraga *et al.*, 2002). For dilute gases these correlations may be neglected. The approximation which uses the product of two one-particle distribution functions is called the *hypothesis of molecular chaos* or *Stoßzahlansatz* according to the historical expression used by Boltzmann.

Correlations in a granular gas occur due to finite-volume effects, when possible colliders are screened by other particles (Chapman and Cowling, 1970). To obtain the rates of collisions rigorously the two-particle distribution function $f_2(\vec{v}_1, \vec{v}_2, \vec{r}_{12}, t)$ at the contact distance between the particles $r_{12} = \sigma$ must be known. An approximation which accounts for finite-volume effects was suggested by Enskog (Chapman and Cowling, 1970), which we use in a more sophisticated form (Resibois and de Leener, 1977):

$$f_2(\vec{v}_1, \vec{v}_2, \sigma, t) \approx g_2(\sigma) f(\vec{v}_1, t) f(\vec{v}_2, t) , \qquad (6.22)$$

where $g_2(\sigma)$ is the contact value of the equilibrium pair correlation function $g_2(r_{12})$, also called *Enskog factor*. This function describes the probability that the distance $|\vec{r}_1 - \vec{r}_2|$ of a pair of particles (which have any velocities) is equal to r_{12}. The same approximation may be used for the inverse collision. For a hard-sphere fluid at equilibrium $g_2(\sigma)$ reads (Carnahan and Starling, 1969)

$$g_2(\sigma) = \frac{(2-\eta)}{2(1-\eta)^3} , \qquad (6.23)$$

where $\eta = \frac{1}{6}\pi n \sigma^3$ is the packing fraction. For non-equilibrium system such as granular gases, this expression is not *a priori* valid. Fortunately, it has been shown by simulations that (6.23) is also very accurate for inelastic hard spheres (Deltour and Barrat, 1997). Physically, the factor $g_2(\sigma)$ accounts for an increased collision frequency due to excluded volume effects.

With the modification (6.22) the Boltzmann equation (6.21) changes to the Boltzmann–Enskog equation

$$\boxed{\frac{\partial}{\partial t} f(\vec{v}_1, t) = g_2(\sigma) I(f, f) ,} \qquad (6.24)$$

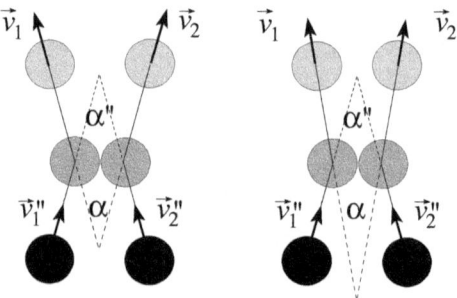

FIG. 6.2. Elastic (left) and inelastic (right) spheres before (black), during (dark grey) and after a collision (light grey). For elastic spheres, the incoming angle equals the outgoing angle ($\alpha = \alpha''$) whereas for inelastic collisions $\alpha < \alpha''$.

where the collision integral $I(f,f)$ has been defined by (6.21). Velocity correlations originate also due to ring collisions (van Noije et al., 1998) which are important for dense systems.

For the case of gases of dissipative particles there is an additional source of correlations: the coefficient of restitution affects only the normal component of the relative velocity of the particles but the tangential velocity remains unchanged. If two particles collide with a (pre-collision) angle α'' the angle after the collision will be smaller $\alpha < \alpha''$ (see Fig. 6.2 and Exercise 2.2). As a consequence subsequent collisions lead to more and more aligned traces of neighbouring particles. It has been shown (van Noije et al., 1997; Brito and Ernst, 1998a) that these correlations give rise to vortices in an otherwise homogeneous granular gas, that is, to macroscopic correlations in the velocity field even in absence of spatial inhomogeneities. This type of structure formation will be described in Chapters 25 and 26.

Since the particles are aligned in macroscopic vortices, such velocity correlations lead to a reduced collision frequency as compared with the collision frequency of a gas of the same average kinetic energy in the absence of vortices. The reduction of the collision frequency leads in its turn to a reduction of the cooling rate, that is, to a retarded decay of temperature even if the gas is still homogeneous (Brito and Ernst, 1998a). The consequences of these correlations on the microscopic level have been considered only recently (Pöschel et al., 2002).

6.4 An important property of the collision integral

For the case of elastic particles for which energy is conserved at impacts, the collision integral has several essential properties (Chapman and Cowling, 1970). Some of them are lost if the collisions are dissipative. The following important property is, however, preserved: given $\langle \psi(t) \rangle \equiv \int d\vec{v}_1 \psi(\vec{v}) f(\vec{v}, t)$ is the average value of some function $\psi(\vec{v})$ and $\Delta\psi(\vec{v}_i) \equiv [\psi(\vec{v}_i') - \psi(\vec{v}_i)]$ denotes the change of $\psi(\vec{v}_i)$ due to a direct collision. Then

$$\boxed{\begin{aligned}\frac{d}{dt}\langle\psi(t)\rangle &= \int d\vec{v}_1 \psi(\vec{v}_1) \frac{\partial}{\partial t} f(\vec{v}_1, t) = g_2(\sigma) \int d\vec{v}_1 \psi(\vec{v}_1) I(f,f) \\ &= \frac{g_2(\sigma)\sigma^2}{2} \int d\vec{v}_1 d\vec{v}_2 \int d\vec{e}\, \Theta(-\vec{v}_{12}\cdot\vec{e}) |\vec{v}_{12}\cdot\vec{e}| \\ &\quad\times f(\vec{v}_1,t) f(\vec{v}_2, t) \Delta[\psi(\vec{v}_1) + \psi(\vec{v}_2)]\end{aligned}}\qquad (6.25)$$

To prove this relation we write

$$\int d\vec{v}_1 \psi(\vec{v}_1) I(f,f)$$
$$= \sigma^2 \int d\vec{v}_1 d\vec{v}_2 \int d\vec{e}\, \Theta(-\vec{v}_{12}\cdot\vec{e}) |\vec{v}_{12}\cdot\vec{e}| \chi f(\vec{v}_1'',t) f(\vec{v}_2'',t) \psi(\vec{v}_1) \qquad (6.26)$$
$$-\sigma^2 \int d\vec{v}_1 d\vec{v}_2 \int d\vec{e}\, \Theta(-\vec{v}_{12}\cdot\vec{e}) |\vec{v}_{12}\cdot\vec{e}| f(\vec{v}_1,t) f(\vec{v}_2, t) \psi(\vec{v}_1)$$

and notice that according to (6.9, 6.14, 6.16),

$$\chi |\vec{v}_{12}\cdot\vec{e}| d\vec{v}_1 d\vec{v}_2 = |\vec{v}_{12}''\cdot\vec{e}| d\vec{v}_1'' d\vec{v}_2'' . \qquad (6.27)$$

Therefore, the first integral in the RHS of (6.26) may be recast into the form

$$\sigma^2 \int d\vec{v}_1'' d\vec{v}_2'' \int d\vec{e}\, \Theta(-\vec{v}_{12}''\cdot\vec{e}) |\vec{v}_{12}''\cdot\vec{e}| f(\vec{v}_1'',t) f(\vec{v}_2'',t) \psi(\vec{v}_1) . \qquad (6.28)$$

Expression (6.28) contains the pre-collision velocities $\vec{v}_1''$, $\vec{v}_2''$ and the present (i.e. after-collision) velocity $\vec{v}_1$. Since $\vec{v}_1''$, $\vec{v}_2''$ are related to $\vec{v}_1$, $\vec{v}_2$ as $\vec{v}_1$, $\vec{v}_2$ to $\vec{v}_1'$, $\vec{v}_2'$, we may change the notation:

$$\sigma^2 \int d\vec{v}_1 d\vec{v}_2 \int d\vec{e}\, \Theta(-\vec{v}_{12}\cdot\vec{e}) |\vec{v}_{12}\cdot\vec{e}| f(\vec{v}_1,t) f(\vec{v}_2,t) \psi(\vec{v}_1') . \qquad (6.29)$$

Expression (6.29) as well as the second term in the RHS of (6.26) reveal complete symmetry with respect to exchange of the particles indices 1 and 2, provided the exchange $1 \leftrightarrow 2$ is accompanied by the change of the variable $\vec{e} \to -\vec{e}$. Therefore,

$$\frac{1}{2}\sigma^2 \int d\vec{v}_1 d\vec{v}_2 \int d\vec{e}\, \Theta(-\vec{v}_{12}\cdot\vec{e}) |\vec{v}_{12}\cdot\vec{e}| f(\vec{v}_1,t) f(\vec{v}_2,t) [\psi(\vec{v}_1') + \psi(\vec{v}_2')] \qquad (6.30)$$

is identical with (6.29). A similar expression can be found for the second term in (6.26). Substituting (6.30) into (6.26) and with the abbreviation $\Delta\psi(\vec{v}_i) \equiv \psi(\vec{v}_i') - \psi(\vec{v}_i)$ we arrive at the relation (6.25) which will be used in many calculations later.

7

SONINE POLYNOMIALS EXPANSION OF THE VELOCITY DISTRIBUTION FUNCTION

The velocity distribution function of a granular gas is different from the Maxwell distribution. It may be represented in the form of a Sonine polynomials expansion. The coefficients of this expansion describe the moments of the velocity distribution function. The first non-trivial Sonine coefficient a_2 is of particular interest for the gas kinetics.

For equilibrium gases whose particles undergo elastic collisions the particle velocities obey the Maxwell distribution (Chapman and Cowling, 1970; Ferziger and Kaper, 1972). The particles of granular gases undergo *inelastic* collisions, which lead to a persistant decrease of energy and, hence, of temperature. Already for this reason the velocity distribution function cannot be independent of time.

If we assume that the collisions are only slightly inelastic ($\varepsilon \lesssim 1$) the granular temperature decays very slowly, so that the system behaves at each time instant like an equilibrium system. The cooling process can be understood as a sequence of equilibrium states and the corresponding process is called *adiabatic cooling*. Since in the most simple case ($\varepsilon = $ const.) all other characteristics of the granular gas (particle mass, density, etc.) do not change with time, the idea of adiabatic cooling suggests that the time dependence of the velocity distribution function occurs only through the time-dependent average velocity, that is, through temperature. This implies a scaled form of the velocity distribution function (Esipov and Pöschel, 1997):

$$f(\vec{v}, t) = \frac{n}{v_T^3(t)} \tilde{f}\left(\frac{\vec{v}}{v_T(t)}\right) = \frac{n}{v_T^3(t)} \tilde{f}(\vec{c}) \tag{7.1}$$

with the scaled velocity $\vec{c} \equiv \vec{v}/v_T(t)$. The thermal velocity $v_T(t)$ is defined by

$$T(t) = \frac{1}{2} m v_T^2(t), \tag{7.2}$$

and temperature is related to the second moment of the velocity distribution,

$$\frac{3}{2} n T(t) = \int d\vec{v}\, \frac{mv^2}{2} f(\vec{v}, t). \tag{7.3}$$

From these arguments we expect that the velocity distribution is close to a Maxwell distribution provided the inelasticity is small ($\varepsilon \lesssim 1$). In this case it is convenient to represent the scaled function $\tilde{f}(\vec{c})$ in form of an expansion (Goldshtein and Shapiro, 1995; van Noije and Ernst, 1998)

$$\tilde{f}(\vec{c}) = \phi(c)\left[1 + \sum_{p=1}^{\infty} a_p S_p(c^2)\right] \tag{7.4}$$

where the leading term

$$\phi(c) \equiv \pi^{-3/2} \exp(-c^2) \tag{7.5}$$

is the Maxwell distribution for the scaled velocity $\vec{c}$. One of the possible choices for the complete set of (orthogonal) functions $\{S_p(x)\}$ are the Sonine polynomials with their corresponding coefficients a_p. The Sonine polynomials constitute a complete set of functions with the orthogonality conditions

$$\int d\vec{c}\,\phi(c) S_p(c^2) S_{p'}(c^2) = \frac{2(p+1/2)!}{\sqrt{\pi}\,p!} \delta_{pp'} \equiv \mathcal{N}_p \delta_{pp'}, \tag{7.6}$$

with $\delta_{pp'}$ being the Kronecker delta. For the dimension $d=3$ the first few Sonine polynomials read

$$\begin{aligned} S_0(x) &= 1, \\ S_1(x) &= -x + \frac{3}{2}, \\ S_2(x) &= \frac{x^2}{2} - \frac{5x}{2} + \frac{15}{8}. \end{aligned} \tag{7.7}$$

Correspondingly, the powers of the scaled velocity $\vec{c}$ can be expressed in terms of the Sonine polynomials, namely,

$$\begin{aligned} 1 &= S_0(c^2) \\ c^2 &= \frac{3}{2} - S_1(c^2) \\ c^4 &= 2S_2(c^2) - 5S_1(c^2) + \frac{15}{4}, \end{aligned} \tag{7.8}$$

etc. These relations are applied to calculate the moments of the distribution function

$$\langle c^p \rangle \equiv \int c^p \tilde{f}(\vec{c})\, d\vec{c}. \tag{7.9}$$

For example, the second moment $\langle c^2 \rangle$ may be obtained by means of (7.8) and the orthogonality relation (7.6):

$$\langle c^2 \rangle = \int d\vec{c}\left[\frac{3}{2} - S_1\left(c^2\right)\right]\phi\left(c\right)\left[1 + \sum_{k=1}^{\infty} a_k S_k\left(c^2\right)\right]$$

$$= \frac{3}{2}\int d\vec{c}\,\phi\left(c\right) S_0\left(c^2\right) S_0\left(c^2\right) + \frac{3}{2}\sum_{k=1}^{\infty} a_k \int d\vec{c}\,\phi\left(c\right) S_0\left(c^2\right) S_k\left(c^2\right)$$

$$- \int d\vec{c}\,\phi\left(c\right) S_1\left(c^2\right) S_0\left(c^2\right) - \sum_{k=1}^{\infty} a_k \int d\vec{c}\,\phi\left(c\right) S_1\left(c^2\right) S_k\left(c^2\right) \quad (7.10)$$

$$= \frac{3}{2} + 0 - 0 - \frac{3}{2}a_1 = \frac{3}{2}\left(1 - a_1\right)$$

where we expressed $1 = c^0$ and c^2 in terms of Sonine polynomials and used the normalization constants $\mathcal{N}_0 = 1$ and $\mathcal{N}_1 = 3/2$. Similar considerations yield the next moment of the distribution function,

$$\langle c^4 \rangle = \frac{15}{4}\left(1 + a_2\right) \quad (7.11)$$

as well as the moments of the Sonine polynomials themselves,

$$\nu_{kp} \equiv \int \phi\left(c\right) S_k\left(c^2\right) c^p d\vec{c}. \quad (7.12)$$

These quantities, especially

$$\nu_{22} = 0 \quad \text{and} \quad \nu_{24} = \frac{15}{4} \quad (7.13)$$

will be used in what follows.

The Sonine polynomials expansion is applicable if the distribution is essentially a Maxwell distribution and if the series of Sonine polynomials in (7.4) converges rapidly. For small dissipation, a satisfying description of the velocity distribution function is achieved by omitting all terms in the expansion (7.4) with $p \geq 3$, that is, assuming that $a_p = 0$ for $p \geq 3$. For large degree of inelasticity, when ε differs noticeably from 1 higher order coefficients become important or, in worst case, the Sonine expansion diverges.

For the sake of completeness we give here the general definition of the Sonine polynomials and some useful relations: The Sonine polynomials that are used in d-dimensional kinetic theory are also known as associated Laguerre polynomials $S_p^{(m)}(x)$ with $m = d/2 - 1$. They are defined by[8]

$$S_p^{(m)}(x) = \sum_{n=0}^{p} \frac{(-1)^n (m+p)!}{(m+n)!(p-n)!n!} x^n \quad (7.14)$$

---

[8]In this book we will mainly address three-dimensional systems and, thus, use the Sonine polynomials with $m = 3/2 - 1 = 1/2$. We omit the upper index, it is shown only in cases where the lack of the index might cause confusion.

and satisfy the orthogonality conditions

$$\int_0^\infty c^{2m+1} \exp\left(-c^2\right) S_p^{(m)}\left(c^2\right) S_{p'}^{(m)}\left(c^2\right) = \delta_{pp'} \frac{(m+p)!}{2p!}. \quad (7.15)$$

The according inverse relations to (7.14) read

$$c^{2l} = \sum_{n=0}^{l} \frac{(-1)^n (m+l)!\, l!}{(m+n)!\,(l-n)!} S_n^{(m)}\left(c^2\right). \quad (7.16)$$

Hence, c^{2l} may be written as a sum of Sonine polynomials $S_n^{(m)}\left(c^2\right)$ with $0 \leq n \leq l$, that is, of $S_l^{(m)}\left(c^2\right)$, $S_{l-1}^{(m)}\left(c^2\right)$, etc. Due to the orthogonality conditions

$$\nu_{kp} = 0 \quad \text{for} \quad 2k > p \quad (7.17)$$

and

$$\nu_{k\,2l} = (-1)^k \frac{(2l+1)!!\, l!}{2^l k!\,(l-k)!} \quad \text{for} \quad k \leq l. \quad (7.18)$$

The moments of the distribution function $\langle c^{2n} \rangle$ may be also expressed in terms of a_k with $0 \leq k \leq n$:

$$\langle c^{2k} \rangle = \frac{(2k+1)!!}{2^k} \left(1 + \sum_{p=1}^{k} (-1)^p \frac{k!}{(k-p)!\,p!} a_p\right). \quad (7.19)$$

Taking into account that the moments of the Maxwell distribution read

$$\langle c^{2k} \rangle_0 = \int c^{2k} \phi(c)\, d\vec{c} = \frac{(2k+1)!!}{2^k}, \quad (7.20)$$

we conclude from (7.19) that the Sonine coefficients characterize the deviation of the moments of the velocity distribution function from the according moments of the Maxwell distribution, that is,

$$a_1 = -\frac{\langle c^2 \rangle - \langle c^2 \rangle_0}{\langle c^2 \rangle_0}, \quad a_2 = \frac{\langle c^4 \rangle - \langle c^4 \rangle_0}{\langle c^4 \rangle_0}, \quad \text{etc.} \quad (7.21)$$

We will apply the Sonine polynomials expansion to solve the Boltzmann equation for a granular gas. The coefficients of the expansion a_p are related to the function $\tilde{f}(\vec{c})$ via

$$a_p = \frac{1}{\mathcal{N}_p} \int d\vec{c}\, S_p\left(c^2\right) \tilde{f}(\vec{c}). \quad (7.22)$$

Obviously, (7.22) may be used to find a_p only if the distribution function $\tilde{f}(\vec{c})$ is known, either from experiments or from simulations. In the next chapter we will show how these coefficients may be directly obtained from the Boltzmann equation. As outlined above, the knowledge about the Sonine coefficients is equivalent to the knowledge about the velocity distribution function itself. Therefore, if we are able to obtain the coefficients we have solved the Boltzmann equation for the dissipative gas.

Since gases of particles that collide with $\varepsilon = $ const. and gases of viscoelastic particles behave qualitatively different, we consider them separately in Chapters 8 and 9, respectively.

8

VELOCITY DISTRIBUTION AND TEMPERATURE OF A GRANULAR GAS FOR THE CASE $\varepsilon =$ CONST.

For $\varepsilon =$ const. the Boltzmann equation may be reduced to a set of two uncoupled equations. One of them describes the evolution of temperature and the other determines the scaled distribution function. We solve this equation using the Sonine polynomials expansion.

8.1 Decomposition of the Boltzmann equation

First we formulate the Boltzmann equation

$$\frac{\partial}{\partial t} f(\vec{v}_1, t) = g_2(\sigma) \sigma^2 \int d\vec{v}_2 \int d\vec{e}\, \Theta(-\vec{v}_{12} \cdot \vec{e}) |\vec{v}_{12} \cdot \vec{e}|$$
$$\times \left\{ \frac{1}{\varepsilon^2} f(\vec{v}_1'', t) f(\vec{v}_2'', t) - f(\vec{v}_1, t) f(\vec{v}_2, t) \right\} \quad (8.1)$$
$$\equiv g_2(\sigma) I(f, f)$$

in terms of the scaled velocity distribution function $\tilde{f}(\vec{c})$. With the Ansatz (7.1) the LHS changes to

$$\frac{\partial}{\partial t} f(\vec{v}_1, t) = \left(-\frac{3n}{v_T^4} \tilde{f}(\vec{c}_1) + \frac{n}{v_T^3} \frac{\partial \tilde{f}}{\partial c_1} \frac{\partial c_1}{\partial v_T} \right) \frac{dv_T}{dt}, \quad (8.2)$$

while the collision integral may be written as

$$I(f, f) = \sigma^2 v_T^3 v_T n^2 v_T^{-6} \int d\vec{c}_2 \int d\vec{e}\, \Theta(-\vec{c}_{12} \cdot \vec{e}) |\vec{c}_{12} \cdot \vec{e}|$$
$$\times \left[\frac{1}{\varepsilon^2} \tilde{f}(\vec{c}_1'') \tilde{f}(\vec{c}_2'') - \tilde{f}(\vec{c}_1) \tilde{f}(\vec{c}_2) \right] \quad (8.3)$$
$$\equiv \sigma^2 n^2 v_T^{-2} \tilde{I}\left(\tilde{f}, \tilde{f} \right).$$

The factor v_T^3 in the first line originates from the differential $d\vec{v} = v_T^3 d\vec{c}$, the factor v_T from the length of the collision cylinder $|\vec{v}_{12} \cdot \vec{e}| = v_T |\vec{c}_{12} \cdot \vec{e}|$, and the factor $n^2 v_T^{-6}$ comes from the product of the two distribution functions (see definition (7.1)). In (8.3) we also introduce the dimensionless collision integral

$\tilde{I}\left(\tilde{f}, \tilde{f}\right)$. The basic property of the collision integral (6.25) may also be written in dimensionless form:

$$\int d\vec{c}\,\psi(\vec{c}_1)\tilde{I}\left(\tilde{f}, \tilde{f}\right) = \frac{1}{2}\int d\vec{c}_1 d\vec{c}_2 \int d\vec{e}\,\Theta\left(-\vec{c}_{12}\cdot\vec{e}\right)|\vec{c}_{12}\cdot\vec{e}| \\ \times \tilde{f}\left(\vec{c}_1\right)\tilde{f}\left(\vec{c}_2\right)\Delta\left[\psi\left(\vec{c}_1\right) + \psi\left(\vec{c}_2\right)\right].$$

(8.4)

Note that the dimensionless collision integral does not depend on time.

Exercise 8.1 *Recast the basic property of the collision integral (6.25) into its dimensionless form (8.4)!*

Using (8.2, 8.3), and the relation $\partial c_1/\partial v_T = -c_1/v_T$ we formulate the Boltzmann equation for the scaled velocity distribution function:

$$-\frac{1}{v_T^2}\frac{dv_T}{dt}\left(3 + c_1\frac{\partial}{\partial c_1}\right)\tilde{f}\left(\vec{c}_1\right) = g_2\left(\sigma\right)\sigma^2 n \tilde{I}\left(\tilde{f}, \tilde{f}\right).$$

(8.5)

The rate of change of the thermal velocity dv_T/dt in (8.5) may be expressed in terms of the temperature decay rate dT/dt. From $T = mv_T^2/2$ we find

$$\frac{1}{v_T^2}\frac{dv_T}{dt} = \frac{1}{2v_T T}\frac{dT}{dt}.$$

(8.6)

The rate of temperature decay may be written, in turn, in terms of the collision integral. From the definition of temperature (7.3) and from relation (6.25) it follows that

$$\frac{d}{dt}\frac{3}{2}nT(t) = \int d\vec{v}_1 \frac{mv_1^2}{2}\frac{\partial}{\partial t}f\left(\vec{v}_1, t\right) = g_2\left(\sigma\right)\int d\vec{v}_1 \frac{mv_1^2}{2}I\left(f, f\right) \\ = g_2\left(\sigma\right)\sigma^2 n^2 v_T \frac{mv_T^2}{2}\int d\vec{c}_1 c_1^2 \tilde{I}\left(\tilde{f}, \tilde{f}\right) \\ = -g_2\left(\sigma\right)\sigma^2 n^2 v_T T \mu_2.$$

(8.7)

Equation (8.3) was used to express $I(f, f)$ in terms of $\tilde{I}(\tilde{f}, \tilde{f})$ and we have also introduced the moments of the dimensionless collision integral

$$\mu_p \equiv -\int d\vec{c}_1 c_1^p \tilde{I}\left(\tilde{f}, \tilde{f}\right).$$

(8.8)

Since the dimensionless collision integral $\tilde{I}(\tilde{f}, \tilde{f})$ is time-independent, the moments μ_p do not depend on time. Hence, the factor

$$\frac{1}{v_T^2}\frac{dv_T}{dt} = \frac{1}{2v_T T}\frac{dT}{dt} = -\frac{1}{3}g_2\left(\sigma\right)\sigma^2 n \mu_2$$

(8.9)

given by (8.6, 8.7) is also time-independent. Thus, we arrive at a time-independent equation for the scaled velocity distribution:

$$\boxed{\frac{\mu_2}{3}\left(3+c_1\frac{\partial}{\partial c_1}\right)\tilde{f}(\vec{c}_1)=\tilde{I}\left(\tilde{f},\tilde{f}\right)} \qquad (8.10)$$

and a time-dependent equation for the evolution of temperature which follows from (8.7):

$$\boxed{\frac{dT}{dt}=-\frac{2}{3}BT\mu_2 \quad \text{with} \quad B=B(t)\equiv v_T(t)\,g_2(\sigma)\,\sigma^2 n\,.} \qquad (8.11)$$

We will use (8.11) also in the form

$$\frac{dT}{dt}=-\zeta T \quad \text{with} \quad \zeta\equiv\frac{2}{3}n\sigma^2 g_2(\sigma)\sqrt{\frac{2T}{m}}\mu_2\,. \qquad (8.12)$$

where ζ is the cooling coefficient for the homogeneous gas.

8.2 The second Sonine coefficient and the moments of the collision integral

The direct solution of the integro-differential equation (8.10) for the scaled function $\tilde{f}(\vec{c})$ is a very complicated mathematical problem. Instead we apply the Sonine expansion method to obtain equations for the Sonine coefficients a_p. These equations are algebraic and may be solved easily (Goldshtein and Shapiro, 1995; van Noije and Ernst, 1998). First we show, however, that the Sonine coefficient a_1 is trivial.

The density of the kinetic energy of a granular gas is

$$\int d\vec{v}\,\frac{mv^2}{2}f(\vec{v},t)=\frac{mv_T^2}{2}n\int c^2\tilde{f}(\vec{c})\,d\vec{c}=\langle c^2\rangle\frac{mv_T^2}{2}n\,, \qquad (8.13)$$

where the velocity is given in units of the thermal velocity $\vec{c}\equiv\vec{v}/v_T$ (see (7.1)). On the other hand from the definitions of temperature (7.3) and thermal velocity v_T (7.2) it follows that

$$\int d\vec{v}\,\frac{mv^2}{2}f(\vec{v},t)=n\frac{m\langle v^2\rangle}{2}=\frac{3}{2}nT=\frac{3}{2}\frac{mv_T^2}{2}n\,. \qquad (8.14)$$

Comparison of (8.13) and (8.14) yields

$$\langle c^2\rangle=\frac{3}{2}\quad\text{and}\quad\langle v^2\rangle=\frac{3}{2}v_T^2\,. \qquad (8.15)$$

Applying (7.10) we obtain

$$\langle c^2\rangle=\frac{3}{2}=\frac{3}{2}(1-a_1) \qquad (8.16)$$

and conclude
$$a_1 = 0. \tag{8.17}$$

Note that this result is general since it follows exclusively from the definitions of temperature and thermal velocity. Thus, the first non-trivial coefficient which describes the deviation of the distribution function from the Maxwell distribution is the coefficient a_2. Therefore, our main concern in this section is to determine the second Sonine coefficient and its relation to the inelasticity of the particles expressed by the coefficient of restitution.

To find equations for a_p we multiply both sides of (8.10) with c_1^p and integrate over $\vec{c}_1$. The RHS then changes to $-\mu_p$ due to the definition (8.8). The first term on the LHS yields $\mu_2 \langle c^p \rangle$ and the second term simplifies after integration by parts:

$$\begin{aligned}
\int c_1^{p+1} \frac{\partial}{\partial c_1} \tilde{f}(\vec{c}_1) d\vec{c}_1 &= \int_0^{2\pi} d\varphi \int_0^{\pi} \sin\theta d\theta \int_0^{\infty} c_1^{p+3} \frac{\partial}{\partial c_1} \tilde{f}(c_1) dc_1 \\
&= 4\pi c_1^{p+3} \tilde{f}(c_1) \Big|_0^{\infty} - 4\pi \int_0^{\infty} dc_1 \tilde{f}(c_1) \frac{\partial}{\partial c_1} c_1^{p+3} \\
&= -4\pi(p+3) \int_0^{\infty} dc_1 \tilde{f}(c_1) c_1^{p+2} \\
&= -(p+3) \langle c^p \rangle .
\end{aligned} \tag{8.18}$$

Here we take into account that for an isotropic gas the distribution function depends only on the absolute value of the velocity, that is, on $|\vec{c}|$, hence, integration over the angles φ, θ yields 4π. We also assume that $\tilde{f}(c_1)$ decreases with c_1 faster than any power of c_1, therefore, the term $c_1^{p+3} \tilde{f}(c_1)$ vanishes in the limit $c_1 = \infty$.[9] Thus, after this transformation the LHS of (8.10) reads $(\mu_2/3)[3 \langle c^p \rangle - (p+3) \langle c^p \rangle]$ and we obtain

$$\frac{1}{3} \mu_2 \, p \, \langle c^p \rangle = \mu_p . \tag{8.19}$$

In Chapter 7 we have shown that the moments $\langle c^p \rangle$ may be expressed in terms of the Sonine coefficients a_p. The moments μ_p of the collision integral may also be expressed in terms of a_p. This follows from the definition (8.8), which shows that μ_p depends on a_p via the distribution function $\tilde{f}(c)$. Thus, the system (8.19) for $p = 2, 4, 6, \ldots$ (odd p yield trivial relations) is an infinite but closed set of equations for the coefficients $a_2, a_3, \ldots$ With the moment $\langle c^2 \rangle$ given by (8.16),

---

[9] The velocity distribution function of dilute hard sphere gases, that is, for $\varepsilon = 1$ decays as $\exp(-c^2)$. Since the coefficient of restitution was assumed to be close to one, we expect that the velocity distribution function of granular gases reveals a similar behaviour. Therefore, we do not expect it to decay algebraically as, for example, for a gas of Maxwellian molecules (Ben-Naim and Krapivsky, 2000; Bobylev et al., 2000), but at least exponentially. There is, however, no strict mathematical justification of the assumption that $\tilde{f}(c_1)$ decreases with c_1 faster than any power of c_1. In Chapter 10 we investigate the tail of the distribution function in detail.

LINEAR APPROXIMATION OF THE SECOND SONINE COEFFICIENT

the equation for $p = 2$ is trivial. For $p = 4$ we obtain the first non-trivial equation with $\langle c^4 \rangle = \frac{15}{4}(1 + a_2)$, see (7.11):

$$5\mu_2 (1 + a_2) - \mu_4 = 0. \tag{8.20}$$

Unfortunately, it is not possible to solve the full set of equations for $p = 2, 4, 6, \ldots$

For small dissipation ($\varepsilon \lesssim 1$) the velocity distribution function is close to the Maxwell distribution. Therefore, for almost elastic particle collisions high-order terms of the expansion (7.4) with $p > 2$ may be neglected. If this assumption is justified, the velocity distribution function is already well approximated by the second Sonine coefficient a_2. In the next section we determine a_2 by means of (8.20). To this end μ_2 and μ_4 are expressed in terms of a_2.

8.3 Linear approximation of the second Sonine coefficient

If we apply the basic property of the collision integral $\tilde{I}(\tilde{f}, \tilde{f})$ written in the dimensionless form (8.4) to $\psi(\vec{c}_1) = c_1^p$ we obtain from the definition of μ_p (8.8)

$$\mu_p = -\frac{1}{2} \int d\vec{c}_1 \int d\vec{c}_2 \int d\vec{e}\, \Theta\left(-\vec{c}_{12} \cdot \vec{e}\right) |\vec{c}_{12} \cdot \vec{e}|\, \tilde{f}(c_1)\, \tilde{f}(c_2)\, \Delta\left(c_1^p + c_2^p\right), \tag{8.21}$$

where $\Delta \psi(\vec{c}_i) \equiv \psi(\vec{c}_i') - \psi(\vec{c}_i)$ denotes the variation of some quantity $\psi(\vec{c}_i)$ due to a direct collision (see (6.25)). We truncate the Sonine polynomials expansion (7.4) after the second term,

$$\tilde{f}(c) \simeq \phi(c)\left[1 + a_2 S_2(c^2)\right] \tag{8.22}$$

and find the coefficients in the form

$$\mu_p = -\frac{1}{2} \int d\vec{c}_1 \int d\vec{c}_2 \int d\vec{e}\, \Theta\left(-\vec{c}_{12} \cdot \vec{e}\right) |\vec{c}_{12} \cdot \vec{e}|\, \phi(c_1)\, \phi(c_2) \tag{8.23}$$
$$\times \left\{1 + a_2 \left[S_2(c_1^2) + S_2(c_2^2)\right] + a_2^2 S_2(c_1^2) S_2(c_2^2)\right\} \Delta\left(c_1^p + c_2^p\right).$$

Integral (8.23) is a representative of a class of integrals which we call *kinetic integrals*. We will be faced, in many of the following calculations, with this type of integrals. Later we will solve them by computational formula manipulation using Maple (see Appendix A). Since it occurs here for the first time we solve it in the traditional way using paper and pencil.

8.3.1 Solution of kinetic integrals

For the computation it is convenient to use the centre of mass velocity $\vec{C}$ and relative velocity $\vec{c}_{12}$:

$$\vec{c}_1 = \vec{C} + \frac{1}{2}\vec{c}_{12}, \qquad \vec{c}_2 = \vec{C} - \frac{1}{2}\vec{c}_{12}. \tag{8.24}$$

It is easy to check that the Jacobian of the transformation (8.24) equals unity. The product of two Maxwell distributions $\phi(\vec{c}_1)\, \phi(\vec{c}_2)$ transforms into a product of a Maxwell distributions for the centre of mass (an effective particle of double mass) and of a Maxwell distribution for the relative motion (an effective particle of half mass)

$$\phi(\vec{c}_1)\phi(\vec{c}_2) \to \frac{1}{(2\pi)^{3/2}} \exp\left(-\frac{1}{2}c_{12}^2\right) \left(\frac{2}{\pi}\right)^{3/2} \exp\left(-2C^2\right) \tag{8.25}$$
$$\equiv \phi(\vec{c}_{12})\phi(\vec{C}).$$

The terms in square brackets in (8.23) transform as follows

$$[S_2(c_1^2) + S_2(c_2^2)] =$$
$$C^4 + (\vec{C}\cdot\vec{c}_{12})^2 + \frac{1}{16}c_{12}^4 + \frac{1}{2}C^2 c_{12}^2 - 5C^2 - \frac{5}{4}c_{12}^2 + \frac{15}{4}. \tag{8.26}$$

To find the quantities $\Delta(c_1^p + c_2^p)$ ($p = 2, 4$) we rewrite the collision law (6.5) in terms of $\vec{C}$ and $\vec{c}_{12}$ too:

$$\begin{aligned}\vec{c}_1' &= \vec{C} + \frac{1}{2}\vec{c}_{12} - \frac{1}{2}(1+\varepsilon)(\vec{c}_{12}\cdot\vec{e})\vec{e} \\ \vec{c}_2' &= \vec{C} - \frac{1}{2}\vec{c}_{12} + \frac{1}{2}(1+\varepsilon)(\vec{c}_{12}\cdot\vec{e})\vec{e}.\end{aligned} \tag{8.27}$$

Then straightforward algebra yields

$$\Delta(c_1^2 + c_2^2) = -\frac{1}{2}(1-\varepsilon^2)(\vec{c}_{12}\cdot\vec{e})^2 \tag{8.28}$$

and

$$\begin{aligned}\Delta(c_1^4 + c_2^4) =\ &2(1+\varepsilon)^2(\vec{c}_{12}\cdot\vec{e})^2(\vec{C}\cdot\vec{e})^2 + \frac{1}{8}(1-\varepsilon^2)^2(\vec{c}_{12}\cdot\vec{e})^4 \\ &-\frac{1}{4}(1-\varepsilon^2)(\vec{c}_{12}\cdot\vec{e})^2 c_{12}^2 - (1-\varepsilon^2)\vec{C}^2(\vec{c}_{12}\cdot\vec{e})^2 \\ &- 4(1+\varepsilon)(\vec{C}\cdot\vec{c}_{12})(\vec{C}\cdot\vec{e})(\vec{c}_{12}\cdot\vec{e}).\end{aligned} \tag{8.29}$$

For the moment we assume that the value of a_2 is small, therefore, contributions to μ_p which are proportional to a_2^2 can be omitted. We will return to these contributions later. Combining the terms in (8.26, 8.28, 8.29) the integrands for μ_2 and μ_4 in (8.23) contain products of the Maxwell distributions for C and c_{12} and the factors C^k, c_{12}^l, $(\vec{c}\cdot\vec{c}_{12})^m$ with different k, l, m and similar other factors. Since this type of integrals will occur frequently we introduce the *Basic Integral* and express μ_2 and μ_4 via these quantities. At this point we introduce a slightly more general integral than necessary for the evaluation of μ_2 and μ_4. The Basic Integral is defined by

$$J_{k,l,m,n,p,\alpha} \equiv \int d\vec{c}_{12} \int d\vec{C} \int d\vec{e}\, \Theta\left(-\vec{c}_{12}\cdot\vec{e}\right) |\vec{c}_{12}\cdot\vec{e}|^{1+\alpha}$$
$$\times \phi(c_{12})\,\phi(C)\,C^k c_{12}^l \left(\vec{C}\cdot\vec{c}_{12}\right)^m \left(\vec{C}\cdot\vec{e}\right)^n (\vec{c}_{12}\cdot\vec{e})^p \quad (8.30)$$

Its general solution in d dimensions is given in Appendix A. For three-dimensional systems for the cases of interest $n=0$, $n=1$ and $n=2$ it reads

$$J_{k,l,m,0,p,\alpha} = \frac{(-1)^p \cdot 8 \cdot 2^{(-k+l+p+\alpha-1)/2}}{(p+\alpha+2)(m+1)} \left[1-(-1)^{m+1}\right]$$
$$\times \Gamma\left(\frac{k+m+3}{2}\right) \Gamma\left(\frac{l+m+p+\alpha+4}{2}\right) \quad (8.31)$$

$$J_{k,l,m,1,p,\alpha} = \frac{(-1)^{p+1} \cdot 4 \cdot 2^{(-k+l+p+\alpha)/2}}{(p+\alpha+3)(m+2)} \left[1-(-1)^m\right]$$
$$\times \Gamma\left(\frac{k+m+4}{2}\right) \Gamma\left(\frac{l+m+p+\alpha+4}{2}\right) \quad (8.32)$$

$$J_{k,l,m,2,p,\alpha} = \frac{(-1)^p \cdot 4 \cdot 2^{(-k+l+p+\alpha-1)/2}}{(p+\alpha+4)(p+\alpha+2)} \left[1-(-1)^{m+1}\right]$$
$$\times \Gamma\left(\frac{k+m+5}{2}\right) \Gamma\left(\frac{l+m+p+\alpha+4}{2}\right) \quad (8.33)$$
$$\times \left(\frac{p+\alpha+1}{m+3} + \frac{1}{m+1}\right)$$

with $\Gamma(x)$ being the Gamma function. The Basic Integral can be evaluated by means of the properties of the Gamma function

$$\Gamma(x+1) = x\Gamma(x) \quad \text{with} \quad \Gamma(1/2) = \sqrt{\pi}. \quad (8.34)$$

For example, $J_{0,0,0,0,2,0}$ reduces to

$$J_{0,0,0,0,2,0} = \frac{(-1)^2 \cdot 8 \cdot 2^{(-0+0+2+0-1)/2}}{(2+0+2)(0+1)} \left[1-(-1)^{0+1}\right]$$
$$\times \Gamma\left(\frac{0+0+3}{2}\right) \Gamma\left(\frac{0+0+2+0+4}{2}\right) \quad (8.35)$$
$$= 4\sqrt{2}\,\Gamma(3/2)\,\Gamma(3) = 4\sqrt{2\pi}.$$

Using (8.26, 8.28), and the definition of the Basic Integral (8.30) the coefficient μ_2 given by (8.23) may be expressed in linear approximation with respect to a_2:

$$\mu_2 = \frac{1}{4}\left(1-\varepsilon^2\right)\left[J_{0,0,0,0,2,0} + a_2\left(J_{4,0,0,0,2,0} + J_{0,0,2,0,2,0} + \frac{1}{16}J_{0,4,0,0,2,0}\right.\right.$$
$$\left.\left.+\frac{1}{2}J_{2,2,0,0,2,0} - 5J_{2,0,0,0,2,0} - \frac{5}{4}J_{0,2,0,0,2,0} + \frac{15}{4}J_{0,0,0,0,2,0}\right)\right]. \quad (8.36)$$

Similar to (8.35) we obtain $J_{4,0,0,0,2,0} = (15/4)\sqrt{2\pi}$, $J_{0,0,2,0,2,0} = 6\sqrt{2\pi}$, etc. Consequently, (8.36) yields

$$\mu_2 = \sqrt{2\pi}\left(1-\varepsilon^2\right)\left(1 + \frac{3}{16}a_2 + \mathcal{O}\left(a_2^2\right)\right). \quad (8.37)$$

Exercise 8.2 *Derive the expression for μ_2, given by (8.23) in terms of Basic Integrals as defined by (8.30)!*

Following the same way μ_4 may be written in terms of Basic Integrals. The expression contains 35 integrals and we do not wish to present them here in detail. After simplification we obtain

$$\mu_4 = 4\sqrt{2\pi}\left[T_1 + a_2 T_2 + \mathcal{O}\left(a_2^2\right)\right] \quad (8.38)$$

with

$$\begin{aligned} T_1 &= \frac{1}{4}\left(1-\varepsilon^2\right)\left(\frac{9}{2}+\varepsilon^2\right) \\ T_2 &= \frac{3}{128}\left(1-\varepsilon^2\right)\left(69+10\,\varepsilon^2\right) + \frac{1}{2}\left(1+\varepsilon\right). \end{aligned} \quad (8.39)$$

From these relations for μ_2 and μ_4 the second Sonine coefficient can be computed by means of (8.20). Discarding terms $\mathcal{O}\left(a^2\right)$, equation (8.20) may be expressed as

$$5\left(1-\varepsilon^2\right)\left(1+\frac{19}{16}a_2\right) = 4\left(T_1 + a_2 T_2\right). \quad (8.40)$$

This linear equation for a_2 may be solved easily. Substituting the expressions (8.39) for T_1 and T_2, we find

$$\boxed{a_2^{\text{NE}} = \frac{16\left(1-\varepsilon\right)\left(1-2\varepsilon^2\right)}{81 - 17\varepsilon + 30\varepsilon^2\left(1-\varepsilon\right)}.} \quad (8.41)$$

This result was obtained by van Noije and Ernst (1998) based on an earlier calculation by Goldshtein and Shapiro (1995).

Figure 8.1 shows the scaled velocity distribution function with the second Sonine coefficient a_2 as obtained by the linear theory (8.41) for different values of the coefficient of restitution ε. The maximum of the velocity distribution for gases of dissipative particles is shifted with respect to the Maxwell distribution.

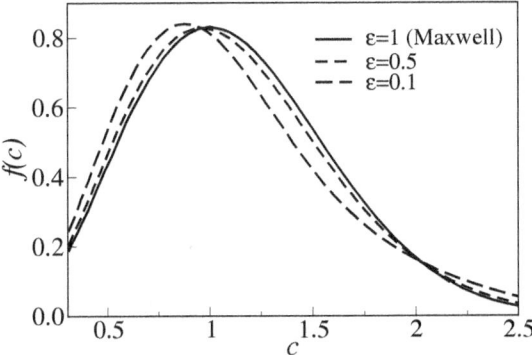

FIG. 8.1. The scaled velocity distribution function for a granular gas of particles which collide with $\varepsilon = $ const. for different values of ε. The Sonine polynomials expansion (7.4) is cut after the second term. For the second Sonine coefficient a_2, the linear result (8.41) is employed.

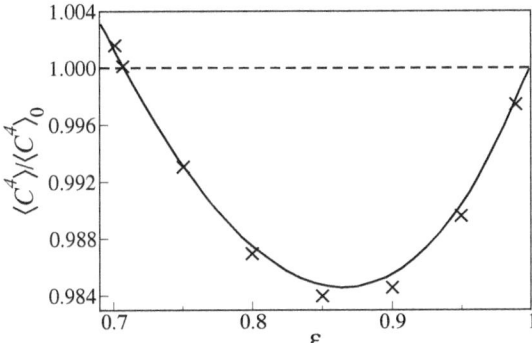

FIG. 8.2. The fourth moment of the scaled velocity distribution function $\langle c^4 \rangle / \langle c^4 \rangle_0 = (4/15) \langle c^4 \rangle$ as a function of the coefficient of restitution ε. For the Maxwell distribution this quantity is one. The points show the results of simulations (data taken from Brey et al., 1996), the line is the result of the linear theory for a_2, equation (8.41).

At the same time the high-velocity tail is more populated. The latter effect will be discussed in Chapter 10.

Brey et al. (1996) have performed direct Monte Carlo simulations (DSMC) of the Boltzmann equation and determined the fourth moment of the velocity distribution. Since a_2 characterizes this quantity, $\langle c^4 \rangle = (15/4)(1 + a_2)$ (see (7.11)), the accuracy of the theory may be checked by comparing the numerical and theoretical values of $\langle c^4 \rangle$. As shown in Fig. 8.2 the linear theory predicts the value of the fourth moment of the velocity distribution function rather accurately for small dissipation.

8.3.2 *Solution of kinetic integrals by means of computer algebra*

The evaluation of even the most simple coefficient μ_2 in linear approximation requires a lot of routine manipulations, although, it is nothing but straightforward algebra. Instead of performing the tedious computation manually, we can leave this problem to a computer. To this end, we wrote a Maple program which processes the kinetic integrals of the type (8.23) automatically. The detailed description of this program is given in Appendix A, here we illustrate its application to the calculation of μ_2 and μ_4 using this program as a black box. Below we give the program worksheet for the computation of μ_2:

```
> restart;
> libname:=libname,'/home/kies/SMGG/':
> with(KineticIntegral);
```
   $[DefDimension, DefJ, DefS, KIinit, getJexpr, unDefJ]$
```
> KIinit();
> DefDimension(3):
```

Dimension = 3

Sonine polynomials defined
```
> DefJ():
```

Basic Integrals will be evaluated
```
> Expr:=Delta2*(-1/2*(1+a2*(S(2,c1p)+S(2,c2p)))):
> getJexpr(0,Expr,0);
```

expressing 16 terms in form of Basic Integrals

$$-\frac{1}{16}\sqrt{\pi}\sqrt{2}\,(\varepsilon - 1)\,(1 + \varepsilon)\,(16 + 3\,a2)$$

```
> unDefJ();
```

Basic Integrals will not be evaluated
```
> getJexpr(0,Expr,0);
```

expressing 16 terms in form of Basic Integrals

$$-\frac{1}{64}(\varepsilon - 1)\,(1 + \varepsilon)(-80\,a2\,J(2, 0, 0, 0, 2, 0) - 20\,a2\,J(0, 2, 0, 0, 2, 0)$$
$$+ 60\,a2\,J(0, 0, 0, 0, 2, 0) + a2\,J(0, 4, 0, 0, 2, 0) + 16\,J(0, 0, 0, 0, 2, 0)$$
$$+ 16\,a2\,J(4, 0, 0, 0, 2, 0) + 8\,a2\,J(2, 2, 0, 0, 2, 0) + 16\,a2\,J(0, 0, 2, 0, 2, 0))$$

The first command **restart** clears the memory from other code which might have been invoked earlier, that is, it initializes the program. The next line extends the path where Maple searches for libraries. In our case this is /home/kies/SMGG/. This directory contains the library **KineticIntegral** which is loaded by the **with()** command. Maple outputs the names of routines which are available with this library: **DefJ, KIinit, DefS, getJexpr, unDefJ** and **DefDimension**. A short explanation is listed below, for a detailed description see Appendix A.

LINEAR APPROXIMATION OF THE SECOND SONINE COEFFICIENT 77

Command	Description
`KIinit()`	Initializes the computation of the kinetic integral and defines some variables and functions
`DefDimension(d)`	defines the dimensionality of the problem, $d = 2$ or $d = 3$
`DefJ()`	defines the Basic Integrals according to (8.31–8.33). Hence, the final expression does not contain Basic Integrals $J_{k,l,m,n,p,\alpha}$ but the according numbers
`unDefJ()`	The Basic Integrals are not evaluated, that is, the result is given in terms of Basic Integrals. This option is of value for testing purposes.

`getJexpr(i,Expr,j)`	Main routine to evaluate the kinetic integral. The parameters are:
	`i` indicates whether the computation is performed for $\varepsilon = $ const. (`i=0`) or for a gas of viscoelastic particles (`i=1`, see Chapter 9)
	`j` indicates the desired output format
	    `j=0`: express the Basic Integral J analytically, using the Gamma function and its properties
	    `j=1`: express J as real numbers
	    `j=2`: present the result as a Taylor expansion using the numerical coefficients ω_0 and ω_1 (see Chapter 9)
	`Expr` the argument of the kinetic integral, for explanation see text

After having loaded the library, the main body of the program starts with the initialization routine `KIinit()`. Since `KIinit()` internally calls `DefJ`, Maple replies `Basic Integrals will be evaluated`, indicating that the result will not contain unevaluated Basic Integrals $J_{k,l,m,n,p,\alpha}$ but the according numbers due to (8.31–8.33). Next we define the core of the kinetic integral. Kinetic integrals have the structure

$$\int d\vec{c}_1 \int d\vec{c}_2 \int d\vec{e}\, \Theta\left(-\vec{c}_{12}\cdot\vec{e}\right) |\vec{c}_{12}\cdot\vec{e}|\, \phi(c_1)\, \phi(c_2)\, \text{Expr}\left(\vec{c}_1, \vec{c}_2\right), \qquad (8.42)$$

where Expr $(\vec{c}_1, \vec{c}_2)$ specifies a mathematical expression in terms of variables given in the table on page 261 (Appendix A). Here we will need only the following variables:

	Mathematical notation	Maple notation
the pre-collision velocities	$\vec{c}_1, \vec{c}_2$	c1p, c2p
the after-collision velocities	$\vec{c}_1', \vec{c}_2'$	c1a, c2a
the Sonine coefficients	a_1, a_2	a1, a2
the Sonine polynomials	$S_i\left(\vec{c}^{\,2}\right)$	S(i,c)
the expressions	$\Delta\left(c_1^2 + c_2^2\right),\ \Delta\left(c_1^4 + c_2^4\right)$	Delta2, Delta4

Note the difference of the mathematical notation and the Maple code for the Sonine polynomials, for example, S(2,c1a) stands for $S_2\left[(\vec{c}')^2\right]$.

With these specifications, expression (8.23) with $p = 2$ for the coefficient μ_2 translates into the Maple input

$$\text{Expr}:=\text{Delta2}*(-1/2*(1+a2*(S(2,c1p)+S(2,c2p)))): , \qquad (8.43)$$

where we again keep only linear terms with respect to a_2. This is contained in the listing given above. The evaluation is then started by getJexpr(0,Expr,0). The first argument indicates that a gas of particles which interact with $\varepsilon =$ const. is considered. The third argument indicates that we seek an analytical solution with exact evaluation of the Basic Integrals, for example, $J_{0,0,0,0,2,0}$ is written as $J_{0,0,0,0,2,0} = 4\sqrt{2\pi}$, but not as $J_{0,0,0,0,2,0} = 10.026513$. In the above example for μ_2 we expect a short result, however, for the case of more complicated expressions the exact pre-factors which arise due to the solution of the Basic Integrals (8.31–8.33) may become rather involved. In these cases, in order to obtain a more compact result it may be advantageous to perform the evaluation with the command getJexpr(0,Expr,1).

Maple replies with expressing 16 terms in form of Basic Integrals. Even for the computation of μ_2, which is one of the most simple cases, the program has to process as much as 16 terms and to find representations in form of Basic Integrals for all of them. In case of more complex kinetic integrals there may appear thousands of terms. The information about the number of terms allows to estimate the duration of the computation, which, certainly, depends on the power of the applied computer.

As the final result Maple outputs

$$-\frac{1}{16}\sqrt{2}\sqrt{\pi}(\varepsilon - 1)(1 + \varepsilon)(16 + 3\,a2) \qquad (8.44)$$

which coincides with the previous result (8.37).

For some purposes, it may be desirable to obtain the result in terms of unevaluated Basic Integrals J. To this end we undefine the Basic Integral by the command unDefJ(). If we then start the calculation by getJexpr(0,Expr,0) Maple outputs the expression shown at the end of the above listing. This expression coincides with (8.36).

The Maple library KineticIntegral.m as well as the program which generates this library, KineticIntergralPackage.mws and all programs which we use in this book are available at URL http://www.oup.co.uk/isbn/0-19-853038-2.

Exercise 8.3 *Derive μ_4 given by (8.38) using Maple!*

8.4 Complete solution for the second Sonine coefficient

To go beyond the linear approximation with respect to a_2 the full equation (8.23) for μ_p is to be solved, including the contribution of the order $\mathcal{O}\left(a_2^2\right)$ which have

COMPLETE SOLUTION FOR THE SECOND SONINE COEFFICIENT

been omitted in the previous section. This has been done by Brilliantov and Pöschel (2000a) in the traditional way using pencil and paper. The coefficient μ_4 in this case is expressed by 328 Basic Integrals. Below we present the complete solution for a_2, obtained by automatic formula manipulation using Maple.

```
> restart;
> libname:=libname,'/home/kies/SMGG/':
> with(KineticIntegral):
> KIinit():
> DefDimension(3):
> DefJ();
```

Basic Integrals will be evaluated

```
> Expr:=Delta2*(-1/2*(1+a2*(S(2,c1p)+S(2,c2p))
> + a2^2*S(2,c1p)*S(2,c2p))):
> mu2:=getJexpr(0,Expr,0);
```

expressing 60 terms in form of Basic Integrals

$$\mu 2 := -\frac{1}{1024}\sqrt{\pi}\sqrt{2}\,(3\,a2 + 32)^2\,(\varepsilon - 1)\,(1 + \varepsilon)$$

```
> Expr:=Delta4*(-1/2*(1+a2*(S(2,c1p)+S(2,c2p))
> + a2^2*S(2,c1p)*S(2,c2p))):
> mu4:=getJexpr(0,Expr,0);
```

expressing 328 terms in form of Basic Integrals

$$\mu 4 := \frac{1}{2048}\sqrt{\pi}\sqrt{2}\,(1+\varepsilon)(-2048\,\varepsilon^3 + 30\,\varepsilon^3\,a2^2 - 1920\,\varepsilon^3\,a2 + 2048\,\varepsilon^2$$
$$- 30\,a2^2\,\varepsilon^2 + 1920\,a2\,\varepsilon^2 - 9\,a2^2\,\varepsilon - 13248\,a2\,\varepsilon - 9216\,\varepsilon + 137\,a2^2$$
$$+ 9216 + 17344\,a2)$$

```
> A2:=solve(5*mu2*(1+a2)-mu4=0,a2):
```

plot results

```
> AA2:=unapply(Re(A2[2]),epsilon):
> A2NE:=epsilon->(16*(1-epsilon)*(1-2*epsilon^2))
> /(81-17*epsilon+30*epsilon^2*(1-epsilon)):
> plot([AA2,A2NE],0..1,linestyle=[1,4],
> legend=["nonlinear theory","linear theory (v. Noije, Ernst)"],
> font=[COURIER,20],thickness=2, labels=[e,a2],color=black);
```

After the initialization we calculate μ_2 and μ_4 applying `getJexpr()`. Using the expressions for these coefficients we solve (8.20) for a_2. This is done in the first 13 lines of the program, which comprise the full solution of the problem which on paper occupies many pages. The rest of the program, which follows after the line `plot results`, is just related to plotting the data and comparing them with the result of the linear theory (8.41). For demonstration in the above program we have used Maple also to plot the results. In the following we will drop plotting commands from the Maple code.

From the program we obtain the moments of the collision integral μ_2 and μ_4 in terms of a_2 which we rewrite in a slightly more convenient notation:

$$\mu_2 = \sqrt{2\pi}\left(1 - \varepsilon^2\right)\left(1 + \frac{3}{16}a_2 + \frac{9}{1024}a_2^2\right) \quad (8.45)$$

$$\mu_4 = 4\sqrt{2\pi}\left(T_1 + a_2 T_2 + a_2^2 T_3\right) \quad (8.46)$$

with

$$
\begin{aligned}
T_1 &= \frac{1}{4}\left(1 - \varepsilon^2\right)\left(\frac{9}{2} + \varepsilon^2\right) \\
T_2 &= \frac{3}{128}\left(1 - \varepsilon^2\right)\left(69 + 10\varepsilon^2\right) + \frac{1}{2}\left(1 + \varepsilon\right) \\
T_3 &= \frac{1}{64}\left(1 + \varepsilon\right) + \frac{1}{8192}\left(1 - \varepsilon^2\right)\left(9 - 30\varepsilon^2\right) \, .
\end{aligned}
\quad (8.47)
$$

This result is exact and represents the complete description of the system on the level of the second Sonine coefficient. Obviously, when substituting μ_2 and μ_4 into (8.20) as it has been done in the program, a cubic equation for a_2 is obtained.

One of the solutions is close to the solution $a_2^{\text{NE}}(\varepsilon)$ (see (8.41)) which has been obtained in linear approximation above. The other solutions diverge in the elastic limit $\varepsilon \to 1$ (Fig. 8.3). Since $a_2 \to 0$ in this limit, it seems most probable that these two roots do not correspond to a physical solution of the Boltzmann equation. We discuss this issue in more detail below.

The solution of the full problem as obtained by the Maple program, may be also written using the solution a_2^{NE} of the linear problem (Brilliantov and Pöschel, 2000a):

$$a_2 = a_2^{\text{NE}}\left(1 - \frac{1005\left(1 - \varepsilon^2\right) - 4096 T_3}{6080\left(1 - \varepsilon^2\right) - 4096 T_2} a_2^{\text{NE}} + \cdots\right), \quad (8.48)$$

where terms of the order $\mathcal{O}\left(\left[a_2^{\text{NE}}\right]^3\right)$ and higher are neglected. In Fig. 8.4 the linear approximation a_2^{NE} and the exact solution of the full problem a_2 as obtained from the Maple program are shown as a function of the coefficient of restitution ε together with the approximation (8.48). The maximal deviation between a_2 and a_2^{NE} is about 10% for small ε and decreases as ε tends to 1.

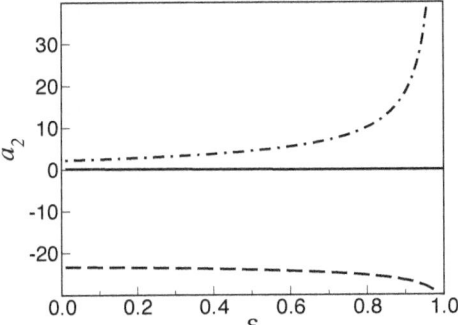

FIG. 8.3. All three solutions of (8.20) for the second Sonine coefficient a_2 over the coefficient of restitution ε.

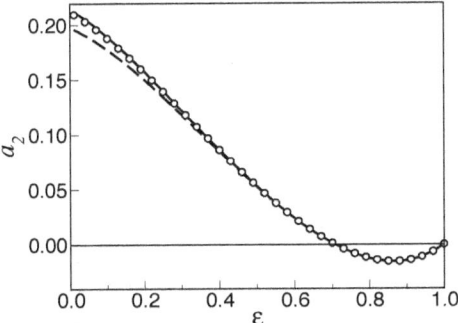

FIG. 8.4. The exact solution for the second Sonine coefficient a_2 as a function of the coefficient of restitution ε (full line). The dashed line shows the solution in linear approximation a_2^{NE} (van Noije and Ernst, 1998) as given in (8.41). Approximation (8.48) is shown by circles.

8.5 Time-dependent scaled velocity distribution function

So far we have discussed the properties of the velocity distribution function in its steady state. This means that all time dependence of $f(\vec{v},t)$ is expressed by the time dependence of the thermal velocity v_T whereas the shape of the distribution function is invariant and may be described by the time-independent scaled velocity distribution function

$$f(\vec{v},t) = \frac{n}{v_T^3(t)} \tilde{f}(\vec{c}) . \tag{8.49}$$

If we consider processes which alter the shape of the distribution function we have to generalize the scaled distribution function and consider its time dependence as well:

$$f(\vec{v},t) = \frac{n}{v_T^3(t)} \tilde{f}(\vec{c},t) , \tag{8.50}$$

Examples of such processes are the relaxation of the distribution function from a certain initial state, for example, from an initial Maxwell distribution, and the relaxation of the distribution function after a perturbation. Another example is the cooling of a gas of viscoelastic particles where the Ansatz (8.49) does not solve the Boltzmann equation as shown in Chapter 9. For such gases the more general scaled velocity distribution function $\tilde{f}(\vec{v}, t)$ is needed.

The Boltzmann equation (6.21) may also be formulated for the scaled function $\tilde{f}(\vec{c}, t)$. For the derivation we follow the same steps as in Section 8.1. We obtain

$$\frac{\partial}{\partial t} f(\vec{v}_1, t) = -\frac{n}{v_T^4}\left(3\tilde{f} + c_1 \frac{\partial \tilde{f}}{\partial c_1}\right)\frac{dv_T}{dt} + \frac{n}{v_T^3}\frac{\partial}{\partial t}\tilde{f}(\vec{c}, t). \quad (8.51)$$

This result coincides with (8.2) for $\tilde{f}(\vec{c})$ (when taking into account $\partial c_1 / \partial v_T = -c_1/v_T$), except for the last term, which originates from the explicit time dependence of $\tilde{f}(\vec{c}, t)$.

This additional term in (8.51) as compared to (8.2) gives rise to a corresponding term in the Boltzmann equation for $\tilde{f}(\vec{c}, t)$ as compared with its analogue (8.10) for $\tilde{f}(\vec{c})$. It reads

$$\frac{\mu_2}{3}\left(3 + c_1 \frac{\partial}{\partial c_1}\right)\tilde{f}(\vec{c}, t) + B^{-1}\frac{\partial}{\partial t}\tilde{f}(\vec{c}, t) = \tilde{I}\left(\tilde{f}, \tilde{f}\right), \quad (8.52)$$

where B has been defined in (8.11). The dimensionless collision integral itself depends now on time via the time-dependent scaled function $\tilde{f}(\vec{c}, t)$. Conseqently, the moments of the collision integral μ_p as well as the quantities $\langle c^p \rangle$ depend on time too.

As previously in Section 8.2 we seek the solution of the Boltzmann equation using the Sonine polynomials expansion

$$\tilde{f}(\vec{c}, t) = \phi(c)\left(1 + \sum_{p=1}^{\infty} a_p(t) S_p(c^2)\right), \quad (8.53)$$

where the Sonine coefficients are now time-dependent. To find equations for $a_p(t)$ we multiply both sides of (8.52) with c_1^p and integrate over $\vec{c}_1$. This is the same procedure which had been applied to derive algebraic equations for the time-independent a_p from (8.10). Since (8.52) differs from (8.10) only by the term $B^{-1}\frac{\partial}{\partial t}\tilde{f}(\vec{c}, t)$, only this term requires further consideration. The transformation of the other terms yields correspondingly $-\frac{1}{3}\mu_2 p\langle c^p \rangle$ and $-\mu_p$. The new term transforms into

$$\int d\vec{c}_1 c_1^p B^{-1} \frac{\partial}{\partial t}\tilde{f}(\vec{c}, t) = B^{-1} \sum_{k=1}^{\infty} \frac{\partial}{\partial t} a_k(t) \int d\vec{c}_1 c_1^p \phi(c_1) S_k(c_1^2)$$
$$= B^{-1} \sum_{k=1}^{\infty} \dot{a}_k \nu_{kp}, \quad (8.54)$$

where we use expansion (8.53) and definition (7.12) of the moments of the Sonine polynomials ν_{kp}. With (8.54), equation (8.52) transforms into a set of equations for $p = 2, 4, \ldots$:

$$\frac{1}{3}\mu_2 p \langle c^p \rangle - B^{-1} \sum_{k=1}^{\infty} \dot{a}_k \nu_{kp} = \mu_p, \tag{8.55}$$

where, according to (7.17), $\nu_{kp} = 0$ if $k > p/2$. The first of them for $p = 2$ is an identity since

$$\begin{aligned} \langle c^2 \rangle &= 3/2, &\text{see (8.16)} \\ a_1 &= 0 \\ \nu_{k2} &= 0 &\text{for all } k > 1, \end{aligned} \tag{8.56}$$

thus, the sum in (8.55) vanishes. For $p = 4$ we obtain with

$$\begin{aligned} \nu_{24} &= 15/4, &\text{see (7.13)} \\ \nu_{k4} &= 0 &\text{for } k > 2 \end{aligned} \tag{8.57}$$

the equation

$$\boxed{\dot{a}_2 - \frac{4}{3} B \mu_2 (1 + a_2) + \frac{4}{15} B \mu_4 = 0} \tag{8.58}$$

Here μ_2 and μ_4 are given by (8.45, 8.46) with the time-dependent coefficient $a_2(t)$. It is assumed that all a_p with $p > 2$ are negligible. In terms of the mean collision time at the initial temperature T_0

$$\tau_c(0)^{-1} \equiv 4\sqrt{\pi} g_2(\sigma) \sigma^2 n \sqrt{\frac{T_0}{m}} \tag{8.59}$$

and with the reduced temperature $u(t) \equiv T(t)/T_0$ the time-dependent coefficient $B(t)$, defined by (8.11), may be written as

$$B(t) = \sqrt{\frac{u(t)}{8\pi}} \tau_c(0)^{-1}. \tag{8.60}$$

With these definitions we recast (8.58) into the form

$$\frac{da_2}{d\tau} = \frac{1}{15} \sqrt{\frac{2}{\pi}} \sqrt{u} F(a_2) \tag{8.61}$$

where $\tau \equiv t/\tau_c(0)$ is the reduced time, measured in collision units and where $F(a_2)$ is a cubic function of a_2:

$$F(a_2) = 5\mu_2 (1 + a_2) - \mu_4. \tag{8.62}$$

This function has three real roots, that is, there are three different time-independent solutions of (8.61) for the second Sonine coefficient a_2. In Chapter 8.7 we will discuss the stability of these solutions.

8.6 Evolution of temperature for $\varepsilon = $ const.

Once the scaled velocity distribution function has approached its stationary state, the velocity distribution function does not change its shape and evolves via the time-dependent thermal velocity $v_T(t)$. The current temperature of the granular gas which determines $v_T(t)$ evolves on a large time scale. Therefore, the evolution of the temperature is not coupled to the time dependence of a_p which evolves on the short time scale $\tau_c(0)$, that is, a_p approaches its stationary value after few collisions per particle.

The time dependence of temperature is determined by (8.11), which we rewrite in terms of the scaled temperature $u(t) \equiv T(t)/T_0$ and with $B(t)$ as given by (8.59):

$$\frac{du}{dt} = -\frac{2}{3\sqrt{8\pi}}\tau_c(0)^{-1}\mu_2 u^{3/2}. \tag{8.63}$$

The constants μ_2 and $\tau_c(0)$ are given by (8.45, 8.59). This first-order differential equation with constant coefficients can be solved with the initial condition $u(0) = 1$ and yields the time-dependent temperature

$$T(t) = \frac{T_0}{(1+t/\tau_0)^2}. \tag{8.64}$$

The characteristic time τ_0 is a function of a_2 via μ_2 given by (8.45):

$$\tau_0^{-1} = \frac{\mu_2}{3\sqrt{8\pi}}\tau_c(0)^{-1} = \frac{1}{6}(1-\varepsilon^2)\left(1 + \frac{3}{16}a_2 + \frac{9}{1024}a_2^2\right)\tau_c(0)^{-1}. \tag{8.65}$$

The coefficient a_2, in its turn, depends on ε according to (8.48, 8.41). The value

$$\gamma = \tau_0^{-1}/\tau_c(0)^{-1} = \gamma_0\left(1 + \frac{3}{16}a_2 + \frac{9}{1024}a_2^2\right), \tag{8.66}$$

with $\gamma_0 = \frac{1}{6}(1-\varepsilon^2)$ is the dimensionless cooling rate of the granular gas (van Noije and Ernst, 1998). This parameter characterizes the temperature decay rate on the collisional time scale. For slightly inelastic particles ($\varepsilon \to 1$) the temperature varies only slightly on this time scale ($\gamma \to 0$).

The functional form of (8.64) is identical with Haff's law (5.5), which has been derived in the very beginning of this chapter using scaling arguments. There we could obtain only a rough estimate for the temperature relaxation time τ_0. Now we derive a more rigorous expression.

It turns out that for small inelasticity $1-\varepsilon^2 \ll 1$ the temperature relaxation time is much larger than mean collision time $\tau_c(0)$, that is,

$$\boxed{\tau_0 \gg \tau_c(0).} \tag{8.67}$$

In this case, the evolution of temperature occurs on a significantly larger time scale than the time scale of particle collision.

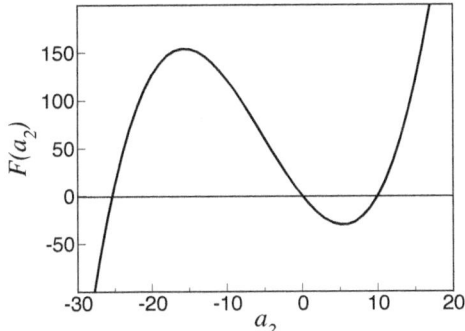

FIG. 8.5. $F(a_2)$ as a function of a_2 for $\varepsilon = 0.8$. The shape of $F(a_2)$ persists for all physical values of the coefficient of restitution, $0 \le \varepsilon \le 1$. Obviously, $F(a_2) = 0$ has three real roots.

8.7 Stability analysis of the Boltzmann equation

In Section 8.4 we have derived a cubic equation for the second Sonine coefficient a_2. It was obtained by substituting μ_2 and μ_4 as given by (8.45, 8.46) into (8.20). This equation was solved by computer algebra and led us to a_2 as a function of the coefficient of restitution ε as drawn in Fig. 8.4. From the plot of the LHS of (8.20), $F(a_2) \equiv 5\mu_2(1 + a_2) - \mu_4$, (see Fig. 8.5) it is obvious that there are three real solutions for a_2.

In the discussion above, we have selected one of the solutions of the cubic equation for a_2 since only this one is in agreement with the condition $\lim_{\varepsilon \to 1} a_2 = 0$. Clearly, the velocity distribution function of a granular gas must approach the Maxwell distribution in the limit of vanishing dissipation, that is, $\lim_{\varepsilon \to 0} \tilde{f}(\vec{c}) = \phi(\vec{c})$. This condition is fulfilled in our approximation (neglecting a_3 and higher) if $\lim_{\varepsilon \to 1} a_2 = 0$. The other two roots, which diverge at $\varepsilon \to 1$ are, however, also valid mathematical solutions of the problem.

It may be speculated whether there exist two additional solutions of the Boltzmann equation for $\varepsilon \ne 1$. If these solutions had a physical meaning, obviously they could not be adequately described by the Sonine polynomials expansion in our approximation. Indeed, the divergence of $\lim_{\varepsilon \to 1} a_2$ must be compensated by higher order terms with $a_3, a_4, \ldots$ which we have neglected here. Given such alternative solutions of the Boltzmann equation exist, they might develop for certain initial conditions at which the granular gas had been prepared. In simulations only the distribution with a_2 given by (8.48) has been observed. The physical meaning of the other two solutions for a_2 so far lacks an interpretation.

Provided there are more than one physically meaningful distribution which solve the Boltzmann equation there arises the question whether the solution (8.48) is stable with respect to small perturbations and what is the domain of attraction of this particular scaled solution in some parameter space. The stability of the scaled velocity distribution (7.1) means that if due to some reasons (fluctuations, external perturbations, etc.) a deviation from the time-independent scaled distribution arose it would quickly relax back to its steady state.

Certainly, the stability problem is very complicated to be solved in general. Therefore, we restrict ourselves to the stability analysis of the scaled distribution $\tilde{f}(\vec{c})$ given by (7.1), with non-zero coefficient a_2 and negligible higher coefficients a_p with $p > 2$. (For this scaled solution our above results for the coefficients μ_2, μ_4 are valid). Moreover, we assume, that small perturbations of the (vanishingly small) coefficients a_p with $p > 2$ do not influence the stability of the distribution, and analyze the stability only with respect to variation of the coefficient a_2.

The stability of the scaled solution requires $dF/da_2 < 0$. This condition is fulfilled only for the middle root (Fig. 8.5). Figure 8.4 shows that this root is a small correction to $a_2^{\rm NE}$ as predicted by the linear theory (van Noije and Ernst, 1998). For any $0 \le \varepsilon \le 1$ the point $a_2 = 0$ belongs to the attractive interval of this stable root. Therefore, a granular gas will relax to its stationary velocity distribution, characterized by a_2 (see (8.48)), if initially the particle velocities obey a Maxwell distribution.

To analyse the relaxation of a_2 to its stable stationary value we discard in $F(a_2)$ terms of the second and third order in a_2,

$$F(a_2) \simeq 5\sqrt{2\pi}\left((1-\varepsilon^2) - \frac{4}{5}T_1\right) + 5\sqrt{2\pi}\left(\frac{19}{16}(1-\varepsilon^2) - \frac{4}{5}T_2\right)a_2 \qquad (8.68)$$
$$\equiv \alpha + \beta a_2,$$

where T_1 and T_2 are given by (8.47). Then (8.61) reads

$$\frac{da_2}{d\tau} = \frac{1}{15}\sqrt{\frac{2}{\pi}}\sqrt{u}\,(\alpha + \beta a_2) = -\sqrt{\frac{2}{\pi}}\frac{|\beta|}{15}\left(a_2 - a_2^{\rm NE}\right)\sqrt{u}, \qquad (8.69)$$

with $a_2^{\rm NE} = -\alpha/\beta$ which follows from (8.20) in linear approximation for $F(a_2)$. The solution of this equation reads

$$a_2(\tau) - a_2^{\rm NE} = [a_2(0) - a_2^{\rm NE}]\exp\left(-\sqrt{\frac{2}{\pi}}\frac{|\beta|}{15}\int_0^\tau \sqrt{u}(\tau')\,d\tau'\right). \qquad (8.70)$$

The exponent in (8.70) decays to zero on the collision time scale. Since the temperature of the granular gas decays on a time scale much larger than the collision time scale, during the initial time interval of a few collisions $u(\tau) \approx 1$. Consequently the relaxation of any (small) perturbation of a_2 to its stationary value occurs on the collision time scale, that is, almost immediately as compared with the time scale of the temperature decay. This means that the relaxation of a perturbation is decoupled from the temperature decay and we may conclude that the scaled solution of the Boltzmann–Enskog equation (8.5) with a_2 given by (8.41), (8.48) and with negligibly small other Sonine coefficients $a_3, a_4, \ldots$ is stable with respect to small perturbations.

8.8 High-order coefficients of the Sonine polynomials expansion

Throughout this Chapter we have assumed that the velocity distribution function is sufficiently described by the second Sonine coefficient a_2. Our calculations have shown that this assumption leads to a rather small value of a_2 for the entire range of ε (Fig. 8.4) which justifies this conjecture. It is worthwile, however, to compare a_2 with the higher coefficients a_3, a_4, etc.

With $p = 2l$ equation (8.55) reads

$$\frac{\mu_2}{3} 2l \left\langle c^{2l} \right\rangle - B^{-1} \sum_{k=1}^{\infty} \dot{a}_k \nu_{k\,2l} = \mu_{2l}\,. \tag{8.71}$$

According to (7.17) all $\nu_{k\,2l} = 0$ for $k > l$, that is, the infinite sum consists of only l non-trivial terms. We express $\left\langle c^{2l} \right\rangle$ in terms of a_k as given by (7.19) and use (7.18) for $\nu_{k\,2l}$ to obtain

$$\sum_{k=1}^{l} \frac{(-1)^k l!}{k!\,(l-k)!} \left[\dot{a}_k - \left(\frac{2l}{3} B \mu_2 \right) a_k \right] + \left(\frac{2l}{3} B \right) \left[\frac{3\sqrt{\pi}}{4 l \Gamma\,(l+3/2)} \mu_{2l} - \mu_2 \right] = 0\,. \tag{8.72}$$

This equation contains only the first l Sonine coefficients and their first-order time derivatives. In a more convenient notation (8.72) reads for $l = 2, 3, 4, \ldots$

$$\begin{aligned}
\dot{a}_2 &= \frac{4}{3} B \mu_2 \left(1 + a_2 \right) - \frac{4}{15} B \mu_4 \\
\dot{a}_3 &= 2 B \mu_2 \left(1 - a_2 + a_3 \right) - \frac{4}{5} B \left(\mu_4 - \frac{2}{21} \mu_6 \right) \\
\dot{a}_4 &= \frac{8}{3} B \mu_2 \left(1 - a_3 + a_4 \right) - \frac{8}{5} B \left(\mu_4 - \frac{4}{21} \mu_6 + \frac{2}{189} \mu_8 \right)
\end{aligned} \tag{8.73}$$

etc.

The first equation coincides with (8.58). Although these equations contain only the first l Sonine coefficients, the moments μ_2, μ_{2l-2}, μ_{2l} depend on the complete set of all Sonine coefficients a_p, thus the set of equations can be closed only by considering the entire (infinite) set (see also Huthmann et al., 2000). To obtain a finite closed set of equations some truncation scheme is needed, for example, $a_n = 0$ for $n > n_0$.

Another approach was proposed by Huthmann et al. (2000) where it was assumed that the high-order Sonine coefficients decay as

$$a_k \sim \lambda^k \tag{8.74}$$

where λ is a small parameter. This implies, for example, that a_2^2 and a_4 are of the same order of magnitude. This assumption simplifies the analysis and allows for perturbative calculations up to any desired order of k. Unfortunately, there is no theoretical justification for this conjecture. In Huthmann et al. (2000) the Sonine

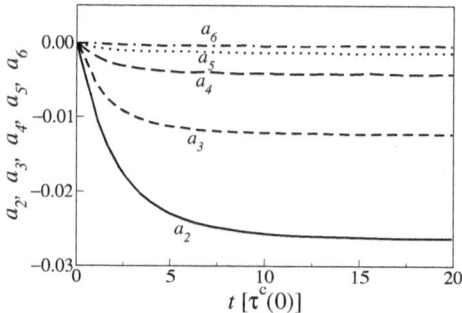

FIG. 8.6. The coefficients $a_2, \ldots a_6$ calculated up to the order $\mathcal{O}\left(\lambda^6\right)$ as a function of time for $\varepsilon = 0.8$ (data taken from Huthmann et al., 2000).

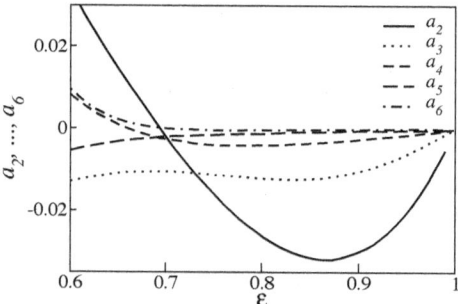

FIG. 8.7. Stationary values of $a_2, \ldots a_6$ calculated up to the order $\mathcal{O}\left(\lambda^6\right)$ as functions of ε (data taken from Huthmann et al., 2000).

coefficients have been computed up to $\mathcal{O}\left(\lambda^6\right)$. The authors performed calculations of μ_p for two-dimensional systems and then solved the system of equations for a_p numerically. Figures 8.6–8.8 illustrate the main results by Huthmann et al. (2000), which may be summarized as follows

- The Sonine coefficients relax to their steady-state values quickly within a time which corresponds to few collisions per particle. This set of Sonine coefficients corresponds to the stable stationary solution of the Boltzmann equation. There exist non-stable stationary solutions too.
- For small inelasticity, the Sonine coefficients are small and decrease rapidly with increasing order. In this case, neglecting all coefficients $a_p, p > 2$ yields rather accurate results. For a_2 the linear theory can be used (see Sec. 8.3), which corresponds to calculations up to the order $\mathcal{O}\left(\lambda^2\right)$.
- For larger inelasticity, the Sonine coefficients a_p with $p > 2$ are not negligible and are of the same order of magnitude as a_2. Moreover, for $\varepsilon \lesssim 0.6$ the coefficient a_2 does not converge with increasing order of λ. This means that for $\varepsilon < 0.6$ it is not possible to expand the velocity distribution function near the Maxwell distribution.

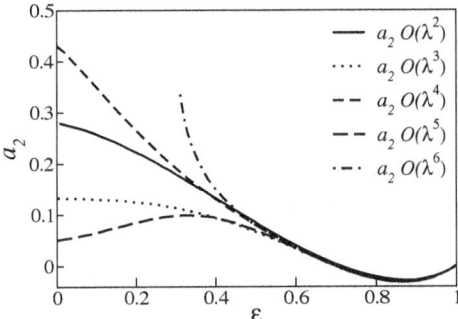

FIG. 8.8. Stationary values of a_2 up to the order $\mathcal{O}\left(\lambda^6\right)$ as a function of ε (data taken from Huthmann et al., 2000).

We wish to note, that the conclusion by Huthmann et al. (2000) that the Sonine polynomials expansion does not converge for large inelasticities does not agree with the molecular dynamics simulations presented in the same paper. Instead, the value of a_2, as obtained by simulations was very close to the prediction of the linear theory, a_2^{NE}, (8.41). Thus, the conjecture (8.74) may be questionable.

We want to remind that the results refer to the simplified case $\varepsilon = $ const. Fortunately, however, for real materials the coefficient of restitution is not a constant but increases with decreasing impact velocity (see Chapter 3), that is, with decreasing temperature. Therefore, even if the initial dissipation was large, it decreases in due course of time to values which allow for the application of the Sonine polynomials expansion.

9

VELOCITY DISTRIBUTION FUNCTION AND TEMPERATURE FOR VISCOELASTIC PARTICLES

For granular gases of viscoelastic particles the shape of the velocity distribution function cannot be described by a time-independent function. Instead, it evolves along with decaying temperature. Its complicated time dependence allows to define the age of a granular gas.

9.1 Why do we expect qualitatively different distribution functions for the cases $\varepsilon = $ const. and $\varepsilon = \varepsilon(g)$?

In the previous section we have analysed the properties of the velocity distribution function for a granular gas consisting of particles that interact through a constant coefficient of restitution. The assumption $\varepsilon = $ const. significantly simplifies the mathematical analysis. We call this type of collisions a simplified collision model. In Section 3.2.1, however, it has been proven that realistic particles cannot interact via $\varepsilon = $ const. Instead, the coefficient of restitution *must* depend on the impact velocity. Thus, for a granular gas of realistic particles, the results of the previous sections need to be reconsidered.

For the most simple physically realistic material properties, that is, for viscoelastic deformation of the colliding particles the coefficient of restitution has been derived as a function of the normal component of the impact velocity $\varepsilon = \varepsilon(g)$ in Sec. 3.2. The central result of this section is the collision law (3.22) which is employed now to derive the velocity distribution function of a more realistic granular gas.

For $\varepsilon = $ const. the Boltzmann equation (6.21) may be reduced to a set of two uncoupled equations: a first-order ordinary differential equation (8.11) for the evolution of the time-dependent temperature $T(t)$ and a time-independent equation (8.10) which describes the shape of the distribution function $\tilde{f}(\vec{c})$. Thus, for $\varepsilon = $ const. there exists a time-independent scaled velocity distribution function $\tilde{f}(\vec{c})$ which solves the Boltzmann equation.

What is the physical reason for this property? Assume that the granular temperature decays at a time scale which is much larger than the mean collision time. This assumption assures that the velocity distribution function is indenpendent of the initial conditions (provided the velocity distribution function is unique). The microscopic configuration of the particles of a granular gas (A) is characterized by their positions $\vec{r}_i$ and their velocities $\vec{v}_i$. In due course of time the particles propagate and collide with each other, which gives rise to an individual path of each grain. If we now initialize another gas (B) with the same configuration, except that all particle velocities are multiplied by a factor α, the particles will move along *identical* pathes with velocities scaled by α. Formally, both systems behave identically except for the definition of the unit of time. This is a consequence of the constant coefficient of restitution.

The main *macroscopic* dynamical property of a homogeneous cooling granular gas – its temperature – is determined by the average collision frequency which is proportional to $\sqrt{T}$. The collision frequency decays according to the temperature decay due to inelastic collisions. If we were to construct a clock which runs slower and slower in agreement with the decay of temperature, that is, keeping the average number of collisions per time unit invariant, we would not observe *any* evolution of the granular gas (including its velocity distribution) as long as it remains in the homogeneous cooling state. The particle velocity $\vec{c}$ in the system with such a special clock is related to the particle velocity in the laboratory system as $\vec{c} \propto \vec{v}/\sqrt{T}$. According to our arguments the velocity distribution function of the former system $\tilde{f}(\vec{c})$ does not depend on time, thus, in our laboratory system the distribution function $\tilde{f}(\vec{v}/\sqrt{T})$ will also be time-independent. This is just the scaled distribution function that characterizes the shape of the velocity distribution. The arguments given above are valid provided the velocity correlations and spatial correlation which develop due to dissipative collisions may be neglected.

For granular gases consisting of viscoelastic particles the situation is different. If we again consider two gases (A) and (B) with scaled velocities, just as in the discussion above, the particles of the gases do not follow the same traces, that is, the systems are not mechanically identical, since the coefficient of restitution for viscoelastic particles depends on the *absolute* relative velocity. The typical coefficient of restitution may be defined by $\varepsilon(v_T)$ where v_T is the time-dependent thermal velocity. Consequently, the gases (A) and (B) are characterized by different coefficients of restitution, which implies different scaled distribution functions. In other words, for such a gas it is not possible to set up a clock as discussed above to obtain a time-invariant system. From this discussion, it follows that for a gas consisting of viscoelastic particles the Boltzmann equation does not separate into an equation for the evolution of temperature and a time-independent equation for the shape of the velocity distribution. Therefore, the shape of the distribution *must* vary over a period of time.

The qualitative evolution of the shape of the distribution function may be predicted by the following reasoning: given almost elastically colliding particles with $\varepsilon(g) \lesssim 1$ where $g \propto \sqrt{T_0}$ and T_0 is the initial temperature. For such ε we can assume that the cooling is quasi-static. At each time instant the typical (thermal) velocity $v_T(t)$ corresponds to an effective value of the coefficient of restitution $\varepsilon(t) = \varepsilon(g(t))$. We can now use the second Sonine coefficient $a_2(\varepsilon)$ as obtained in Sec. 8.2 for $\varepsilon = $ const. to characterize the shape of the velocity distribution function. We assume that in the quasi-static regime the current value of a_2 is determined by the current value of the effective coefficient of restitution $\varepsilon(t)$ in the same way as it is shown in Fig. 8.4. As the gas cools down the effective coefficient of restitution increases and tends to 1. Thus, we expect that for $\varepsilon(g) \lesssim 1$ the absolute value of the (negative) coefficient a_2 decreases and tends to zero.

Assume now that the same qualitative analysis is valid for smaller ε too. Then according to Fig. 8.4 the second Sonine coefficient has initially a positive sign, that is, the velocity distribution is shiftet to the left with respect to the Maxwell distribution (see Fig. 8.1). As the gas cools, the thermal velocity decreases which corresponds to a gradual increase of the effective ε. Eventually, ε reaches the value which corresponds to $a_2 = 0$. At this instant the velocity distribution coincides with a Maxwell distribution. While further cooling a_2 becomes negative; correspondingly, the velocity distribution function shifts to the right with respect to the Maxwell distribution. With decaying temperature the particles behave more and more elastically and, therefore, a_2 decreases further. This process continues until $\varepsilon \approx 0.87$ when a_2 reaches its minumum (see Fig. 8.4). From this moment on a_2 increases (its absolute value decreases) with decaying temperature, that is, the distribution again approaches the Maxwell distribution.

This qualitative description of the scenario is based on the results of the preceding section which have been obtained with the simplifying assumption $\varepsilon = \text{const}$. Although the above discussion for the time dependence of the second Sonine coefficients leads to the qualitatively correct result (cf. Fig. 9.5), a quantitative description certainly needs a careful mathematical analysis (Brilliantov and Pöschel, 2000d, 2001) which will be given in the following sections.

9.2 Collision integral for a gas of viscoelastic particles

The collision integral

$$I(f,f) = \sigma^2 \int d\vec{v}_2 \int d\vec{e}\, \Theta(-\vec{v}_{12} \cdot \vec{e}) |\vec{v}_{12} \cdot \vec{e}| \qquad (9.1)$$
$$\times \{\chi f(\vec{v}_1'', t) f(\vec{v}_2'', t) - f(\vec{v}_1, t) f(\vec{v}_2, t)\},$$

which has been defined in (6.21) contains a factor χ. As discussed in Sec. 6.2, this factor equals the product of the Jacobian of the transformation $(\vec{v}_1'', \vec{v}_2'') \to (\vec{v}_1, \vec{v}_2)$ and the ratio of the lengths of the collision cylinders of the inverse and the direct collisions (6.16):

$$\chi \equiv \frac{|\vec{g}''|}{|\vec{g}|} \frac{\mathcal{D}(\vec{v}_1'', \vec{v}_2'')}{\mathcal{D}(\vec{v}_1, \vec{v}_2)}, \qquad (9.2)$$

(recall that $\vec{v}_1''$, $\vec{v}_2''$ and $\vec{g}''$ are the pre-collision velocities and the normal relative velocity of the inverse collision). For the case $\varepsilon = \text{const}$. we had obtained $\chi = 1/\varepsilon^2$, see (6.17).

For viscoelastic particles, the coefficient of restitution for the inverse collision reads according to (3.22)

$$\varepsilon(g'') = 1 - C_1 A \kappa^{2/5} (g'')^{1/5} + C_2 A^2 \kappa^{4/5} (g'')^{2/5} \mp \cdots \qquad (9.3)$$

It is convenient to write $g'' \equiv |\vec{g}''|$ as a function of $g \equiv |\vec{g}|$. This may be done iteratively using (6.9)

$$g'' = \frac{g}{\varepsilon(g'')} = \frac{g}{\varepsilon\left(\frac{g}{\varepsilon(g'')}\right)} = \cdots, \qquad (9.4)$$

or explicitly

$$g'' = \frac{g}{1 - \dfrac{C_1 A \kappa^{2/5} g^{1/5}}{\left(1 - \dfrac{C_1 A \kappa^{2/5} g^{1/5}}{\cdots}\right)^{1/5}} + \dfrac{C_2 A^2 \kappa^{4/5} g^{2/5}}{\left(1 - \dfrac{C_1 A \kappa^{2/5} g^{1/5}}{\cdots}\right)^{2/5}}}. \qquad (9.5)$$

We expand this expression with respect to the small dissipative parameter A, use the relation $C_2 = \frac{3}{5} C_1^2$ from (3.45) and obtain the ratio of the lengths of the collision cylinders

$$\frac{g''}{g} = 1 + C_1 A \kappa^{2/5} g^{1/5} + \frac{3}{5}\left(C_1 A \kappa^{2/5}\right)^2 g^{2/5} \mp \cdots \qquad (9.6)$$

To find the Jacobian of the transformation $(\vec{v}_1'', \vec{v}_2'') \to (\vec{v}_1, \vec{v}_2)$, we recast the collision law (6.10) for the inverse collision into the form

$$\vec{v}_1'' = \vec{v}_1 + \frac{1}{2}(\vec{g}'' - \vec{g}), \qquad \vec{v}_2'' = \vec{v}_2 - \frac{1}{2}(\vec{g}'' - \vec{g}), \qquad (9.7)$$

which follows from (6.8, 6.9). We now choose the coordinate axis z along the unit vector $\vec{e}$. Thus,

$$\begin{aligned} g_z &= v_{1,z} - v_{2,z} = g \\ g_z'' &= v_{1,z}'' - v_{2,z}'' = -g''. \end{aligned} \qquad (9.8)$$

The negative sign in the second equation is caused by the fact that before the collision the particles approach each other, that is, $\vec{v}_{12}'' \cdot \vec{e} < 0$, whereas after the collision they depart from each other, that is, $\vec{v}_{12} \cdot \vec{e} > 0$. In components (9.7) reads

$$\begin{aligned} v_{1,x}'' &= v_{1,x} & v_{2,x}'' &= v_{2,x} \\ v_{1,y}'' &= v_{1,y} & v_{2,y}'' &= v_{2,y} \\ v_{1,z}'' &= v_{1,z} - \frac{1}{2}(g'' + g) & v_{2,z}'' &= v_{2,z} + \frac{1}{2}(g'' + g). \end{aligned} \qquad (9.9)$$

The relative velocity g depends on $v_{1,z}$ and $v_{2,z}$ according to (9.8) and g'' depends on g via (9.6) and, thus, also on $v_{1,z}, v_{2,z}$. Therefore, (9.9) explicitly expresses all components of the velocities $\vec{v}_1''$, $\vec{v}_2''$ in terms of components of $\vec{v}_1$ and $\vec{v}_2$. Straightforward calculation yields the Jacobian of this transformation

$$\frac{\mathcal{D}(\vec{v}_1'', \vec{v}_2'')}{\mathcal{D}(\vec{v}_1, \vec{v}_2)} = \left[1 + \frac{6}{5} C_1 A \kappa^{2/5} g^{1/5} + \frac{21}{25} \left(C_1 A \kappa^{2/5} \right)^2 g^{2/5} \mp \cdots \right]. \qquad (9.10)$$

Multiplying (9.10) with g''/g from (9.6) and collecting terms of the same order of A, we find the factor χ:

$$\chi = 1 + \frac{11}{5} C_1 A \kappa^{2/5} |\vec{v}_{12} \cdot \vec{e}|^{1/5} + \frac{66}{25} C_1^2 A^2 \kappa^{4/5} |\vec{v}_{12} \cdot \vec{e}|^{2/5} + \cdots \qquad (9.11)$$

In terms of the scaled velocities $\vec{v}_1 = v_T \vec{c}_1$ and $\vec{v}_2 = v_T \vec{c}_2$

$$\tilde{\chi} = 1 + \frac{11}{5} C_1 \delta' |\vec{c}_{12} \cdot \vec{e}|^{1/5} + \frac{66}{25} C_1^2 \delta'^2 |\vec{c}_{12} \cdot \vec{e}|^{2/5} + \cdots \qquad (9.12)$$

is obtained with

$$\delta'(t) \equiv A \kappa^{2/5} \left(\frac{2T(t)}{m} \right)^{1/10} \equiv \delta \left(\frac{2T(t)}{T_0} \right)^{1/10} \qquad (9.13)$$

The factor $\tilde{\chi}$ depends not only on the scaled velocity $\vec{c}_{12}$ but also explicitly on time via δ'. In (9.13) we also define the small parameter

$$\delta \equiv A \kappa^{2/5} \left(\frac{T_0}{m} \right)^{1/10} \qquad (9.14)$$

with T_0 being the initial temperature. Because of the time dependence of χ we use the generalized scaled velocity distribution function (8.50)

$$f(\vec{v}, t) = \frac{n}{v_T^3(t)} \tilde{f}(\vec{c}, t). \qquad (9.15)$$

Correspondingly, the dimensionless collision integral depends now explicitly on time:

$$\tilde{I}(\tilde{f}, \tilde{f}) = \int d\vec{c}_2 \int d\vec{e} \, \Theta(-\vec{c}_{12} \cdot \vec{e}) |\vec{c}_{12} \cdot \vec{e}| \\ \times \left[\tilde{\chi} \tilde{f}(\vec{c}_1'', t) \tilde{f}(\vec{c}_2'', t) - \tilde{f}(\vec{c}_1, t) \tilde{f}(\vec{c}_2, t) \right]. \qquad (9.16)$$

9.3 Moments of the collision integral for viscoelastic particles

We assume that the dissipation is small and that it is sufficient to keep only the first non-vanishing coefficient a_2 in the Sonine polynomials expansion for the velocity distribution function. Then we can use (8.58)

$$\dot{a}_2 - \frac{4}{3} B\mu_2 (1 + a_2) + \frac{4}{15} B\mu_4 = 0 \qquad (9.17)$$

to determine the evolution of a_2: Indeed, in the derivation of (8.58) we did not presume any particular model for the coefficient of restitution. The moments μ_p in (9.17) are given by (8.21)

$$\mu_p = -\frac{1}{2} \int d\vec{c}_1 \int d\vec{c}_2 \int d\vec{e}\, \Theta \left(-\vec{c}_{12} \cdot \vec{e}\right) |\vec{c}_{12} \cdot \vec{e}|\, \tilde{f}(c_1, t)\, \tilde{f}(c_2, t) \Delta \left(c_1^p + c_2^p\right). \qquad (9.18)$$

The calculation of the moments μ_p for a gas of viscoelastic particles is very similar to the case of a constant coefficient of restitution. An important difference arises, however, due to the dependence of ε on the impact velocity, which reads in terms of the scaled variables as

$$\varepsilon = 1 - C_1 \delta'(t) |\vec{c}_{12} \cdot \vec{e}|^{1/5} + \frac{3}{5} C_1^2 \delta'^{\,2}(t) |\vec{c}_{12} \cdot \vec{e}|^{2/5} + \cdots \qquad (9.19)$$

The terms $\Delta (c_1^p + c_2^p)$ with $p = 2, 4$ depend explicitly on ε due to (8.28, 8.29) which implies an additional velocity dependence via the factors $(1 + \varepsilon)$, $(1 - \varepsilon^2)$, etc. Nevertheless, the structure of the integrands remains essentially the same. Therefore, the result for μ_p may be still expressed in terms of the Basic Integrals $J_{k,l,m,n,p,\alpha}$ given by (8.30), that is, we can apply the Maple program which was explained in Section 8.3 (see also Appendix A). We expand the moments formally, keeping terms up to the second order in δ' and a_2:

$$\mu_2 = \left(\mathcal{A}_1 + \mathcal{A}_2 a_2 + \mathcal{A}_3 a_2^2\right) + \delta' \left(\mathcal{A}_4 + \mathcal{A}_5 a_2 + \mathcal{A}_6 a_2^2\right) \\ - \delta'^{\,2} \left(\mathcal{A}_7 + \mathcal{A}_8 a_2 + \mathcal{A}_9 a_2^2\right) \qquad (9.20)$$

and

$$\mu_4 = \left(\mathcal{B}_1 + \mathcal{B}_2 a_2 + \mathcal{B}_3 a_2^2\right) + \delta' \left(\mathcal{B}_4 + \mathcal{B}_5 a_2 + \mathcal{B}_6 a_2^2\right) \\ - \delta'^{\,2} \left(\mathcal{B}_7 + \mathcal{B}_8 a_2 + \mathcal{B}_9 a_2^2\right), \qquad (9.21)$$

where $\mathcal{A}_n$ and $\mathcal{B}_n$ are pure numbers. The result is expressed with the abbreviation

$$\begin{aligned} \omega_0 &\equiv 2\sqrt{2\pi}\, 2^{1/10} \Gamma\left(\frac{21}{10}\right) C_1 \approx 6.485 \\ \omega_1 &\equiv \sqrt{2\pi}\, 2^{1/5} \Gamma\left(\frac{16}{5}\right) C_1^2 \approx 9.285\,. \end{aligned} \qquad (9.22)$$

The Maple program for the calculation of μ_2 and μ_4 is given below. It coincides with the program for the case $\varepsilon = $ const. given on page 79, except for the line getJexpr(1,Expr,2). The first parameter in this expression indicates that the calculation is done with the coefficient of restitution for viscoelastic particles and the third argument means that the output is given in terms of ω_0 and ω_1 due to (9.22).

```
> restart;
> libname:=libname,'/home/kies/SMGG/':
> with(KineticIntegral):
> KIinit():
> DefDimension(3):
```

Dimension = 3

Sonine polynomials defined
```
> DefJ():
```

Basic Integrals will be evaluated
```
> Expr:=Delta2*(-1/2*(1+a2*(S(2,c1p)+S(2,c2p)) +
> a2^2*S(2,c1p)*S(2,c2p))):
> mu2:=getJexpr(1,Expr,2);
```

expressing 120 terms in form of Basic Integrals

$$\mu 2 := \frac{1}{2500} \omega 0 \left(600\, a2 + 21\, a2^2 + 2500\right) dprime -$$
$$\frac{1}{640000} \omega 1 \left(4641\, a2^2 + 190400\, a2 + 640000\right) dprime^2 + O(dprime^3)$$

```
> Expr:=Delta4*(-1/2*(1+a2*(S(2,c1p)+S(2,c2p)) +
> a2^2*S(2,c1p)*S(2,c2p))):
> mu4:=getJexpr(1,Expr,2);
```

expressing 626 terms in form of Basic Integrals

$$\mu 4 := \frac{1}{8} a2 \sqrt{\pi} \sqrt{2} \left(32 + a2\right) - \frac{7}{12500} \omega 0 \left(-10000 - 12900\, a2\right.$$
$$\left. + 81\, a2^2\right) dprime + \frac{7}{70400000} \omega 1 \left(637119\, a2^2 - 77440000\right.$$
$$\left. - 109022400\, a2\right) dprime^2 + O(dprime^3)$$

In a more convenient representation, the coefficients of (9.20, 9.21) read

$$\mathcal{A}_1 = 0, \qquad \mathcal{A}_2 = 0, \qquad \mathcal{A}_3 = 0,$$
$$\mathcal{A}_4 = \omega_0, \qquad \mathcal{A}_5 = \frac{6}{25}\omega_0, \qquad \mathcal{A}_6 = \frac{21}{2500}\omega_0, \qquad (9.23)$$
$$\mathcal{A}_7 = \omega_1, \qquad \mathcal{A}_8 = \frac{119}{400}\omega_1, \qquad \mathcal{A}_9 = \frac{4641}{640000}\omega_1$$

and

$$\mathcal{B}_1 = 0, \qquad \mathcal{B}_2 = 4\sqrt{2\pi}, \qquad \mathcal{B}_3 = \frac{1}{8}\sqrt{2\pi},$$

$$\mathcal{B}_4 = \frac{28}{5}\omega_0, \qquad \mathcal{B}_5 = \frac{903}{125}\omega_0, \qquad \mathcal{B}_6 = -\frac{567}{12500}\omega_0, \qquad (9.24)$$

$$\mathcal{B}_7 = \frac{77}{10}\omega_1, \qquad \mathcal{B}_8 = \frac{476973}{44000}\omega_1, \qquad \mathcal{B}_9 = -\frac{4459833}{70400000}\omega_1,$$

where ω_0 and ω_1 are given by (9.22).

Thus, (9.17, 8.11) together with (9.13, 9.20, 9.21) represent a closed set of equations for the evolution of temperature and the coefficient a_2:

$$\frac{dT}{dt} = -\frac{2}{3}BT\mu_2 \quad \text{with} \quad B(t) \equiv v_T(t)\, g_2(\sigma)\, \sigma^2 n$$

$$\dot{a}_2 = \frac{4}{3} B\mu_2 (1 + a_2) - \frac{4}{15} B\mu_4$$

$$\delta'(t) = \delta \left(\frac{2T(t)}{T_0} \right)^{1/10} \qquad (9.25)$$

$$\mu_2 = (\mathcal{A}_1 + \mathcal{A}_2 a_2 + \mathcal{A}_3 a_2^2) + \delta'(\mathcal{A}_4 + \mathcal{A}_5 a_2 + \mathcal{A}_6 a_2^2)$$
$$\quad - \delta'^2 (\mathcal{A}_7 + \mathcal{A}_8 a_2 + \mathcal{A}_9 a_2^2)$$

$$\mu_4 = (\mathcal{B}_1 + \mathcal{B}_2 a_2 + \mathcal{B}_3 a_2^2) + \delta'(\mathcal{B}_4 + \mathcal{B}_5 a_2 + \mathcal{B}_6 a_2^2)$$
$$\quad - \delta'^2 (\mathcal{B}_7 + \mathcal{B}_8 a_2 + \mathcal{B}_9 a_2^2).$$

9.4 Equations for temperature and for the shape of the velocity distribution function

Inserting the last three relations of the set (9.25) into its first equations for $a_2(t)$ and $T(t)$ and using the coefficients $\mathcal{A}$ and $\mathcal{B}$ as obtained in the previous section we arrive at a set of equations for a_2 and for the reduced temperature $u(t) \equiv T(t)/T_0$:

$$\dot{u} + \tau_0^{-1} \varphi_1(a_2)\, u^{8/5} - \delta \tau_0^{-1} q_1 \varphi_2(a_2)\, u^{17/10} = 0 \qquad (9.26)$$

$$\dot{a}_2 - r_0\sqrt{u}\,\mu_2 (1 + a_2) + \frac{1}{5} r_0 \sqrt{u}\,\mu_4 = 0, \qquad (9.27)$$

where we introduce the short-hand notations

$$\varphi_1(a_2) \equiv \frac{5}{3} + \frac{2}{5} a_2 + \frac{7}{500} a_2^2$$
$$\varphi_2(a_2) \equiv \frac{5}{3} + \frac{119}{240} a_2 + \frac{1547}{128000} a_2^2 \qquad (9.28)$$

and the characteristic time

$$\tau_0^{-1} = \frac{16\, q_0}{5} \tau_c(0)^{-1} \delta = \frac{16\, q_0}{5} 4\sqrt{\pi} g_2(\sigma)\, \sigma^2 n \sqrt{T_0/m}\, \delta. \qquad (9.29)$$

The constants q_0, q_1 and r_0 are given by

$$q_0 \equiv \frac{2^{1/10}\omega_0}{16\sqrt{2\pi}} \approx 0.173$$

$$r_0 \equiv \frac{2}{3\sqrt{2\pi}} \tau_c(0)^{-1} \qquad (9.30)$$

$$q_1 \equiv 2^{1/10} \frac{\omega_1}{\omega_0} \approx 1.534,$$

with $\omega_{0/1}$ being defined by (9.22). Note that the characteristic time τ_0 is much larger than the initial mean collision time $\tau_c(0)$ since $\delta^{-1} \gg 1$ (see (9.13)).

We seek the solution of the set of equations (9.26, 9.27) as expansions in terms of the small dissipative parameter δ:

$$u = u_0 + \delta\, u_1 + \delta^2\, u_2 + \cdots \qquad (9.31)$$

$$a_2 = a_{20} + \delta\, a_{21} + \delta^2\, a_{22} + \cdots \qquad (9.32)$$

Substituting (9.31, 9.32, 9.20, 9.21) into (9.26, 9.27), these equations can be solved perturbatively for each order of δ. The sets of equations in zeroth and first order will be given explicitly in the following sections.

9.5 Velocity distribution and temperature in $\mathcal{O}(\delta^0)$

Collecting terms of the order $\mathcal{O}(\delta^0)$ we obtain from (9.26, 9.27)

$$\dot{u}_0 + \tau_0^{-1} \left(\frac{5}{3} + \frac{2}{5} a_{20} + \frac{7}{500} a_{20}^2 \right) u_0^{8/5} = 0 \qquad (9.33)$$

$$\dot{a}_{20} + r_1\sqrt{u_0} \left(a_{20} + \frac{1}{32} a_{20}^2 \right) = 0 \qquad (9.34)$$

with

$$r_1 \equiv \frac{1}{5} r_0 \mathcal{B}_2 = \frac{8}{15} \tau_c(0)^{-1}. \qquad (9.35)$$

Moreover, we used the definition of r_0, given by (9.30) and the zeroth order coefficients $\mathcal{B}_2$ and $\mathcal{B}_3$ in the expansion of $\mu_4(\delta)$ as given by (9.24). Changing variables

$$t \to \tau = r_1 \int_0^t dt'\, \sqrt{u_0(t')} \qquad (9.36)$$

in (9.34), the solution of this (Riccati) equation is found to be

$$\boxed{a_{20}(\tau) = \frac{a_{20}(0)}{\left[1 + \frac{1}{32} a_{20}(0)\right] e^\tau - \frac{1}{32} a_{20}(0)}.} \qquad (9.37)$$

From (9.33) we see that the characteristic time scale for $u_0(t)$ is $\tau_0 \gg \tau_c(0)$, therefore, during the first stage of the gas evolution when $t \sim \tau_c(0) \ll \tau_0$ we can approximate $u(t) = T(t)/T_0 \approx 1$ and, thus, $\tau \approx r_1 t$. Moreover, if the initial deviation from the Maxwell distribution is small, i.e., $a_{20}(0)/32 \ll 1$, we approximate for this time interval

$$a_{20}(t) \approx a_{20}(0) \exp\left(-\frac{8}{15}\frac{t}{\tau_c(0)}\right). \tag{9.38}$$

For large $t \sim \tau_0 \gg \tau_c(0)$ the Sonine coefficient $a_{20}(t)$ vanishes which corresponds to the relaxation of an initially non-Maxwellian velocity distribution to the Maxwell distribution. This relaxation occurs within few collisions per particle, similarly as the relaxation in molecular gases.

We now assume that the initial velocities obey a Maxwell distribution, that is, $a_{20}(0) = 0$ for $t = 0$. Then any deviation from the Maxwell distribution originates from the inelasticity of the particles. For the case $a_{20}(0) = 0$ [and thus $a_{20}(t) = 0$, see (9.37)] equation (9.33) reads

$$\dot{u}_0 + \frac{5}{3}\tau_0^{-1} u_0^{8/5} = 0 \tag{9.39}$$

with the solution (Brilliantov and Pöschel, 2000d)

$$\boxed{u_0(t) = \frac{T(t)}{T_0} = \left(1 + \frac{t}{\tau_0}\right)^{-5/3}.} \tag{9.40}$$

This result, except for the constant τ_0, coincides with the time dependence of the temperature obtained previously using scaling arguments (Schwager and Pöschel, 1998), see (5.8).

9.6 Velocity distribution and temperature in $\mathcal{O}(\delta^1)$

In $\mathcal{O}(\delta^1)$ from (9.26, 9.27) we obtain

$$\dot{u}_1 + \frac{8}{3\tau_0} u_0^{3/5} u_1 + \frac{2}{5\tau_0} u_0^{8/5} a_{21} - \frac{5}{3\tau_0} q_1 u_0^{17/10} = 0 \tag{9.41}$$

$$\dot{a}_{21} + r_1\sqrt{u_0}\, a_{21} + r_2 u_0^{3/5} = 0 \tag{9.42}$$

with

$$r_2 \equiv \frac{4}{15}\frac{2^{1/10}}{\sqrt{8\pi}}\frac{\mathcal{B}_4 - 5\mathcal{A}_4}{\tau_c(0)}. \tag{9.43}$$

For $t \ll \tau_0$ we have $u_0 \approx 1$ and (9.42) reduces to

$$\dot{a}_{21} + r_1 a_{21} = -r_2 \tag{9.44}$$

with the solution

$$a_{21}(t) = -\frac{r_2}{r_1}\left[1 - \exp(-r_1 t)\right] = -h\left[1 - \exp\left(-\frac{8}{15}\frac{t}{\tau_c(0)}\right)\right]. \qquad (9.45)$$

The constant h reads

$$h \equiv \frac{r_2}{r_1} = \frac{3}{10}\Gamma\left(\frac{21}{10}\right) 2^{1/5} C_1 \approx 0.415, \qquad (9.46)$$

where we use the definitions of r_1, r_2 and the values of $\mathcal{A}_k$ and $\mathcal{B}_k$ given above. As it follows from (9.45), after a transient time of the order of few collisions per particle, that is, for $\tau_c(0) < t \ll \tau_0$, the coefficient $a_2(t)$ saturates at

$$a_2 = -h\delta \approx -0.415\,\delta. \qquad (9.47)$$

After this transient time a_2 starts to evolve very slowly with regard to the time scale of the collision $\sim \tau_c(0)$. This part of the evolution of a_2 is of the same nature as the initial fast relaxation of a_2 to its stationary value for granular gases with constant ε. The difference is that for $\varepsilon = $ const. the evolution of $a_2(t)$ is completed after this relaxation, whereas for gases of viscoelastic particles a complicated evolution of the shape of the distribution function on the slow time scale τ_0 just began.

For $t \gg \tau_0$ the scaled temperature varies as $u_0 \propto (t/\tau_0)^{-5/3}$ according to (9.40). Therefore, (9.42) reads

$$\dot{a}_{21} + r_1 \left(\frac{t}{\tau_0}\right)^{-5/6} a_{21} = -r_2 \left(\frac{t}{\tau_0}\right)^{-1}. \qquad (9.48)$$

Let us use the power-law Ansatz

$$a_{21}(t) = A\left(\frac{t}{\tau_0}\right)^{-\nu} \qquad (9.49)$$

and perform the asymptotic analysis of (9.48) to determine the exponent ν and the pre-factor A. Inserting the Ansatz into (9.48) we obtain

$$-\nu A \tau_0^{-1}\left(\frac{\tau_0}{t}\right)^{\nu+1} + r_1 A\left(\frac{\tau_0}{t}\right)^{\nu+5/6} = -r_2\left(\frac{\tau_0}{t}\right). \qquad (9.50)$$

In the limit $(\tau_0/t) \ll 1$ the first term of the LHS is negligible as compared to the second one since $(\tau_0/t)^{\nu+1} \ll (\tau_0/t)^{\nu+5/6}$. Neglecting this term we find $\nu = 1/6$ and $A = -r_2/r_1$. Thus for $t \gg \tau_0$ we obtain

$$a_{21}(t) = -\frac{r_2}{r_1}\left(\frac{t}{\tau_0}\right)^{-1/6} = -h\left(\frac{t}{\tau_0}\right)^{-1/6}. \qquad (9.51)$$

The formal solution of the linear first-order differential equation (9.42) reads

$$a_{21}(t) = -6\tau_0 r_2 \exp\left[-6\tau_0 r_1 \left(1 + \frac{t}{\tau_0}\right)^{1/6}\right] \int_{6\tau_0 r_1}^{6\tau_0 r_1 (1+t/\tau_0)^{1/6}} \frac{e^x}{x} dx. \quad (9.52)$$

With the definitions of r_1, r_2 and τ_0 the pre-factors are

$$6\tau_0 r_1 = \frac{1}{q_0 \delta} \qquad 6\tau_0 r_2 = \frac{12}{5}\frac{1}{\delta}. \quad (9.53)$$

Therefore, in linear approximation with respect to δ we find for $a_2(t) = \delta\, a_{21}(t)$

$$a_2(t) = -\frac{12}{5} w(t)^{-1} \{\mathrm{Li}\,[w(t)] - \mathrm{Li}\,[w(0)]\} \quad (9.54)$$

with

$$w(t) \equiv \exp\left[\frac{1}{q_0 \delta}\left(1 + \frac{t}{\tau_0}\right)^{1/6}\right] \quad (9.55)$$

and with the definition of the logarithmic integral

$$\mathrm{Li}\,(x) \equiv \int_0^x 1/\ln(t)\, dt. \quad (9.56)$$

It is not difficult to show that the general expression (9.54) comprises both limits, (9.45) for $t \ll \tau_0$ and (9.51) for $t \gg \tau_0$.

We have not been able to solve (9.41) for the scaled temperature $u_1(t)$ rigorously, however, its solution for $t \gg \tau_0$ can be obtained. Substituting the asymptotic expressions $u_0(t) \simeq (t/\tau_0)^{-5/3}$ and $a_{21}(t) \simeq -h(t/\tau_0)^{-1/6}$ into (9.41) this equation turns into

$$\dot{u}_1 + \frac{8}{3}\left(\frac{t}{\tau_0}\right)^{-1} u_1 = \left(\frac{2}{5}h + \frac{5}{3}q_1\right)\left(\frac{t}{\tau_0}\right)^{-17/6}. \quad (9.57)$$

Using again the power-law Ansatz $u_1(t) \sim (t/\tau_0)^\alpha$ and performing the asymptotic analysis just in the same way as it has been done for (9.48) we obtain the exponent $\alpha = -11/6$ as well as the corresponding prefactor. The result for $t \gg \tau_0$ reads

$$u_1(t) = \left(\frac{12}{25}h + 2q_1\right)\left(\frac{t}{\tau_0}\right)^{-11/6} \approx 3.268 \left(\frac{t}{\tau_0}\right)^{-11/6}, \quad (9.58)$$

where the above results for the constants h and q_1 have been used. Combining (9.58, 9.40) we obtain the time dependence of the temperature for $t \gg \tau_0$ in linear approximation with respect to δ:

$$\frac{T(t)}{T_0} \simeq \left(\frac{t}{\tau_0}\right)^{-5/3} + 3.268\, \delta \left(\frac{t}{\tau_0}\right)^{-11/6} \quad (9.59)$$

In Figs 9.1 and 9.2 the time dependence of the coefficient $a_2(t)$ of the Sonine polynomials expansion and of the temperature of the granular gas are shown.

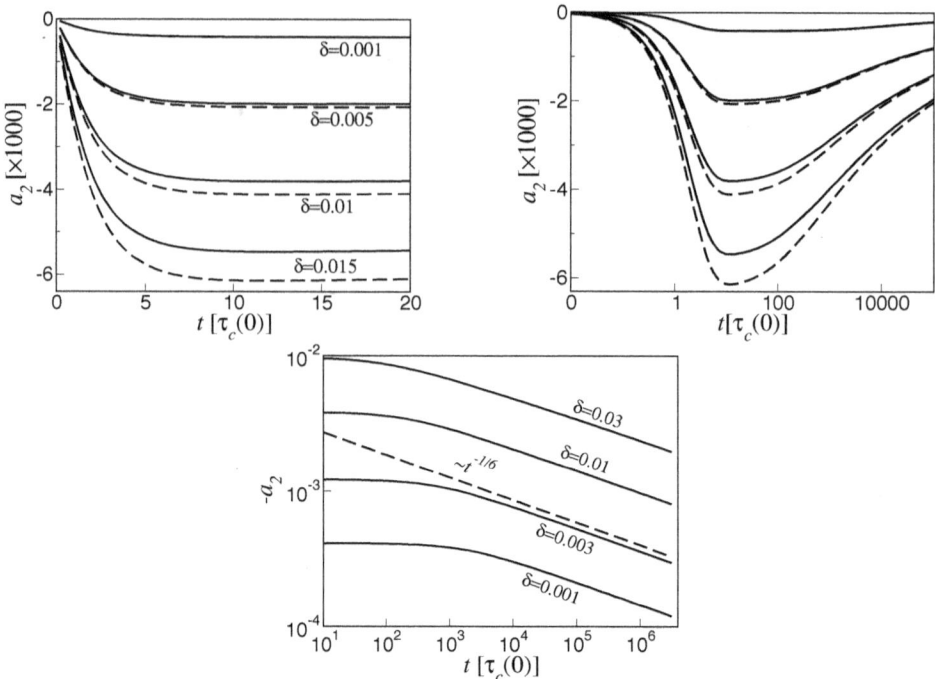

FIG. 9.1. Time dependence of the second Sonine coefficient. Top: a_2 for $\delta = 0.001, 0.005, 0.01, 0.015$ (solid lines) together with the linear approximation (dashed lines). Bottom: $-a_2(t)$ over time (log-scale) for $\delta = 0.03, 0.01, 0.003, 0.001$ together with the power-law asymptotics $\sim t^{-1/6}$ (dashed line).

The analytical results are compared with the numerical solution of the system (9.26, 9.27). As it follows from the figures the analytical theory reproduces fairly well the numerical results for the case of small dissipation δ.

In Section 9.1 we have discussed the evolution of the velocity distribution of a granular gas qualitatively, just by physical reasoning. Based on the more rigorous quantitative analysis Fig. 9.1 reveals now the complicated scenario of the evolution of a granular gas of viscoelastic particles. First, the gas leaves the initial Maxwell distribution which is expressed by $a_2 \neq 0$. The deviation from the Maxwell distribution grows quickly and saturates after a few collisions per particle at the time scale $\sim \tau_c(0)$. At this instant the deviation from the Maxwell distribution is maximal, $a_2 \approx -0.4\delta$ (see (9.47) and Fig. 9.1, top). After this maximal deviation is reached, the second stage of the evolution starts, where a_2 decays to zero on the slow time scale $\sim \tau_0 \sim \delta^{-1}\tau_c(0) \gg \tau_c(0)$, which corresponds to the time scale of the temperature evolution (Fig. 9.1, middle). In this regime, the decay of the coefficient $a_2(t)$ occurs according to a power law $\sim t^{-1/6}$ (Fig. 9.1, bottom). Provided the granular gas stays homogeneous for

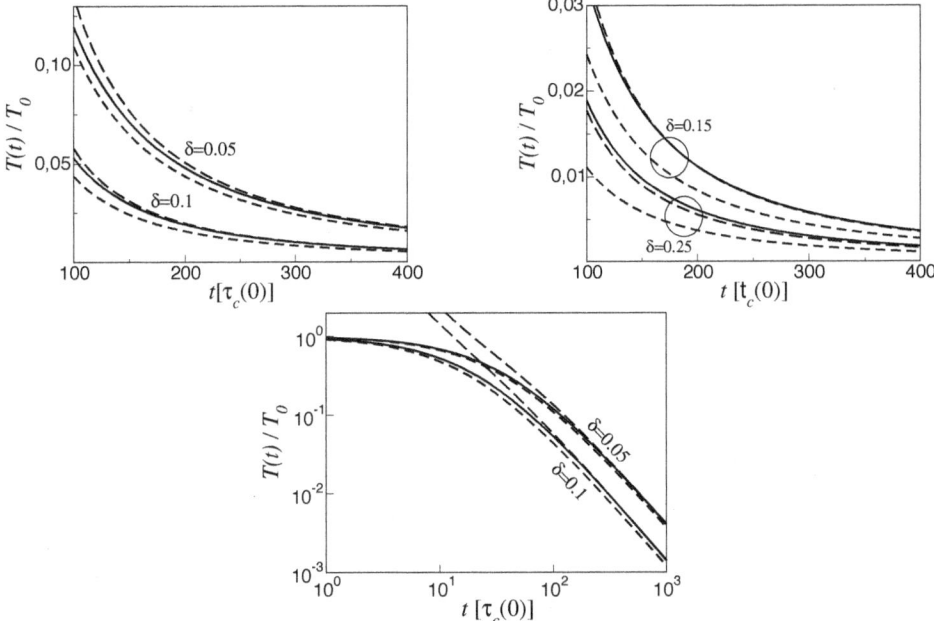

FIG. 9.2. Evolution of the reduced temperature $u(t) = T(t)/T_0$. Solid lines: numerical solution, short-dashed: $u_0(t) = (1 + t/\tau_0)^{-5/3}$ (zeroth order theory), long-dashed: $u(t) = u_0(t) + \delta u_1(t)$ (first-order theory).

long enough time, asymptotically the Maxwell distribution is approached.

Figure 9.2 illustrates the significance of the first-order correction $u_1(t)$ of the time-evolution of temperature. This correction becomes more significant for larger dissipation parameter δ (see Fig. 9.2 top, right). For large times the results of the first-order theory (with $u_1(t)$ incorporated) practically coincide with the numerical results, while the zeroth order theory (without $u_1(t)$) reveals noticeable deviations (Fig. 9.2, bottom).

9.7 Beyond the linear theory

For larger values of δ the linear theory fails. Unfortunately, the equations obtained for the second-order approximation $\mathcal{O}(\delta^2)$ are too complicated to be treated analytically, however, they can still be solved numerically (see Fig. 9.3). As compared to the case of small δ, an additional intermediate regime of the evolution of the velocity distribution is observed. The first (fast) stage of evolution takes place, as before, on the time scale of few collisions per particle (Fig. 9.3). For $\delta \gtrsim 0.15$ the initial (transient) maximal value of a_2 is positive. Then, on the second stage, which continues for $10-100$ collision times, a_2 changes its sign and reaches a maximal negative deviation. Finally, on the third (slow) stage, $a_2(t)$ relaxes to zero on the slow time scale τ_0, just as for small δ. In Fig. 9.3 we show

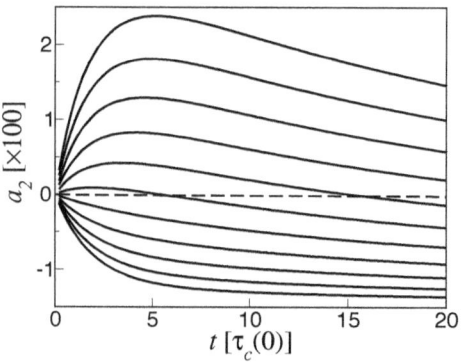

FIG. 9.3. Time dependence of the second Sonine coefficient for $\delta = 0.1, 0.11, 0.12, \ldots, 0.20$ (bottom to top).

the first stage of the evolution of $a_2(t)$ for systems with large δ.

Figure 9.4 shows the numerical solution of (9.26, 9.27) for the second Sonine coefficient $a_2(t)$ as a function of time. The different stages of the evolution of the velocity distribution function can be clearly distinguished. The shape of the curve $a_2(t)$ agrees with the qualitative discussion given in Section 9.1 which was based on the results for $\varepsilon = $ const.

When drawing the results with logarithmic time axis (Fig. 9.5) the complicated evolution of the first non-vanishing Sonine coefficient a_2 becomes obvious. Thus, we conclude that for the case of larger dissipative parameter δ the evolution of the velocity distribution function exhibits a complicated non-monotonic behaviour which consists of different regimes. Physically, this behaviour is caused by an intrinsic time scale which describes the viscoelastic collisions. This time scale couples the evolution of the shape of the velocity distribution function with the evolution of temperature.

9.8 Age of granular gases

Given at time t_1 a granular gas of simplified particles ($\varepsilon = $ const.) at temperature $T_1 = T(t_1)$. Let their velocities be distributed according to (8.22) with the second Sonine coefficient (8.48). The further evolution of the gas for $t > t_1$ is then described by

$$T(t) = \frac{T_1}{\left(1 + \dfrac{t - t_1}{\tau_0}\right)^2} \qquad (9.60)$$

$$a_2 = \text{const.}, \qquad (9.61)$$

where τ_0 may be expressed in terms of T_1. The temperature T_1 in its turn may be considered as a result of the evolution of the granular gas which was initialized at time $t_0 < t_1$ with temperature $T_0 > T_1$. Knowing the functions $T(t)$ and a_2 for

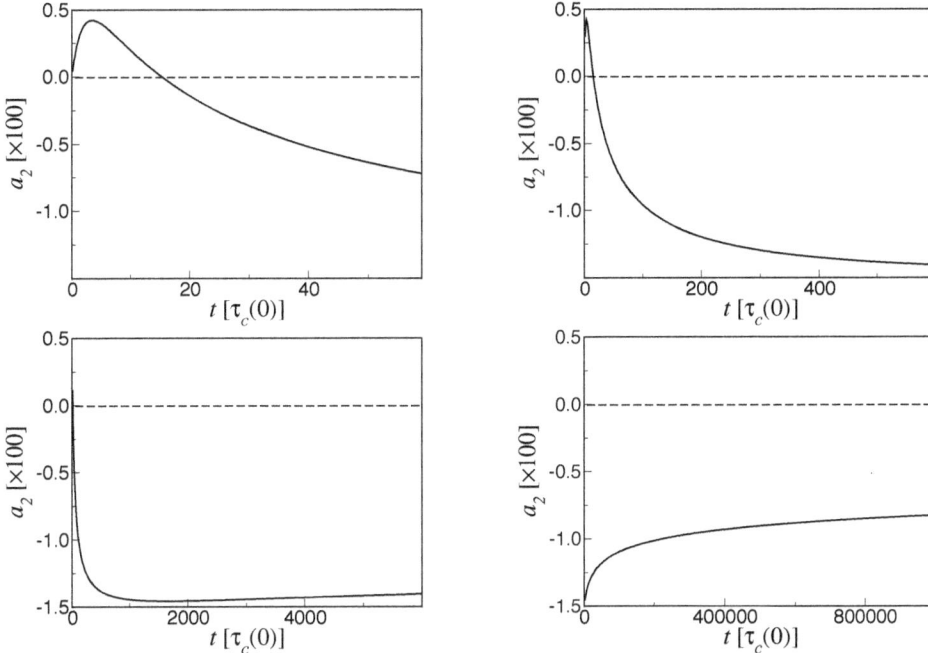

FIG. 9.4. The second Sonine coefficient a_2 for $\delta = 0.16$ over time. The numerical solution of (9.26, 9.27) shows all stages of the evolution.

$t > t_1$, we neither know anything about the initial temperature $T(t_0)$ nor about the initial time t_0. An infinite number of combinations of T_0 and t_0 may be used in (9.60) instead of T_1 and t_1, giving the same dependence $T(t)$ for $t > t_1$. This means that at time t_1 the gas does not carry any information about its starting point t_0. If we call the time span between the initialization of the gas and the present time its *age*, there are no means to determine the age of the gas by recording the particle velocities.

This differs drastically for a gas of viscoelastic particles. The evolution of its temperature is described by

$$T(t) = \frac{T_1}{\left(1 + \dfrac{t - t_1}{\tau_0}\right)^{5/3}}, \qquad (9.62)$$

which is not qualitatively different from (9.60). This dependence can be extended to earlier times just as described above. The second Sonine coefficient, however, is not a constant now, but reveals a complicated time behaviour as drawn in Fig. 9.5. If we knew from measurements of the particle velocities a piece of this curve in a certain interval $t_1 < t < t_2$ we could (under some conditions to be discussed below) reconstruct the entire curve. From Fig. 9.5 we see that there is, obviously,

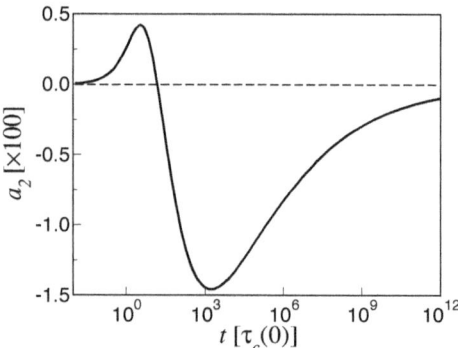

FIG. 9.5. The plot of $a_2(t)$ for $\delta = 0.16$ over logarithmic time illustrates the complicated evolution of the shape of the velocity distribution function characterised by the second Sonine coefficient.

no physically meaningful way to extend the curve further to the left than shown in the figure. This means that a granular gas of viscoelastic particles has an age.

We wish to formulate the statements of this paragraph more carefully: Assume we know, for example, from the nature of the physical process which gave rise to the granular gas, that the initial velocities of its particles obey a Maxwell distribution. Then the evolution of the velocity distribution function, quantified by the first non-trivial Sonine coefficient a_2, is described by the results of the preceding section. We have seen that this quantity evolves characteristically in time. Assume that we know the size and material properties of the granular gas particles and that we are able to measure the number density n of the gas and its velocity distribution $f(\vec{v}, t)$. From $f(\vec{v}, t)$ we can obtain the temperature (from the second moment of f) and the second Sonine coefficient a_2 (from the fourth moment of f). In the previous analysis the initial mean collision time $\tau_c(0)$ was used as the unit of time, which is determined by the initial temperature T_0. We also used the reduced temperature $u(t) \equiv T(t)/T_0$.

The main idea of determining the age of the gas is to solve the evolution equations backwards, provided the gas was all the time in the state of homogeneous cooling. Suppose that we know from measurements the temperature and a_2 at time $t = t_1$. Then we can use $T_1 = T(t_1)$ as the reference temperature and respectively $\tau_c(t_1)$ as the unit of time. Hence, we reformulate the above evolution equations (9.25–9.27) using the substitutes

$$t \to -t; \qquad T_0 \to T_1; \qquad a_2(0) = 0 \to a_2(t_1). \tag{9.63}$$

The corresponding equations then read

$$-\dot{u} + \tau_0^{-1} \varphi_1(a_2) u^{8/5} - \delta \tau_0^{-1} q_1 \varphi_2(a_2) u^{17/10} = 0 \tag{9.64}$$

$$-\dot{a}_2 - r_0 \sqrt{u} \mu_2 (1 + a_2) + \frac{1}{5} r_0 \sqrt{u} \mu_4 = 0, \tag{9.65}$$

where

$$u(t) \equiv \frac{T(t)}{T_1} \; ; \qquad \tau_0^{-1} = \frac{16\, q_0}{5} 4\sqrt{\pi} g_2(\sigma) \sigma^2 n \sqrt{\frac{T_1}{m}} \delta, \qquad (9.66)$$

and with

$$\delta'(t) = \delta\, (u(t))^{1/10} \; ; \qquad \delta = A\kappa^{2/5} \left(\frac{T_1}{m}\right)^{1/10}. \qquad (9.67)$$

The remaining quantities are defined in (9.28, 9.30). The initial conditions for (9.64, 9.65) are $u = 1$ and $a_2 = a_2(t_1)$. Solving these equations numerically we obtain a persistent *increase* of temperature and the respective backward evolution of a_2.

Assume that the condition $a_2(t_0) = 0$ is fulfilled at time t_0, then the time

$$(t_1 - t_0) \left[4\sqrt{\pi} g_2(\sigma) \sigma^2 n \sqrt{\frac{T_1}{m}} \right]^{-1} \qquad (9.68)$$

is the approximative age of the granular gas, which has the initial temperature $T(t_0)$. Using only the second Sonine coefficient, we cannot discriminate between two possible initialization times of the gas, t_0^1 and t_0^2 ($t_0^1 < t_0^2$), where the condition $a_2 = 0$ is satisfied (see Fig. 9.5). Fortunately, as shown in Fig. 9.4 (top, left) the times where $a_2 = 0$ differ by about 10 mean collision times only, while the age of the granular gas may be orders of magnitude larger. Thus, the evolution of the shape of the distribution function allows the measurement of the age of the granular gas up to a good accuracy.

In practice, the experimental data for the temperature and a_2 contain some noise. Therefore, more accurate results may be obtained if, instead of using only one initial set T_1, $a_2(t_1)$, a part of a evolution curve is traced for both quantities. Then solving the equations, one can vary the initial set T_1, $a_2(t_1)$ (within the uncertainty due to the noise) in order to obtain the best coincidence with the observed parts of the curves.

In this section we have shown how to determine the age of a granular gas by a single measurement of a_2 and T at a certain time t_1. To determine the age from these data we need additional information about the particles (radii and masses), the particle material properties, and the gas density. We wish to mention that the age can also be determined without these data if we know from measurements the functions $a_2(t)$ and $T(t)$ in a certain interval $t_1 \leq t \leq t_2$.

10
HIGH-ENERGY TAIL OF THE VELOCITY DISTRIBUTION FUNCTION

The velocity distribution of particles whose velocities significantly exceed the thermal velocity cannot be described by a Maxwell distribution. This part of the velocity distribution is described by an exponential function which decays significantly slower than a Maxwell distribution.

10.1 Overpopulation of the high-velocity tail

As we have discussed in the previous sections, the velocity distribution function for granular gases is roughly a Maxwell distribution. The deviations from the Maxwell distribution have been characterized by the first non-vanishing term of the Sonine polynomials expansion. Actually, this is true only for the main part of the velocity distribution. For fast particles, whose velocities significantly exceed the thermal velocity the velocity distribution function differs drastically from the Maxwell distribution: The number of particles of a granular gas moving at velocity $v \in (v_1, v_1 + \Delta v_1)$ with $v_1 \gg \sqrt{T}$ is much larger than according to the Maxwell distribution of the same temperature. This effect is called *overpopulation of the high-velocity tail* and has been first predicted by Esipov and Pöschel (1997). They have shown that for large scaled velocities, $c \gg 1$, the velocity distribution function decays exponentialy, $\tilde{f}(c) \propto \exp(-\text{const} \cdot c)$. In contrast, the Maxwell distribution decays as $\exp(-c^2)$, that is, much faster. Therefore, for $v \gg v_T$ the Maxwell distribution is not valid even as an approximation. The particular value of the constant describing the exponential tail has been found later by van Noije and Ernst (1998). The predicted overpopulation of the tail has been confirmed by Brey *et al.* (1996) using the Direct Simulation Monte Carlo method and by Huthmann *et al.* (2000) using molecular dynamics Simulations.

The overpopulation of the high-velocity tail is an interesting and important phenomenon which is specific for granular gases. This behaviour can be understood if one notices that the number of particles of the reduced velocity $c = v/v_T$ is determined by a balance of *three* processes: (i) losses due to collisions of particles at velocity c (thus changing their velocity to some c'), (ii) gains due to particles resulting at c after a collision, and (iii) it varies without collisions due to decaying thermal velocity v_T of a cooling gas, that is, although the particle's velocity v stays the same its reduced velocity $c = v/v_T$ changes due to temperature decay. For gases of elastic particles process (iii) is irrelevant and the balance of the processes (i) and (ii) yields the Maxwell distribution. For dissipative gases of particles which collide with $\varepsilon =$ const., the process (ii) for the high velocity tail may be neglected as compared to (i). Process (iii) causes, in this case, an increase of the number of particles in the high velocity tail with $c \gg 1$. The resulting balance of (i) and (iii) yields a steady-state exponential overpopulation of the high-energy tail (Esipov and Pöschel, 1997).

While for the main part of the distribution this effect is less important, for $v/\sqrt{T} \gg 1$ it dominates the distribution. This is due to the fact that for any given temperature the distribution function decays steeply with increasing velocity, hence, a small difference of the temperature causes a significant change of the popolation of high-velocity intervals. As a result the high-velocity tail of the distribution function becomes overpopulated (Esipov and Pöschel, 1997).

10.2 High-velocity tail for $\varepsilon =$ const.

Consider the collision integral $\tilde{I}(\tilde{f},\tilde{f})$ given by (8.3). For large velocity ($c_1 \gg 1$) the main contribution to the collision integral corresponds to collisions with other particles whose velocities c_2 are relatively close to the average velocity, that is, $c_2 \approx 1 \ll c_1$. Thus, for those pairs of velocities $\vec{c}_{12}$ in $\tilde{I}(\tilde{f},\tilde{f})$ may be replaced by $\vec{c}_1$. Writing the collision integral as difference of the gain and the loss term, $\tilde{I}_g - \tilde{I}_l$, the loss term can be approximated by

$$\tilde{I}_l = -\int d\vec{c}_2 \int d\vec{e}\, \Theta(-\vec{c}_{12}\cdot\vec{e})\,|\vec{c}_{12}\cdot\vec{e}|\,\tilde{f}(\vec{c}_1)\,\tilde{f}(\vec{c}_2)$$
$$\approx -c_1 \tilde{f}(c_1) \int d\vec{e}\,\Theta(-\vec{c}_1\cdot\vec{e})\,|\hat{c}_1\cdot\vec{e}|\int d\vec{c}_2 \tilde{f}(\vec{c}_2) = -\pi c_1 \tilde{f}(c_1)$$
(10.1)

where we introduce the unit vector $\hat{c}_1 = \vec{c}_1/c_1$, use the normalization $\int \tilde{f}(\vec{c}_2)\,d\vec{c}_2 = 1$ and the integral

$$\int d\vec{e}\,\Theta(-\vec{c}_1\cdot\vec{e})\,|\hat{c}_1\cdot\vec{e}| = \int_0^{2\pi} d\varphi \int_{\pi/2}^{\pi} \sin\theta\cos\theta\, d\theta = \pi.$$
(10.2)

(In (10.2) the z-axis is chosen along the vector $\vec{c}_1$ and φ and θ are correspondingly the polar and azimuthal angles of the unit vector $\vec{e}$.) Contributions to the gain term come mainly from collisions of fast particles of even larger velocities than c_1 which occur only rarely in the system. Therefore, the gain term $\tilde{I}_g$ in the collision integral may be neglected as compared with the loss term $\tilde{I}_l$. The inequality $\tilde{I}_g \ll \tilde{I}_l$ will be checked below. Consequently, the collision integral may be approximated by

$$\tilde{I}(\tilde{f},\tilde{f}) \approx \tilde{I}_l(\tilde{f},\tilde{f}) \approx -\pi c_1 \tilde{f}(\vec{c}_1)$$
(10.3)

which allows to recast (8.10) for the scaled function $\tilde{f}(\vec{c})$ into

$$\mu_2 \tilde{f}(\vec{c}_1) + \frac{1}{3}\mu_2 c_1 \frac{d}{dc_1}\tilde{f}(\vec{c}_1) = -\pi c_1 \tilde{f}(\vec{c}_1)\,.$$
(10.4)

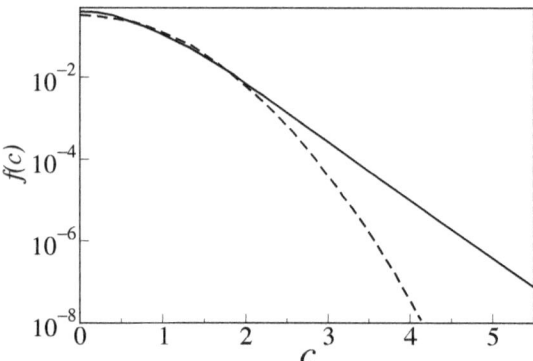

FIG. 10.1. Scaled velocity distribution $\tilde{f}(c)$ as a function of the dimensionless velocity $c = v/v_T(t)$, where $v_T(t)$ is the thermal velocity at time t. The coefficient of restitution is $\varepsilon = 0.1$. The solid line shows the results of a direct Monte Carlo simulation (DSMC) in two dimensions and the dashed line is the corresponding two-dimensional Maxwell distribution $\phi = \pi^{-1} \exp(-c^2)$. The data have been taken from Brey et al. (1999a).

For $c_1 \gg 1$ the first term on the LHS of (10.4) is negligible with respect to the other terms since they contain c_1 as a factor. Omitting this term we obtain an equation for the scaled velocity distribution function for large velocities:

$$\frac{d\tilde{f}}{dc} = -\frac{3\pi}{\mu_2}\tilde{f} \tag{10.5}$$

with the solution

$$\tilde{f}(c) = \mathcal{A}\exp\left(-\frac{3\pi}{\mu_2}c\right). \tag{10.6}$$

The integration constant $\mathcal{A}$ cannot be determined on this level of consideration while μ_2 is given by (8.45). The overpopulation with respect to the Maxwell distribution $\sim \exp(-c^2)$ occurs for $c > 3\pi/\mu_2$, or since μ_2 implies a factor $(1-\varepsilon^2)$, the overpopulation takes place for $c \gtrsim 1/(1-\varepsilon^2)$.

Figures 10.1 and 10.2 show the velocity distribution as obtained by Direct Monte Carlo (DMC) simulations of the Boltzmann equation (Brey et al., 1999a). For velocities $c \sim 1$ the Maxwell distribution represents an adequate description, while for $c \gtrsim 3$ a qualitatively different behaviour is observed (Fig. 10.1) which confirms the exponential overpopulation of the tail. Figure 10.2 compares numerical and theoretical results for the constant in the exponential distribution $f(c) \propto \exp(-\text{const.}\, c)$. The agreement between the numerical and theoretical results is rather good.

We still have to prove that $\tilde{I}_g \ll \tilde{I}_l$ for $c_1 \gg 1$ which has been used above to obtain (10.3). For $c_1 \gg 1$ we can approximate $\vec{c}_{12} \approx \vec{c}_1$ and the law of the inverse collision (6.10) reads

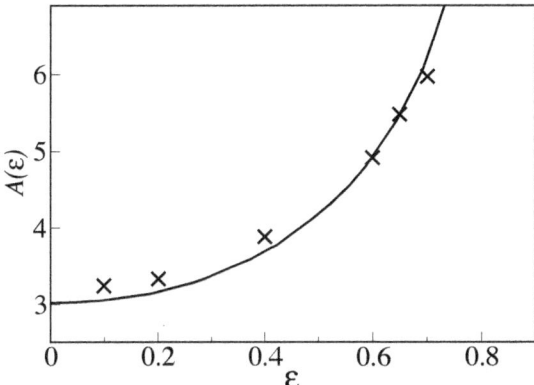

FIG. 10.2. Slope of the logarithm of the high-energy tail of the velocity distribution as a function of ε (data taken from Brey et al., 1999a). Crosses: DSMC (two-dimensional) with $\varepsilon = 0.1$. Line: theoretical prediction for the two-dimensional gas (11.8). The value of μ_2 is taken in linear approximation with respect to a_2 (see (11.4)), $\mu_2 = \sqrt{\pi/2}\left(1 - \varepsilon^2\right)\left(1 + \frac{3}{16}a_2^{NE}\right)$, where a_2^{NE} is defined by (11.7).

$$\vec{c}_1'' = \vec{c}_1 - \frac{1}{2}\left(1 + \frac{1}{\varepsilon}\right)(\vec{c}_1 \cdot \vec{e})\vec{e}$$
$$\vec{c}_2'' = \vec{c}_2 + \frac{1}{2}\left(1 + \frac{1}{\varepsilon}\right)(\vec{c}_1 \cdot \vec{e})\vec{e}.$$
(10.7)

For most of the collisions (except for a small number of grazing collisions where $\vec{c}_1$ is almost perpendicular to $\vec{e}$) $|\vec{c}_1 \cdot \vec{e}| \sim c_1 \gg 1$ and c_2 in (10.7) can be neglected. With these approximations (10.7) turns into

$$c_1'' = c_1\sqrt{1 - \frac{1}{4}\left(1 + \frac{1}{\varepsilon}\right)\left(3 - \frac{1}{\varepsilon}\right)\left(\hat{\vec{c}}_1 \cdot \vec{e}\right)^2}$$
$$c_2'' = \frac{1}{2}\left(1 + \frac{1}{\varepsilon}\right)c_1\left|\hat{\vec{c}}_1 \cdot \vec{e}\right|.$$
(10.8)

Now we show that $\tilde{f}(c) \propto \exp(-Ac)$ is a consistent solution for $c \gg 1$. Let us compare the factor $\tilde{f}(\vec{c}_1'')\tilde{f}(\vec{c}_2'')$ in $\tilde{I}_g$ with the corresponding factor $\tilde{f}(\vec{c}_1)\tilde{f}(\vec{c}_2)$ in $\tilde{I}_l$ for large velocities. If the exponential distribution is correct, the ratio $\tilde{I}_g/\tilde{I}_l$ reads

$$\frac{\tilde{I}_g}{\tilde{I}_l} \propto \frac{\tilde{f}(\vec{c}_1'')\tilde{f}(\vec{c}_2'')}{\tilde{f}(\vec{c}_1)\tilde{f}(\vec{c}_2)} \propto \exp\left[-A\left(c_1'' + c_2'' - c_1\right)\right].$$
(10.9)

Using (10.8) it may be proved that the exponent in (10.9) is proportional to c_1 and is always negative for $\varepsilon < 1$ (again with the exception of grazing collisions where it vanishes). To prove this we notice that for non-grazing collisions (i.e.

when $|\vec{c}_1 \cdot \vec{e}| \gg 1$) equation (10.7) implies $\vec{c}_1 = \vec{c}_1'' + \vec{c}_2''$. Then with the general inequality $\left|\vec{a} + \vec{b}\right| \leq |\vec{a}| + \left|\vec{b}\right|$ we obtain

$$c_1'' + c_2'' - c_1 = |\vec{c}_1''| + |\vec{c}_2''| - |\vec{c}_1| \geq |\vec{c}_1'' + \vec{c}_2''| - |\vec{c}_1| = 0. \qquad (10.10)$$

The last two equations (10.9, 10.10) show that $\tilde{I}_g / \tilde{I}_l \ll 1$ for $c_1 \gg 1$ for non-grazing collisions. Those collisions require separate discussion: For grazing collisions $|\vec{c}_1 \cdot \vec{e}| = c_1 \cos\theta \approx 1$ and $\cos\theta \sim 1/c_1 \ll 1$. Thus $\theta \approx \pi/2$, that is, $\theta = \pi/2 - \Delta\theta$. Then, since $\cos(\pi/2 - \Delta\theta) = \sin(\Delta\theta) \sim \Delta\theta \sim 1/c_1$, we conclude that for grazing collisions $\Delta\theta$ belongs to a small angular interval Ω_{gr} of the order $\mathcal{O}(1/c_1)$. Outside this interval the ratio $\tilde{I}_g / \tilde{I}_l$ as a function of c_1 is exponentially small. Inside the interval Ω_{gr} this ratio is approximatively one. Thus, only grazing collisions with angles between vectors $\vec{c}_1$ and $\vec{e}$ within the angular interval Ω_{gr} contribute significantly to the gain term. This contribution may be estimated (ommiting factors of the order one) by

$$\tilde{I}_g \sim \int_{\Omega_{\text{gr}}} d\theta \, c_1 \cos\theta \tilde{f}(\vec{c}_1) \simeq \Omega_{\text{gr}} \tilde{f}(\vec{c}_1) \simeq \tilde{f}(\vec{c}_1) / c_1, \qquad (10.11)$$

where we take into account $c_1 \cos\theta \approx 1$ in the interval Ω_{gr} of size $\sim 1/c_1$. With the above estimate for the loss term, $\tilde{I}_l \sim c_1 \tilde{f}(\vec{c}_1)$, we obtain the ratio of the gain and the loss terms in the collision integral for large c_1:

$$\tilde{I}_g / \tilde{I}_l \sim 1/c_1^2 \ll 1. \qquad (10.12)$$

10.3 High-velocity tail for viscoelastic particles

The above estimates of the loss term and of the ratio of the gain and loss terms in the collision integral are not limited to any particular collision model, therefore, (10.1, 10.12) are also valid for the case of viscoelastic particles. Similarly as for the case $\varepsilon = \text{const.}$ we write the kinetic equation (8.52) for $c \gg 1$

$$\frac{\mu_2}{3} c \frac{\partial}{\partial c} \tilde{f}(\vec{c}, t) + B^{-1} \frac{\partial}{\partial t} \tilde{f}(\vec{c}, t) \approx -\pi c \tilde{f}(\vec{c}, t) \qquad (10.13)$$

First, we evaluate the second term of the LHS. Using the time derivative of (8.53) (neglecting high-order terms, i.e., $a_p = 0$ for $p > 2$)

$$\frac{\partial \tilde{f}}{\partial t} = \dot{a}_2 \phi(c) S_2(c^2) \qquad (10.14)$$

and $\dot{a}_2$ as given by (9.17), we estimate

$$B^{-1} \frac{\partial \tilde{f}}{\partial t} = \frac{4}{3} \left[\mu_2 (1 + a_2) - \frac{1}{5} \mu_4 \right] \phi(c) S_2(c^2) \propto c^4 \exp(-c^2), \qquad (10.15)$$

where we take into account that according to the definition of the Sonine polynomials (7.7), $S_2(c^2) \propto c^4$ for $c \gg 1$ and that μ_2, μ_4 and a_2 are independent of c. Similarly, with $\tilde{f} \propto c^4 \exp(-c^2)$ for $c \gg 1$ we obtain estimates for the first term on the LHS of (10.13) and for the RHS of this equation:

$$c\frac{\partial \tilde{f}}{\partial c} \propto [4c^4 \exp(-c^2) - 2c^6 \exp(-c^2)] \propto c^6 \exp(-c^2) \quad (10.16)$$

$$c\tilde{f} \propto c^5 \exp(-c^2). \quad (10.17)$$

From (10.15–10.17) follows that the terms in the LHS of (10.13) scale as $\propto c^6 \exp(-c^2)$ and $\propto c^4 \exp(-c^2)$, while in the RHS as $\propto c^5 \exp(-c^2)$. Although these terms have the same factor $\exp(-c^2)$, the terms are of different order with respect to c, that is, the approximation is inconsistent. Similarly, it may be shown that inconsistencies arise at any order of the Sonine polynomials expansion. Indeed, using the expansion (8.53) up to order n, yields the estimates $\propto c^{2n+2} \exp(-c^2)$ and $\propto c^{2n} \exp(-c^2)$ for the LHS of (10.13), however, for the RHS $\propto c^{2n+1} \exp(-c^2)$ is obtained.

Nevertheless, the exponential Ansatz

$$\tilde{f}(\vec{c}, t) \propto \exp[-\varphi(t) c] \quad (10.18)$$

for the solution of the kinetic equation (10.13) turns out to be self-consistent for $c \gg 1$. To prove this, we substitute (10.18) into (10.13). The function $\varphi(t)$ in (10.18) must then satisfy

$$\dot{\varphi} + \frac{1}{3}\mu_2 B \varphi = \pi B, \quad (10.19)$$

where the time dependence of B is given by (8.11). From (9.20) follows that μ_2 depends on time via $\delta'(t) \propto \delta$ and $a_2(t)$.[10] As we shown in the previous section, the value of $a_2(t)$ is of the order δ. Hence, in linear approximation with respect to δ:

$$\mu_2(t) \simeq \delta'(t) \omega_0, \quad (10.20)$$

with $\mathcal{A}_{1,2,3} = 0$ and $\mathcal{A}_4 = \omega_0$, given by (9.23).

Using the definition of δ' (9.13) and expression (8.60) for $B(t)$ we write

$$\mu_2(t) B(t) = \delta \left(\frac{2T(t)}{T_0}\right)^{1/10} \omega_0 \frac{\tau_c(0)^{-1}}{\sqrt{8\pi}} \sqrt{\frac{T(t)}{T_0}} = \frac{5}{2}\tau_0^{-1} u^{3/5}(t)$$

$$B(t) = \frac{\tau_c(0)^{-1}}{\sqrt{8\pi}} \sqrt{u}(t), \quad (10.21)$$

---

[10] The moment of the collision integral μ_2 depends on the distribution function which for $c \gg 1$ deviates noticeably from the equivalent value as obtained from the Sonine polynomials expansion, which is valid for $c \sim 1$. According to the exponentially fast decay of the distribution function, however, this interval of velocities contributes hardly to the moments of the collision integral. Therefore, the relation (9.20) for μ_2 can be used which has been obtained previously without taking into account the overpopulation of the high-energy tail.

where we use the definition (9.29) for τ_0 and the relation (9.30) between q_0 and ω_0. In the linear approximation with respect to δ we use (9.40) for the time dependence of the temperature. Then (10.19) reads

$$\dot{\varphi} + \frac{5}{6\tau_0}\left(1 + \frac{t}{\tau_0}\right)^{-1}\varphi = \sqrt{\frac{\pi}{8}}\tau_c(0)^{-1}\left(1 + \frac{t}{\tau_0}\right)^{-5/6}. \qquad (10.22)$$

Substituting the Ansatz $\varphi \propto (1 + t/\tau_0)^\nu$ we find the exponent $\nu = 1/6$ and the prefactor, $\sqrt{\pi/8}\,\tau_0/\tau_c(0)$. Using (9.29) for $\tau_0/\tau_c(0)$ we arrive at the final result

$$\varphi(t) = b\delta^{-1}\left(1 + \frac{t}{\tau_0}\right)^{1/6}, \qquad (10.23)$$

where

$$b = \sqrt{\frac{\pi}{2}}\left(\frac{5}{32q_0}\right) = \frac{5^{7/5}}{2^{5/2}\Gamma(3/5)} \approx 1.13. \qquad (10.24)$$

Thus, we obtain the velocity distribution function for $c \gg 1$:

$$\tilde{f}(\vec{c}, t) \sim \exp\left[-\frac{b}{\delta}c\left(1 + \frac{t}{\tau_0}\right)^{1/6}\right]. \qquad (10.25)$$

Note that this expression is valid only for times $t \gg \tau_c(0)$, when the deviations from the Maxwell distribution are already well developed; it is not applicable for the transient time $t \sim \tau_c(0)$.

Equation (10.25) shows that the overpopulation of the high-velocity tail decreases on the same time scale $\sim \tau_0$ as temperature. This is different from the case of a granular gas with $\varepsilon = \text{const.}$: in the latter system, the overpopulation of the tail persists, while for gases of viscoelastic particles it decays as the system evolves.

Exercise 10.1 *Find the solution of (10.22), that is, find the exponent ν and the prefactor for the Ansatz $\varphi \propto (1 + t/\tau_0)^\nu$!*

11

TWO-DIMENSIONAL GRANULAR GASES

For the sake of simplicity of the notation we have considered up to now only three-dimensional granular gases, which is certainly the most realistic case. However, under some specific experimental conditions granular gases behave as being effectively two-dimensional. Moreover, computer experiments are frequently performed for two-dimensional systems due to the limitation of computer power. Therefore, we give the corresponding generalized expressions. All physical ideas, the mathematical tools, and even the methods of the analysis for the description of d-dimensional systems are almost identical with those for three-dimensional gases.[11] Thus, we will mention here only the differences in the definitions and give the final results.

The generalized definition of temperature reads

$$\int \frac{1}{2} m v^2 f(\vec{v}, t) \, d\vec{v} = \frac{d}{2} n T(t) , \qquad (11.1)$$

where the thermal velocity $v_T(t)$ is defined as before, $T(t) = m v_T^2(t)/2$. The scaled distribution function is now defined by

$$f(\vec{v}, t) = \frac{n}{v_T^d(t)} \tilde{f}(\vec{c}) , \qquad (11.2)$$

where $\vec{c} = \vec{v}/v_T(t)$. The generalized Sonine polynomials expansion for a d-dimensional gas has still the form (7.4), but the Maxwell distribution is now $\phi(c) = \pi^{-d/2} \exp(-c^2)$, while the Sonine polynomials for general dimensionality d differ from (7.7):

$$\begin{aligned} S_0(x) &= 1 , \\ S_1(x) &= -x + \frac{d}{2} , \\ S_2(x) &= \frac{1}{2} x^2 - \frac{1}{2}(d+2) x + \frac{1}{8}(d+2) d . \end{aligned} \qquad (11.3)$$

The moments of the dimensionless collision integral μ_p are still defined by (8.8) and in linear approximation with respect to a_2 they read (van Noije and Ernst, 1998)

---

[11] This does not apply for the case $d = 1$. One-dimensional gases are special, for a gas of identical elastic particles even *any* initial velocity distribution is stable (Du et al., 1995).

$$\mu_2 = \frac{\pi^{d/2}}{\sqrt{2\pi}\Gamma(d/2)}\left(1-\varepsilon^2\right)\left[1+\frac{3}{16}a_2+\mathcal{O}\left(a_2^2\right)\right] \quad (11.4)$$

$$\mu_4 = \frac{4\pi^{d/2}}{\sqrt{2\pi}\Gamma(d/2)}\left[T_1 + a_2 T_2 + \mathcal{O}\left(a_2^2\right)\right] \quad (11.5)$$

with

$$\begin{aligned} T_1 &= \frac{1}{4}\left(1-\varepsilon^2\right)\left(d+\frac{3}{2}+\varepsilon^2\right) \\ T_2 &= \frac{3}{128}\left(1-\varepsilon^2\right)\left(10d+39+10\varepsilon^2\right) + \frac{1}{4}\left(1+\varepsilon\right)(d-1) \,. \end{aligned} \quad (11.6)$$

These results lead to the second Sonine coefficient a_2 in linear approximation with respect to a_2:

$$a_2^{NE} = \frac{16\left(1-\varepsilon\right)\left(1-2\varepsilon^2\right)}{9+24d+8\varepsilon d-41\varepsilon+30\varepsilon^2\left(1-\varepsilon\right)}\,, \quad (11.7)$$

and to the expression for the velocity distribution function for very large velocities, $c \gg 1$ (tail overpopulation),

$$\tilde{f}(c) \simeq \exp\left[-\frac{\pi^{(d-1)/2}d}{\Gamma\left(\frac{d+1}{2}\right)\mu_2}c\right]. \quad (11.8)$$

The time dependence of temperature for d-dimensional granular gas obeys Haff's law

$$T(t) = \frac{T_0}{(1+t/\tau_0)^2} \quad (11.9)$$

with

$$\tau_0^{-1} = \frac{1}{2d}\left(1-\varepsilon^2\right)\left(1+\frac{3}{16}a_2\right)\tau_c(0)^{-1}, \quad (11.10)$$

where the initial mean collision time reads

$$\tau_c(0)(t)^{-1} = \frac{2\pi^{(d-1)/2}}{\Gamma\left(\frac{d}{2}\right)}g_2(\sigma)n\sigma^{d-1}\sqrt{\frac{T(0)}{m}}. \quad (11.11)$$

The contact value of the pair-correlation function $g_2(\sigma)$ for two-dimensional systems of hard discs reads (Hansen and McDonald, 1986):

$$g_2(\sigma) = \frac{1-(7/16)\eta}{(1-\eta)^2}, \quad (11.12)$$

with the volume fraction in two dimensions $\eta = \frac{1}{4}\pi\sigma^2 n$.

For two-dimensional granular gases of viscoelastic spheres, the corresponding results read

$$\mu_2 = (\mathcal{A}_1 + \mathcal{A}_2 a_2 + \mathcal{A}_3 a_2^2) + \delta' (\mathcal{A}_4 + \mathcal{A}_5 a_2 + \mathcal{A}_6 a_2^2) \\ - \delta'^2 (\mathcal{A}_7 + \mathcal{A}_8 a_2 + \mathcal{A}_9 a_2^2) \quad (11.13)$$

and

$$\mu_4 = (\mathcal{B}_1 + \mathcal{B}_2 a_2 + \mathcal{B}_3 a_2^2) + \delta' (\mathcal{B}_4 + \mathcal{B}_5 a_2 + \mathcal{B}_6 a_2^2) \\ - \delta'^2 (\mathcal{B}_7 + \mathcal{B}_8 a_2 + \mathcal{B}_9 a_2^2) , \quad (11.14)$$

with

$$\mathcal{A}_1 = 0, \quad \mathcal{A}_2 = 0, \quad \mathcal{A}_3 = 0,$$
$$\mathcal{A}_4 = \frac{1}{2}\omega_0, \quad \mathcal{A}_5 = \frac{3}{25}\omega_0, \quad \mathcal{A}_6 = \frac{21}{5000}\omega_0, \quad (11.15)$$
$$\mathcal{A}_7 = \frac{1}{2}\omega_1, \quad \mathcal{A}_8 = \frac{119}{800}\omega_1, \quad \mathcal{A}_9 = \frac{4641}{1280000}\omega_1 ;$$

$$\mathcal{B}_1 = 0, \quad \mathcal{B}_2 = \sqrt{2\pi}, \quad \mathcal{B}_3 = \frac{1}{32}\sqrt{2\pi},$$
$$\mathcal{B}_4 = \frac{23}{10}\omega_0, \quad \mathcal{B}_5 = \frac{1671}{500}\omega_0, \quad \mathcal{B}_6 = -\frac{369}{50000}\omega_0, \quad (11.16)$$
$$\mathcal{B}_7 = \frac{67}{20}\omega_1, \quad \mathcal{B}_8 = \frac{438483}{88000}\omega_1, \quad \mathcal{B}_9 = -\frac{2895543}{140800000}\omega_1 .$$

These results may be obtained using the Maple program given on page 96 with DefDimension(2). The evolution of temperature obeys the same law as for the three-dimensional case:

$$T(t) = \frac{T_0}{(1+t/\tau_0)^{5/3}}, \quad (11.17)$$

with the characteristic time

$$\tau_0^{-1} = \frac{24}{5} q_0 \tau_c(0)^{-1} \delta, \quad (11.18)$$

where the initial mean collision time $\left[2n\sigma g_2(\sigma)\sqrt{\pi T(0)/m}\right]^{-1}$ follows from (11.11) for $d = 2$. Correspondingly in linear approximation with respect to δ the evolution of a_2 is described by

$$a_2(t) = -3w(t)^{-1} \{\text{Li}[w(t)] - \text{Li}[w(0)]\} \quad (11.19)$$

with

$$w(t) \equiv \exp\left[\frac{5}{8q_0\delta}\left(1 + \frac{t}{\tau_0}\right)^{1/6}\right] \quad (11.20)$$

and with the definition of the logarithmic integral on page 101.

Finally, the velocity distribution function for $c \gg 1$ has the same form as for three-dimensional gases

$$\tilde{f}(\vec{c},t) \sim \exp\left[-\frac{b}{\delta}c\left(1+\frac{t}{\tau_0}\right)^{1/6}\right] \quad (11.21)$$

with

$$b = \frac{5}{12\sqrt{2\pi}q_0} \approx 0.959 \quad (11.22)$$

SUMMARY

We have analysed the most important characteristics of the granular gas in the homogeneous cooling state – its velocity distribution function. The take-home message is:

1. The homogeneous cooling state may be characterized by the granular temperature $T(t)$, which permanently decreases due to dissipative collisions according to a power law:

$$T(t) = \frac{T(0)}{(1 + t/\tau_0)^\alpha},$$

   where τ_0 is the characteristic time, $\alpha = 2$ for the simplified collision model $\varepsilon = \text{const.}$ and $\alpha = 5/3$ for a gas of viscoelastic particles.

2. The velocity distribution function is close to the Maxwell distribution. Its shape is characterised by the scaled velocity distribution function

$$f(\vec{v}, t) = \frac{n}{v_T^3(t)} \tilde{f}(\vec{c}), \qquad \vec{c} = \vec{v}/v_T(t),$$

   where $v_T(t) = \sqrt{2T(t)/m}$ is the thermal velocity. The deviation of the scaled function from the Maxwell distribution $\phi(c)$ is described by the Sonine polynomials expansion

$$\tilde{f}(\vec{c}) = \phi(c)\left(1 + \sum_{p=1}^\infty a_p S_p(c^2)\right)$$

   with $a_1 = 0$, that is, the first non-trivial coefficient is a_2.

3. For the case of the simplified collision model $\varepsilon = \text{const.}$ the shape of the scaled function persists, hence, the Sonine coefficients are constants. For small dissipation the first non-trivial Sonine coefficient a_2 characterizes the velocity distribution. In linear approximation it is given by

$$a_2 = \frac{16(1-\varepsilon)(1-2\varepsilon^2)}{81 - 17\varepsilon + 30\varepsilon^2(1-\varepsilon)}.$$

4. For a gas of viscoelastic particles the shape of the scaled distribution function evolves in the due of time. This evolution may be characterized by the time dependence of the second Sonine coefficient $a_2(t)$. Contrary to the case $\varepsilon = \text{const.}$ the evolution of temperature and of $a_2(t)$ are coupled. The complicated evolution of the velocity distribution function allows to define the age of a granular gas.

5. Although the main part of the velocity distribution is close to the Maxwell distribution, the high-energy tail ($c \gg 1$) obeys a different function:

$$\tilde{f}(\vec{c}) \propto \exp(-Ac) .$$

For the case $\varepsilon =$ const. the coefficient A is a constant too, while for viscoelastic particles A depends on time. Since the above exponential distribution decays slower than the Maxwell distribution, this effect is also called *overpopulation of the tail* of the velocity distribution function.

III. Single-Particle Transport – Self-Diffusion and Brownian Motion

Brownian motion of tracer particles in a granular gas in the homogeneous cooling state and the corresponding coefficient of self-diffusion are considered. The inelastic nature of particle collisions gives rise to an enhanced spreading of the particles as compared with diffusion in molecular gases of the same temperature. The process of diffusion differs qualitatively for simplified granular particles ($\varepsilon = $ const.) and viscoelastic particles. Since viscoelastic particles are much more mobile we suspect that clustering in gases of such particles may be suppressed.

We introduce the basic relations describing the diffusion process and calculate the self diffusion coefficient using the pseudo-Liouville operator technique. This method allows the resolution of some mathematical difficulties that occur when standard methods of statistical mechanics are applied to hard sphere gases.

12

DIFFUSION AND SELF-DIFFUSION

The basic concepts of diffusion and self-diffusion are discussed. We derive relations between the mean square displacement of a diffusing particle, the diffusion coefficient and the velocity-time correlation function. These relations are generalized to the case of non-equilibrium granular gases.

12.1 Transport in granular gases

Transport processes in their conventional meaning imply macroscopic currents of mass, momentum, energy, etc. which occur correspondingly due to the presence of (even small) gradients of density, average velocity, temperature, etc. According to their definition, there are no such transport phenomena in the homogeneous cooling state.

In spite of the lack of macroscopic currents, individual-particle transport still exists in the homogeneous cooling state. The gas may consist of uniformly distributed particles (host particles) and of sparsely scattered guest particles, which differ in size, mass, material properties, etc. The stochastic motion of the guest particles, caused by collisions with the host particles is called *Brownian motion*. In isotropic systems, the average displacement of the guest particles is zero (since there is no preferred direction for the particles to move). The mean square displacement, however, increases with time, so that over a period of time the Brownian particle shifts more and more from its initial location. This kind of motion, which can occur even without macroscopic gradients, is called *diffusion*.

The concepts of diffusion and Brownian motion have been introduced for molecular systems in thermodynamic equilibrium, where particles suffer *elastic* collisions. For the description of granular gases, which are non-equilibrium systems, these concepts have to be generalized.

Consider a gas which consists of N_B uniformly distributed host particles B and of $N_A \ll N_B$ sparsely distributed guest particles A. The total number density $n = N/V = (N_A + N_B)/V \simeq N_B/V$ is supposed to be homogeneous, while the local concentration $n_A(\vec{r}, t) \equiv \Delta N_A(\vec{r}, t)/\Delta V$ is not necessarily uniform (here $\Delta N_A(\vec{r}, t)$ denotes the number of particles A at time t in a small volume ΔV located at point $\vec{r}$). If the concentration of the guest particles is not homogeneous, a diffusion flux of particles A arises, which is directed opposite to the concentration gradient. This process is described by

$$\vec{J}_A(\vec{r}) = -D_A \vec{\nabla} n_A(\vec{r}), \qquad (12.1)$$

where $\vec{J}_A(\vec{r})$ is the current of the guest particles A at $\vec{r}$ against the density gradient $\vec{\nabla} n_A(\vec{r})$ and D_A is the diffusion coefficient. With the continuity equation

$$\frac{\partial n_A(\vec{r})}{\partial t} + \vec{\nabla} \vec{J}_A(\vec{r}) = 0 \qquad (12.2)$$

we obtain the diffusion equation

$$\frac{\partial n_A(\vec{r})}{\partial t} = D_A \vec{\nabla}^2 n_A(\vec{r}) \,. \qquad (12.3)$$

If the particles A and B are mechanically identical, however, somehow distinguishable (e.g. by colour), the process is called self-diffusion. In this case, the system is uniform and the particles A are called *tracers*.

12.2 Diffusion coefficient and mean square displacement

The coefficient of diffusion (and self-diffusion)[12] is closely related to the mean square displacement of tracers with time. Assume at time $t = 0$ the tracer particles are located at $\vec{r} = 0$. The mean square displacement of particles at time t is then

$$\left\langle [\vec{r}(t)]^2 \right\rangle \equiv \frac{1}{N_A} \int d\vec{r}\, [\vec{r}(t)]^2\, n_A(\vec{r},t) \,, \qquad (12.4)$$

where $N_A = \int d\vec{r}\, n_A(\vec{r},t) = $ const. is the total number of tracers. Now we multiply both sides of (12.3) by $\vec{r}^2/N_A$ and integrate over $d\vec{r}$. With (12.4) the LHS reads

$$\frac{\partial}{\partial t} \frac{1}{N_A} \int d\vec{r}\, r^2 n_A = \frac{d}{dt} \left\langle [\vec{r}(t)]^2 \right\rangle \,. \qquad (12.5)$$

For the RHS we apply Green's theorem (Abramowitz and Stegun, 1965) to the functions n_A and r^2:

$$\int d\vec{r} \left(r^2 \vec{\nabla}^2 n_A - n_A \vec{\nabla}^2 r^2 \right) = \int_S dS\, \vec{l} \cdot \left(r^2 \frac{\partial n_A}{\partial \vec{l}} - n_A \frac{\partial r^2}{\partial \vec{l}} \right) , \qquad (12.6)$$

where S denotes the surface which confines the system of interest and $\vec{l}$ is the normal of this surface. If we put the surface S to infinity, where the concentration of the host particles n_A vanishes, the surface integral in (12.6) becomes zero. Hence, we obtain for the RHS of (12.3)

$$D_A \frac{1}{N_A} \int d\vec{r}\, r^2 \vec{\nabla}^2 n_A = D_A \frac{1}{N_A} \int d\vec{r}\, n_A \vec{\nabla}^2 r^2 = 6 D_A \,, \qquad (12.7)$$

where we use the relation $\vec{\nabla}^2 r^2 = \vec{\nabla} \cdot \left(\vec{\nabla} r^2 \right) = \vec{\nabla} \cdot (2\vec{r}) = 6$ and the normalization condition for n_A. Combining (12.5) and (12.7) we obtain a relation between

---

[12] In the following we will call both the coefficients of diffusion and self-diffusion as *diffusion coefficient*.

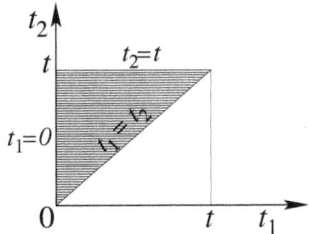 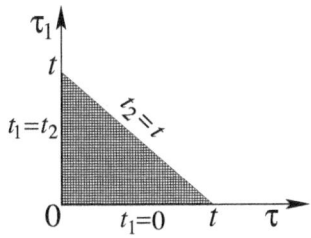

FIG. 12.1. Transformation of the integration variables t_1, t_2 into the variables τ, τ_1. The integration is performed with the condition $t_2 \geq t_1$; the correspondence between the lines confining the integration domains in both sets of variables, t_1, t_2 and τ, τ_1 is indicated.

the mean square displacement and the diffusion coefficient for three-dimensional systems

$$\frac{d}{dt}\left\langle [\vec{r}(t)]^2 \right\rangle = 6D_A \quad \text{or} \quad \left\langle [\vec{r}(t)]^2 \right\rangle = 6D_A t \qquad (12.8)$$

12.3 Diffusion coefficient and velocity-time correlation function

Using the kinematic relation $\vec{r}(t) = \int_0^t \vec{v}(t_1)\,dt_1$ (again for $\vec{r} = 0$ at $t = 0$) we write

$$\left\langle [\vec{r}(t)]^2 \right\rangle = \left\langle \int_0^t \vec{v}(t_1)\,dt_1 \cdot \int_0^t \vec{v}(t_2)\,dt_2 \right\rangle \qquad (12.9)$$

and encounter with the velocity autocorrelation function

$$K_v(t_1, t_2) \equiv \langle \vec{v}(t_1) \cdot \vec{v}(t_2) \rangle . \qquad (12.10)$$

The obvious relation $K_v(t_1, t_2) = K_v(t_2, t_1)$ allows us to rewrite (12.9) with the condition $t_2 \geq t_1$ imposed:

$$\left\langle [\vec{r}(t)]^2 \right\rangle = 2\int_0^t dt_1 \int_{t_1}^t dt_2 K_v(t_2, t_1) . \qquad (12.11)$$

For gases in equilibrium, the velocity autocorrelation function depends on the time lag $\tau \equiv t_2 - t_1$ only and decays with a characteristic time τ_v. First, we assume that the gas is in equilibrium, $K_v(t_2, t_1) = K_v(\tau)$, and transform the variables, $t_1, t_2 \to \tau, \tau_1$ with $\tau_1 = t_1$. The Jacobian of this transformation equals unity and the change of the integration area is illustrated in Fig. 12.1. In the new variables we obtain

$$\int_0^t dt_1 \int_{t_1}^t dt_2 K_v(t_2, t_1) = \int_0^t d\tau \int_0^{t-\tau} d\tau_1 K_v(\tau) = \int_0^t d\tau K_v(\tau)(t - \tau) \qquad (12.12)$$

and, correspondingly, the mean square displacement reads

$$\left\langle [\vec{r}(t)]^2 \right\rangle = 2t \int_0^t d\tau \, \langle \vec{v}(0) \cdot \vec{v}(\tau) \rangle \left(1 - \frac{\tau}{t}\right). \tag{12.13}$$

For longer time periods as compared with the characteristic time, that is, for $t \gg \tau_v$, we may neglect τ/t as compared with 1. With (12.8):

$$\left\langle [\vec{r}(t)]^2 \right\rangle = 2t \int_0^t d\tau \, \langle \vec{v}(0) \cdot \vec{v}(\tau) \rangle = 6 D_A t, \tag{12.14}$$

which implies that at $t \gg \tau_v$, the diffusion coefficient has the general expression

$$\boxed{D_A = \frac{1}{3} \int_0^\infty \langle \vec{v}(0) \cdot \vec{v}(t) \rangle \, dt} \tag{12.15}$$

Equation (12.15) expresses the transport coefficient D_A as the time-integral of the velocity correlation function. This is the simplest fluctuation–dissipation relation, which relates a fluctuating quantity (particle velocity) to a dissipative quantity (diffusion coefficient). Similar expressions exist for other transport coefficients (viscosity, thermal conductivity, etc.) and the corresponding fluctuating quantities (Resibois and de Leener, 1977).

We recall that all the above discussion refers to the case of equilibrium gases. Granular gases are *a priori* non-equilibrium systems, nevertheless, the concept of the diffusion coefficient may be generalized for such systems. Obviously, this refers only to dilute systems (Esipov and Pöschel, 1997) where the particles are highly mobile. Whereas the diffusion coefficient D for equilibrium systems is just a constant, the time dependence of temperature of a cooling granular gas causes the diffusion coefficient to be time-dependent as well. Therefore, the natural generalization of the diffusion coefficient for non-equilibrium systems is the diffusivity

$$\boxed{\left\langle (\Delta r(t))^2 \right\rangle = 6 \int^t D(t') \, dt'} \tag{12.16}$$

The brackets $\langle \cdots \rangle$ denote averaging over the non-equilibrium ensemble, whose evolution is described by the time-dependent N-particle distribution function $\rho(\vec{r}_1, \ldots \vec{r}_N, \vec{v}_1, \ldots \vec{v}_N, t)$ (for simplicity, we have retained the same notation as for the equilibrium average). We will show that for non-equilibrium systems, such as granular gases, the diffusivity can also be expressed in terms of the velocity-time correlation function. To this end we need an appropriate tool to describe the detailed particle dynamics for the case of dissipative systems. The technique of the pseudo-Liouville and binary-collision operators occurs to be very convenient for this purpose. Let us consider this in more detail.

13

PSEUDO-LIOUVILLE AND BINARY COLLISION OPERATORS IN DISSIPATIVE GAS DYNAMICS

For hard spheres, the interaction potential diverges, that is, an infinite force arises when two particles are brought into contact. This causes serious problems when the standard tools of the kinetic theory are applied to hard-sphere fluids. To bypass this difficulty the elegant technique of the pseudo-Liouville operator has been developed. We illustrate its application to dissipative gases.

13.1 Liouville operator in classical mechanics

Consider the evolution of a dynamical variable $A(t)$ which depends on time only through the time dependence of the coordinates and velocities of the particles

$$A(t) = A(\{\vec{r}_i(t), \vec{v}_i(t)\}), \qquad (13.1)$$

where $\{\vec{r}_i(t), \vec{v}_i(t)\} \equiv \vec{r}_1(t), \vec{v}_1(t), \vec{r}_2(t), \vec{v}_2(t), \ldots$ Then its time derivative reads

$$\frac{dA}{dt} = \sum_i \left(\dot{\vec{r}}_i \frac{\partial A}{\partial \vec{r}_i} + \dot{\vec{v}}_i \frac{\partial A}{\partial \vec{v}_i} \right). \qquad (13.2)$$

If there are no dissipative forces and if the particles interact via a smooth (singularity-free) potential the time derivatives of the coordinates $\vec{r}_i$ and momenta $\vec{p}_i = m\vec{v}_i$ of the particles are expressed via the Hamiltonian $\mathcal{H}$:

$$\dot{\vec{r}}_i = \frac{\partial \mathcal{H}}{\partial \vec{p}_i}, \qquad \dot{\vec{p}}_i = -\frac{\partial \mathcal{H}}{\partial \vec{r}_i}. \qquad (13.3)$$

The derivative $\dot{A}$ may then be expressed by

$$\frac{d}{dt} A = \{A, \mathcal{H}\} = \mathcal{L} A, \qquad (13.4)$$

which defines the Poisson brackets

$$\{A, \mathcal{H}\} \equiv \sum_i \left(\frac{\partial \mathcal{H}}{\partial \vec{p}_i} \cdot \frac{\partial}{\partial \vec{r}_i} - \frac{\partial \mathcal{H}}{\partial \vec{r}_i} \cdot \frac{\partial}{\partial \vec{p}_i} \right) A \qquad (13.5)$$

and the Liouville operator[13] $\mathcal{L}$. For the Hamiltonian of N identical particles

---

[13] In the standard definition of $\mathcal{L}$ (e.g. Resibois and de Leener, 1977), the imaginary unit $\sqrt{-1}$ is used. This assures that the operator $\mathcal{L}$ is self-adjoint (Resibois and de Leener, 1977). To avoid complicated notations we skip this factor; it will not affect the results which we derive here.

$$\mathcal{H} = \sum_i \frac{\vec{p}_i^{\,2}}{2m} + \sum_{i<j} \Phi\left(\vec{r}_i - \vec{r}_j\right) \tag{13.6}$$

with $\Phi\left(\vec{r}_i - \vec{r}_j\right)$ being the interaction potential, the Liouville operator reads

$$\begin{aligned}\mathcal{L} &= \sum_i \vec{v}_i \cdot \frac{\partial}{\partial \vec{r}_i} + \sum_{i<j} \frac{1}{m}\frac{\partial \Phi\left(\vec{r}_{ij}\right)}{\partial \vec{r}_j} \cdot \left(\frac{\partial}{\partial \vec{v}_i} - \frac{\partial}{\partial \vec{v}_j}\right) \\ &\equiv \sum_i \mathcal{L}_i^0 + \sum_{i<j} \mathcal{L}'_{ij}\,. \end{aligned} \tag{13.7}$$

Note, that although the free-streaming component of the Liouville operator

$$\mathcal{L}_i^0 \equiv \vec{v}_i \cdot \frac{\partial}{\partial \vec{r}_i} \tag{13.8}$$

is well defined independently of the interaction potential, the interaction component

$$\mathcal{L}'_{ij} \equiv -\frac{1}{m}\frac{\partial \Phi\left(\vec{r}_{ij}\right)}{\partial \vec{r}_{ij}} \cdot \left(\frac{\partial}{\partial \vec{v}_i} - \frac{\partial}{\partial \vec{v}_j}\right) \tag{13.9}$$

is defined only for smooth potentials. Hard spheres of diameter σ interact, however, with the potential

$$\Phi_{\text{hc}}\left(\vec{r}_{ij}\right) = \begin{cases} 0 & \text{for} \quad |\vec{r}_{ij}| > \sigma \\ \infty & \text{for} \quad |\vec{r}_{ij}| \leq \sigma \end{cases} \tag{13.10}$$

and, contrary to the case of soft spheres, suffer instantaneous collisions. For hard-sphere systems the interaction component of the Liouville operator $\mathcal{L}'_{ij}$ is not well defined due to the singularity of the interaction potential at $r_{ij} = \sigma$ (the value of $\partial \Phi/\partial \vec{r}_i$ is not defined at this point). Hence, for hard-sphere system the above approach cannot be applied.

Nevertheless, since the dynamics of the hard-sphere system is still completely deterministic because the after-collision velocities are uniquely determined by the collision rule, an operator can be constructed, which will preserve the formal structure of the Liouville equation. This operator is called *pseudo-Liouville* operator and may describe both elastic and inelastic hard-core collisions. The pseudo-Liouville operator was first introduced by Ernst *et al.* (1969) and Ernst and Dorfman (1972) to describe systems of elastic hard-core particles. The idea occurred to be very fruitful, since the approach may be easily generalized to account for internal degrees of freedom (rotation) of colliding particles (e.g. Chandler, 1975; Berne, 1977; Brilliantov and Revokatov, 1984) as well as for dissipative collisions. For this reason, the pseudo-Liouville operator is widely used in the theory of granular gases (e.g. Luding *et al.*, 1998; van Noije *et al.*, 1998). The rigorous derivation of the pseudo-Liouville operator is rather lengthy and technical (see, e.g. Ernst *et al.*, 1969; Resibois and de Leener, 1977 for details). Here we give a simplified non-rigorous derivation which, we hope, elucidates the main physical idea of the rigorous approach.

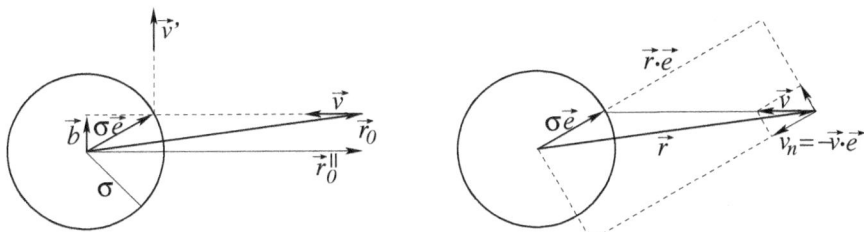

FIG. 13.1. Scattering of a point particle by a spherical hard-core potential. Left: The vector $\sigma\vec{e}$ specifies the collision point and $\vec{b} = \vec{r}_0 - \vec{r}_0^{\parallel}$ determines the impact parameter $b = |\vec{b}|$. Right: The velocity component $v_n = -\vec{v}\cdot\vec{e}$ refers to the normal motion and $\vec{r}\cdot\vec{e}$ is the distance in normal direction which cannot be smaller than σ.

13.2 Derivation of the binary-collision operator

Consider the scattering of a point particle by a hard-core scatterer of radius σ located at the origin (Fig. 13.1). Let $\vec{r}_0$ be the initial radius-vector of the particle, $\vec{v}$ its initial velocity, which does not change before the collision occurs at time t^* and $\vec{v}'$ the velocity after the collision. The vector $\sigma\vec{e}$ specifies the point where the particle hits the spherical surface ($\vec{e}$ is a unit vector). Then the radius-vector of the particle reads

$$\vec{r}(t) = \begin{cases} \vec{r}_0 + \vec{v}t & \text{for} \quad t < t^* \\ \sigma\vec{e} = \vec{r}_0 + \vec{v}t^* & \text{for} \quad t = t^* \\ \sigma\vec{e} + \vec{v}'(t - t^*) & \text{for} \quad t > t^* \end{cases} \quad (13.11)$$

Hence, any function $A(t)$, which depends on time via the time-dependent dynamical variables $\vec{r}(t), \vec{v}(t)$ may be expressed by

$$A(t) = \Theta(t^* - t) A(\vec{r}_0 + \vec{v}t, \vec{v}) + \Theta(t - t^*) A(\sigma\vec{e} + \vec{v}'(t - t^*), \vec{v}'), \quad (13.12)$$

where $\Theta(x)$ is the Heaviside function (6.19). Obviously, the first term in (13.12) vanishes after the collision and thus, describes the pre-collision motion. The second term vanishes before the collision and, therefore, refers to the after-collision motion. Equation (13.12) is not general enough, since it explicitly requires the occurrence of the collision at time t^*. For arbitrary $\vec{r}_0$ and $\vec{v}$, however, the particles do not necessarily collide. To specify the collision conditions we introduce the vector $\vec{b}$,

$$\vec{b} = \vec{r}_0 - \vec{r}_0^{\parallel} \qquad \text{with} \qquad \vec{r}_0^{\parallel} = -\vec{r}_0 \cdot \vec{v}/v, \quad (13.13)$$

which determines the impact parameter of the collision $b = |\vec{b}|$. Certainly, the collision takes place only if two conditions are fulfilled (see Fig. 13.1):

1. The impact parameter b is smaller than the radius of the scatterer σ.
2. The particle approaches the scatterer, which requires $\vec{v} \cdot \vec{r}_0 < 0$.

Therefore, the expression for A in (13.12) must contain the factor

$$\Theta(\sigma - b)\Theta(-\vec{v}\cdot\vec{r}_0) \tag{13.14}$$

which assures the collision to occur. Conversely, the factor

$$1 - \Theta(\sigma - b)\Theta(-\vec{v}\cdot\vec{r}_0) \tag{13.15}$$

assures that the collision does not occur. In the former case (13.12) is applicable, while in the latter case the particle moves freely along a straight line with velocity v and, hence,

$$A(t) = A(\vec{r}_0 + \vec{v}t, \vec{v}) . \tag{13.16}$$

With (13.12–13.16) the general relation for the evolution of A reads

$$\begin{aligned}A(t) =& \Theta(\sigma - b)\Theta(-\vec{v}\cdot\vec{r}_0) \\ & \times [\Theta(t^* - t) A(\vec{r}_0 + \vec{v}t, \vec{v}) + \Theta(t - t^*) A(\sigma\vec{e} + \vec{v}'(t-t^*), \vec{v}')] \\ & + [1 - \Theta(\sigma - b)\Theta(-\vec{v}\cdot\vec{r}_0)] A(\vec{r}_0 + \vec{v}t, \vec{v}) .\end{aligned} \tag{13.17}$$

Using $\frac{d}{dx}\Theta(x) = \delta(x)$ we obtain for the time derivative of A

$$\begin{aligned}\frac{dA}{dt} =& \Theta(\sigma - b)\Theta(-\vec{v}\cdot\vec{r}_0)\left[-\delta(t^* - t) A(\vec{r}_0 + \vec{v}t, \vec{v}) + \Theta(t^* - t)\,\vec{v}\cdot\frac{\partial A}{\partial\vec{r}}\right.\\ &\left. + \delta(t - t^*) A(\sigma\vec{e} + \vec{v}'(t-t^*), \vec{v}') + \Theta(t - t^*)\,\vec{v}'\cdot\frac{\partial A}{\partial\vec{r}}\right]\\ & + [1 - \Theta(\sigma - b)\Theta(-\vec{v}\cdot\vec{r}_0)]\,\vec{v}\cdot\frac{\partial A}{\partial\vec{r}} .\end{aligned} \tag{13.18}$$

The application of the free-streaming operator $\vec{v}(t)\cdot\partial/\partial\vec{r}$ on A yields

$$\begin{aligned}\vec{v}(t)\cdot\frac{\partial A(t)}{\partial\vec{r}} =& \Theta(\sigma - b)\Theta(-\vec{v}\cdot\vec{r}_0)\left[\Theta(t^* - t)\,\vec{v}\cdot\frac{\partial A}{\partial\vec{r}} + \Theta(t - t^*)\,\vec{v}'\cdot\frac{\partial A}{\partial\vec{r}}\right]\\ & + [1 - \Theta(\sigma - b)\Theta(-\vec{v}\cdot\vec{r}_0)]\,\vec{v}\cdot\frac{\partial A}{\partial\vec{r}} .\end{aligned} \tag{13.19}$$

Subtracting (13.19) from (13.18) we obtain

$$\frac{dA}{dt} - \vec{v}\cdot\frac{\partial A(t)}{\partial\vec{r}} = \Theta(\sigma - b)\Theta(-\vec{v}\cdot\vec{r}_0)\,\delta(t - t^*)\,[A(\sigma\vec{e}, \vec{v}') - A(\sigma\vec{e}, \vec{v})] , \tag{13.20}$$

where we use (13.11) and take into account that the δ-function is even. The factor $\delta(t - t^*)$ implies that all time-dependent terms should be taken at $t = t^*$.

On the LHS of (13.20) we recognize the free-streaming part of the Liouville operator and the RHS is the interaction part. We recast the RHS into the form

$\mathcal{L}'A(\vec{r},\vec{v})$. From (13.11) follows $\vec{r} - \sigma\vec{e} = \vec{v}(t - t^*)$, which after multiplying by $\vec{e}$ may be transformed into

$$t - t^* = \frac{\vec{r} \cdot \vec{e} - \sigma}{(\vec{v} \cdot \vec{e})}. \tag{13.21}$$

Using the property of the δ-function,

$$\delta[f(x)] = \frac{\delta(x - x_0)}{|f'(x_0)|} \quad \text{where} \quad f(x_0) = 0 \tag{13.22}$$

we can write

$$\delta(t - t^*) = \delta\left(\frac{\vec{r} \cdot \vec{e} - \sigma}{\vec{v} \cdot \vec{e}}\right) = |\vec{v} \cdot \vec{e}|\,\delta(\vec{r} \cdot \vec{e} - \sigma). \tag{13.23}$$

Note that according to the collision geometry (see Fig. 13.1) the unit vector $\vec{e}$ specifies the direction of the normal motion of the particle with respect to the scatterer ($v_n = -\vec{v} \cdot \vec{e}$ is the normal component of the velocity), thus, $\vec{r} \cdot \vec{e} \geq \sigma$. The condition imposed by $\delta(\vec{r} \cdot \vec{e} - \sigma)$ in (13.23) is satisfied exclusively in the point

$$\vec{r} = \sigma\vec{e} \quad \text{or} \quad \hat{\vec{r}} \equiv \frac{\vec{r}}{|\vec{r}|} = \vec{e}; \quad |\vec{r}| = \sigma. \tag{13.24}$$

If this condition is satisfied, the condition $\Theta(\sigma - b)$ for the impact parameter is fulfilled automatically and may be omitted. The condition $\Theta(-\vec{v} \cdot \vec{r}_0)$ may be substituted by the equivalent $\Theta(-\vec{v} \cdot \vec{e})$ as illustrated in Fig. 13.1. Then the RHS of (13.20) may be written in the form

$$|\vec{v} \cdot \vec{e}|\,\Theta(-\vec{v} \cdot \vec{e})\,\delta(\vec{r} \cdot \vec{e} - \sigma)\,[A(\vec{r},\vec{v}') - A(\vec{r},\vec{v})]$$
$$= |\vec{v} \cdot \vec{e}|\,\Theta(-\vec{v} \cdot \vec{e})\,\delta(|\vec{r}| - \sigma)\left(\hat{b}^{\vec{e}} - 1\right)A(\vec{r},\vec{v}), \tag{13.25}$$

where we introduce the operator $\hat{b}^{\vec{e}}$ (specified by the unit vector $\vec{e}$) which, acting on a function of dynamical variables, replaces the velocity just before a collision by the velocity just after the collision. We obtain the final form of (13.20):

$$\boxed{\frac{dA}{dt} = \left(\vec{v} \cdot \frac{\partial}{\partial \vec{r}} + \mathcal{L}'\right)A(\vec{r},\vec{v}) = \mathcal{L}A(\vec{r},\vec{v})} \tag{13.26}$$

with the operator

$$\boxed{\mathcal{L}' = \hat{T} = \left|\vec{v} \cdot \hat{\vec{r}}\right|\Theta\left(-\vec{v} \cdot \hat{\vec{r}}\right)\delta(|\vec{r}| - \sigma)\left(\hat{b}^{\hat{\vec{r}}} - 1\right)} \tag{13.27}$$

The operator $\mathcal{L}$ for systems of particles with hard-core interaction is called pseudo-Liouville operator and $\hat{T}$ is called collision operator. This form of $\hat{T}$ has been used by Luding et al. (1998).

The physical meaning of the factors of the operator $\hat{T}$ is obvious:

$\delta(|\vec{r}| - \sigma)$ assures that the collision occurs at the point of contact,

$\Theta(-\vec{v} \cdot \hat{\vec{r}})$ selects the appropriate direction of the velocity,

$|\vec{v} \cdot \hat{\vec{r}}|$ characterizes the normal component of the relative velocity and determines the collision frequency

$(\hat{b}^{\vec{r}} - 1)$ describes how the variable A changes due to the collision.

With the property of the δ-function

$$\delta(\vec{r} - \sigma \vec{e}) = \frac{1}{\sigma^2} \delta(|\vec{r}| - \sigma) \delta(\hat{\vec{r}} - \vec{e}) \tag{13.28}$$

the operator $\hat{T}$ may be written in a form that emphasizes its similarity with the collision integral in the Boltzmann equation:

$$\hat{T} = \sigma^2 \int d\vec{e} \, \Theta(-\vec{v} \cdot \vec{e}) |\vec{v} \cdot \vec{e}| \delta(\vec{r} - \sigma \vec{e}) (\hat{b}^{\vec{e}} - 1). \tag{13.29}$$

Consider now a system of two identical hard-core particles of diameter σ in the absence of external forces. Its dynamics may be described by the centre of mass motion, characterized by $\vec{r}_{cm}$ and $\vec{v}_{cm}$ and by the relative motion, characterized by $\vec{r}_{12}$ and $\vec{v}_{12}$:

$$\vec{r}_{cm} = \frac{1}{2}\vec{r}_1 + \frac{1}{2}\vec{r}_2, \qquad \vec{v}_{cm} = \frac{1}{2}\vec{v}_1 + \frac{1}{2}\vec{v}_2. \tag{13.30}$$

The time derivative of some variable $A(\vec{r}_1, \vec{r}_2, \vec{v}_1, \vec{v}_2)$ reads

$$\frac{dA}{dt} = \left(\dot{\vec{r}}_{12} \frac{\partial A}{\partial \vec{r}_{12}} + \dot{\vec{v}}_{12} \frac{\partial A}{\partial \vec{v}_{12}} \right) + \dot{\vec{r}}_{cm} \frac{\partial A}{\partial \vec{r}_{cm}} + \dot{\vec{v}}_{cm} \frac{\partial A}{\partial \vec{v}_{cm}}. \tag{13.31}$$

The terms in brackets refer to the relative motion and describe the scattering of an effective point particle by a hard-core scatterer of radius σ; this is exactly the same problem as addressed above. Therefore, we can recast (13.31) using the new notations (13.26):

$$\frac{dA}{dt} = \left(\vec{v}_{12} \frac{\partial A}{\partial \vec{r}_{12}} + \hat{T}_{12} A \right) + \vec{v}_{cm} \frac{\partial A}{\partial \vec{r}_{cm}} = (\mathcal{L}_1^0 + \mathcal{L}_2^0) + \hat{T}_{12} A, \tag{13.32}$$

where $\hat{T}_{12}$ is the *binary collision operator*, since it describes the collision of two particles (1 and 2 in our case). It is defined by (13.27) or (13.29) with the vectors $\vec{r}$ and $\vec{v}$ substituted by $\vec{r}_{12}$ and $\vec{v}_{12}$, respectively. In (13.32) we take into account $\dot{\vec{v}}_{cm} = 0$ if there are no external forces, use the definition (13.8) and the identity

$$\vec{v}_{12} \frac{\partial}{\partial \vec{r}_{12}} + \vec{v}_{cm} \frac{\partial}{\partial \vec{r}_{cm}} = \vec{v}_1 \frac{\partial}{\partial \vec{r}_1} + \vec{v}_2 \frac{\partial}{\partial \vec{r}_2}. \tag{13.33}$$

The generalization for the case of force-free many-particle systems is straightforward. Indeed, due to the instantaneous character of hard-core collisions, at any time instant there may occur at most one collision, provided the system is finite. Thus, all interactions between particles occur via a series of successive (binary) collisions, in which only one pair of particles is involved. Therefore, we conclude that the dynamics of a many-particle hard sphere system is described by

$$\frac{dA}{dt} = \mathcal{L}A, \qquad \mathcal{L} = \sum_j \vec{v}_j \cdot \frac{\partial}{\partial \vec{r}_j} + \sum_{i<j} \hat{T}_{ij} \qquad (13.34)$$

The first sum in the definition of the pseudo-Liouville operator $\mathcal{L}$ corresponds to the free streaming of the particles, while the second sum corresponds to the interaction of the particles, described by the binary collision operator:

$$\hat{T}_{ij} = \sigma^2 \int d\vec{e}\, \Theta(-\vec{v}_{ij} \cdot \vec{e}) |\vec{v}_{ij} \cdot \vec{e}| \delta(\vec{r}_{ij} - \sigma\vec{e}) \left[\hat{b}_{ij}^{\vec{e}} - 1\right] \qquad (13.35)$$

Here $\vec{v}_{ij} \cdot \vec{e} = (\vec{v}_i - \vec{v}_j) \cdot \vec{e}$ is the normal component of the relative velocity of the colliding pair, which (multiplied by the infinitesimal time dt) defines the length of the collision cylinder. The Heaviside step-function $\Theta(x)$ selects approaching particles and the δ-function determines the unit vector $\vec{e}$, which specifies the collision. The operator $\hat{b}_{ij}^{\vec{e}}$ is defined by

$$\hat{b}_{ij}^{\vec{e}} f(\vec{r}_i, \vec{r}_j, \vec{v}_i, \vec{v}_j \cdots) = f(\vec{r}_i, \vec{r}_j, \vec{v}_i', \vec{v}_j' \cdots) \qquad (13.36)$$

where f is some function of dynamical variables. The after-collision velocities of the colliding pair $\vec{v}_i'$ and $\vec{v}_j'$ are given in terms of their pre-collision values $\vec{v}_i$, $\vec{v}_j$ and of the unit vector $\vec{e} = \vec{r}_{ij}/r_{ij}$ at the collision instant. Note that according to the structure of the binary collision operator, the contribution from a given pair ij to the time derivative dA/dt at time t is non-zero only if these particles are in contact at time t.

In a rigorous derivation of the pseudo-Liouville operator in the definition of $\hat{T}_{ij}$ appears a pre-factor which prevents successive collisions of the same pair of particles (Ernst et al., 1969; Resibois and de Leener, 1977). This pre-factor is omitted here since it does not affect the subsequent analysis but complicates the notation.

From the definition of the binary collision operator and the collision rules (6.5) follow the properties of the collision operator:

$$\begin{aligned} \hat{T}_{ij}\vec{v}_j &= -\hat{T}_{ij}\vec{v}_i \\ \hat{T}_{ji}\vec{v}_i &= \hat{T}_{ij}\vec{v}_i \end{aligned} \qquad (13.37)$$

Exercise 13.1 *Derive (13.37) using the definition of the collision operator (13.35) and the collision rules (6.5)!*

13.3 Application of the pseudo-Liouville operator to standard problems

13.3.1 Time dependence of temperature

The formalism of the pseudo-Liouville operator allows to perform various calculations in a very elegant way, in particular those which refer to the one-particle dynamics. Let us consider some representative examples.

From (13.34) and the definition of the temperature (6.4) for a homogeneous gas follows

$$\frac{3}{2}\frac{dT}{dt} = \frac{m}{2}\frac{d}{dt}\langle v_1^2\rangle_t = \frac{m}{2}\langle \mathcal{L} v_1^2\rangle_t, \qquad (13.38)$$

where $\langle \cdots \rangle_t$ denotes averaging using the many-particle distribution function $\rho(t) \equiv \rho(\vec{r}_1, \ldots, \vec{r}_N, \vec{v}_1, \ldots, \vec{v}_N, t)$ at time t. Taking into account $\mathcal{L}_i^0 v_1^2 = 0$ and the identity of the particles we obtain

$$\frac{3}{2}\frac{dT}{dt} = \frac{m}{2}\left\langle \left(\sum_i \mathcal{L}_i^0 + \sum_{i<j} \hat{T}_{ij}\right) v_1^2 \right\rangle_t = (N-1)\frac{m}{2}\langle \hat{T}_{12} v_1^2\rangle_t. \qquad (13.39)$$

The evaluation of $\langle \hat{T}_{12} v_1^2\rangle_t$ is straightforward, but since we will frequently meet quantities of this type below, we give here the detailed solution. Writing explicitly the averaging with $\rho(t)$ and using

$$\left(\hat{b}_{ij}^{\vec{e}} - 1\right) v_1^2 = (v_1')^2 - v_1^2 \qquad (13.40)$$

which follows from the definition of the binary collision operator (13.35), we recast (13.39) into the form

$$\frac{3}{2}\frac{dT}{dt} = \frac{m}{2}(N-1)\int d\vec{r}_1 \cdots d\vec{r}_N \int d\vec{v}_1 \cdots d\vec{v}_N \rho(t) \\ \times \sigma^2 \int d\vec{e}\,\Theta(-\vec{v}_{12}\cdot\vec{e})\,|\vec{v}_{12}\cdot\vec{e}|\,\delta(\vec{r}_{12} - \sigma\vec{e})\left[(v_1')^2 - v_1^2\right]. \qquad (13.41)$$

Now we use the definition of the two-particle correlation function (Resibois and de Leener, 1977)

$$N(N-1)\int d\vec{r}_3 \cdots d\vec{r}_N \int d\vec{v}_3 \cdots d\vec{v}_N \rho(t) \equiv f_2(\vec{r}_1, \vec{r}_2, \vec{v}_1, \vec{v}_2, t) \qquad (13.42)$$

and the assumption that the coordinate part of $f_2(\vec{r}_1, \vec{r}_2, \vec{v}_1, \vec{v}_2, t)$ factorizes from the velocity part. Moreover, we assume that the velocity part may be written as a product of one-particle distribution functions. This approximation corresponds to the hypothesis of molecular chaos. For a uniform isotropic system

$$f_2(\vec{r}_1, \vec{r}_2, \vec{v}_1, \vec{v}_2, t) = g_2(r_{12}) f(\vec{v}_1, t) f(\vec{v}_2, t), \qquad (13.43)$$

where $r_{12} = |\vec{r}_{12}|$ and $g_2(r_{12})$ is the pair-correlation function. With (13.42, 13.43) we rewrite (13.41) as

$$\frac{3}{2}\frac{dT}{dt} = \frac{m}{2}\left\{\frac{1}{N}\int d\vec{r}_1 d\vec{r}_2\, g_2(r_{12})\,\delta(\vec{r}_{12} - \sigma\vec{e})\right\}$$
$$\times \sigma^2 \int d\vec{v}_1 d\vec{v}_2 \int d\vec{e}\,\Theta(-\vec{v}_{12}\cdot\vec{e})|\vec{v}_{12}\cdot\vec{e}|\, f(\vec{v}_1,t)\,f(\vec{v}_2,t)\left[(v'_1)^2 - v_1^2\right]. \quad (13.44)$$

To find the factor in curled brackets we transform the coordinates $\vec{r}_1, \vec{r}_2 \to \vec{r}_1, \vec{r}_{12}$ (the Jacobian of this transformation equals unity) and integrate over $\vec{r}_{12}, \vec{r}_1$:

$$\frac{1}{N}\int d\vec{r}_1 d\vec{r}_2 g_2(r_{12})\,\delta(\vec{r}_{12} - \sigma\vec{e}) = \int d\vec{r}_{12} g_2(r_{12})\,\delta(\vec{r}_{12} - \sigma\vec{e})\frac{1}{N}\int d\vec{r}_1$$
$$= g_2(\sigma)\frac{V}{N} = n^{-1}g_2(\sigma). \quad (13.45)$$

The other factor in (13.44) may be transformed as

$$\sigma^2\int d\vec{v}_1 d\vec{v}_2 \int d\vec{e}\,\Theta(-\vec{v}_{12}\cdot\vec{e})|\vec{v}_{12}\cdot\vec{e}|\,f(\vec{v}_1,t)\,f(\vec{v}_2,t)\left[(v'_1)^2 - v_1^2\right]$$
$$= \frac{1}{2}\sigma^2 v_T^3 n^2\int d\vec{c}_1 d\vec{c}_2 \int d\vec{e}\,\Theta(-\vec{c}_{12}\cdot\vec{e})|\vec{c}_{12}\cdot\vec{e}|\,\tilde{f}(\vec{c}_1,t)\,\tilde{f}(\vec{c}_2,t)\,\Delta(c_1^2 + c_2^2)$$
$$= -\sigma^2 v_T^3 n^2 \mu_2, \quad (13.46)$$

where we use the symmetry with respect to exchange of the particle indices $1 \leftrightarrow 2$, the scaled velocity distribution function, $f(\vec{v},t) = (n/v_T^3)\,\tilde{f}(\vec{c},t)$, equation (8.50) and the definition (8.21) of the variable μ_2. Combining (13.45) and (13.46) we obtain the RHS of (13.41), which finally takes the form

$$\frac{3}{2}\frac{dT}{dt} = -v_T g_2(\sigma)\sigma^2 nT\mu_2 = -BT\mu_2. \quad (13.47)$$

This expression coincides with the previously obtained equation (8.11).

Exercise 13.2 *Prove (13.39)!*

13.3.2 Time-correlation function of dynamical variables

With the assumption of uncorrelated successive binary collisions, that is, of molecular chaos, the pseudo-Liouville operator formalism allows for an elegant calculation of time-correlation functions. The time derivative of A can be written as

$$\frac{dA}{dt} = \mathcal{L}A \quad (13.48)$$

where the pseudo-Liouville operator $\mathcal{L}$ is given by (13.34). Formal integration yields for $t > t'$:

$$A(t) = e^{\mathcal{L}(t-t')} A\left[\{\vec{r}_i(t'), \vec{v}_i(t')\}\right]. \tag{13.49}$$

The time-correlation function then reads

$$\langle A(t') A(t) \rangle = \int d\Gamma \rho(t') A(t') e^{\mathcal{L}(t-t')} A(t'), \tag{13.50}$$

where $\int d\Gamma$ denotes integration over all degrees of freedom. The distribution function $\rho(t')$ depends on temperature T, particle number density n, etc., which change on a time scale $t \gg \tau_c$, where τ_c is the mean collision time.

Due to the assumption of molecular chaos, the velocities of the particles are not correlated which is equivalent with the assumption that the collisions do not depend on preceeding collisions. With this assumption any function of the dynamical variables that does not depend on the position of the particles corresponds to a Markov process, that is, the function changes at the collision instants without memory. The time-correlation function of a Markov process is an exponentially decaying function (for a rigorous mathematical formulation of this statement, see Resibois and de Leener, 1977). Therefore, the time correlation function can be expressed as (see e.g. Chandler, 1975; Berne, 1977; Brilliantov and Revokatov, 1984)

$$\langle A(t') A(t) \rangle = \langle A^2 \rangle_{t'} \exp\left[-\frac{|t-t'|}{\tau_A(t')}\right], \quad t > t', \tag{13.51}$$

where $\langle \cdots \rangle_{t'}$ denotes averaging with the distribution function taken at time t'. To write the time-correlation function in the form (13.51) we assume that the relaxation time $\tau_A(t')$ is much shorter than the time which characterizes the change of $\langle A^2 \rangle_{t'}$; the latter quantity evolves on the same time scale as the distribution function $\rho(t')$. This assumption implies an *adiabatic* approximation for $\langle A^2 \rangle_{t'}$ on the time scale $\tau_A(t')$. In Section 14.5 we consider the velocity correlation time beyond this approximation.

In adiabatic approximation (13.51) the inverse of the relaxation time, τ_A^{-1}, equals the initial slope of the time-correlation function $\langle A(t') A(t) \rangle$ (Chandler, 1975; Berne, 1977; Brilliantov and Revokatov, 1984). It may be found from the time derivative of this function, taken at $t = t'$. Equations (13.50, 13.51) then yield

$$-\tau_A^{-1}(t') = \int d\Gamma \rho(t') \frac{A\mathcal{L}A}{\langle A^2 \rangle_{t'}} = \frac{\langle A\mathcal{L}A \rangle_{t'}}{\langle A^2 \rangle_{t'}}. \tag{13.52}$$

The relaxation time $\tau_A(t')$, which depends on time via the distribution function $\rho(t')$, changes on the slow time scale.

14

COEFFICIENT OF SELF-DIFFUSION

The coefficient of self-diffusion for granular gases with the simplified collision model $\varepsilon = $ const. differs drastically from the coefficient for granular gases of viscoelastic particles. While for the former case the mean square displacement depends logarithmically on time, for the latter case it obeys a power law.

14.1 Velocity correlation time

In the previous section we have discussed a method to calculate the evolution of a general function of dynamic variables $A(t)$ and its time-correlation function. Now we specify $A(t)$ to be the velocity of a tagged particle, say $\vec{v}_1(t)$, and calculate the velocity autocorrelation function and the self-diffusion coefficient. With $3T(t) = m \langle v^2 \rangle_t$ equations (13.51, 13.52), which correspond to the adiabatic approximation, read

$$\langle \vec{v}_1(t') \cdot \vec{v}_1(t) \rangle = \frac{3T(t')}{m} \exp\left[-\frac{|t-t'|}{\tau_v(t')}\right] \tag{14.1}$$

$$-\tau_v^{-1}(t') = (N-1) \frac{\langle \vec{v}_1 \cdot \hat{T}_{12} \vec{v}_1 \rangle_{t'}}{\langle \vec{v}_1 \cdot \vec{v}_1 \rangle_{t'}}. \tag{14.2}$$

To obtain (14.2) we use (13.34, 13.35) and take into account $\mathcal{L}_0 \vec{v}_1 = 0$, $\hat{T}_{ij} \vec{v}_1 = 0$ for $i \neq 1$, $i < j$ and the identity of the particles. From the properties of the binary collision operator (13.37) follows the relation

$$\langle \vec{v}_1 \cdot \hat{T}_{12} \vec{v}_1 \rangle = \frac{1}{2} \langle \vec{v}_{12} \cdot \hat{T}_{12} \vec{v}_1 \rangle, \tag{14.3}$$

Exercise 14.1 *Prove the property (14.3) of the binary collision operator!*

Using the definition of the binary collision operator (13.35) together with (13.36) for the operator $\hat{b}_{ij}^{\vec{e}}$, which has the property

$$\vec{v}_{12} \cdot \left(\hat{b}_{ij}^{\vec{e}} - 1\right) \vec{v}_1 = \vec{v}_{12} \cdot (\vec{v}_1' - \vec{v}_1) = -\frac{1}{2}(1+\varepsilon)(\vec{v}_{12} \cdot \vec{e})^2, \tag{14.4}$$

we obtain the velocity correlation time

$$\tau_v^{-1}(t) = -\left[\frac{3T(t)}{m}\right]^{-1}(N-1)\frac{1}{2}\left\langle \vec{v}_{12}\cdot\hat{T}_{12}\vec{v}_1\right\rangle$$

$$= \frac{1}{4}\left[\frac{3T(t)}{m}\right]^{-1}(N-1)\int d\vec{r}_1\cdots d\vec{r}_N\int d\vec{v}_1\cdots d\vec{v}_N\rho(t) \qquad (14.5)$$

$$\times \sigma^2\int d\vec{e}\,\Theta\left(-\vec{v}_{12}\cdot\vec{e}\right)|\vec{v}_{12}\cdot\vec{e}|\,\delta\left(\vec{r}_{12}-\sigma\vec{e}\right)(1+\varepsilon)\left(\vec{v}_{12}\cdot\vec{e}\right)^2.$$

This relation is very similar to (13.41) and, therefore, may be evaluated analogously. Employing again the approximation (13.43), which corresponds to the hypothesis of molecular chaos, yields the result

$$\tau_v^{-1}(t) = \frac{1}{4}\left[\frac{3T(t)}{m}\right]^{-1}g_2(\sigma)n^{-1}\sigma^2\int d\vec{v}_1 d\vec{v}_2$$

$$\times\int d\vec{e}\,\Theta\left(-\vec{v}_{12}\cdot\vec{e}\right)|\vec{v}_{12}\cdot\vec{e}|\,f(\vec{v}_1,t)f(\vec{v}_2,t)(1+\varepsilon)\left(\vec{v}_{12}\cdot\vec{e}\right)^2. \qquad (14.6)$$

Exercise 14.2 *Derive (14.6)!*

It is convenient to express the velocity correlation time τ_v using the scaled distribution function $\tilde{f}$:

$$\tau_v^{-1}(t) = \frac{1}{6}v_T(t)g_2(\sigma)\sigma^2 n$$

$$\times\int d\vec{c}_1 d\vec{c}_2\int d\vec{e}\,\Theta\left(-\vec{c}_{12}\cdot\vec{e}\right)|\vec{c}_{12}\cdot\vec{e}|\,\tilde{f}(\vec{c}_1,t)\tilde{f}(\vec{c}_2,t)(1+\varepsilon)\left(\vec{c}_{12}\cdot\vec{e}\right)^2, \qquad (14.7)$$

where we use the definition of $\tilde{f}$, (8.50) and $T(t) = \frac{1}{2}mv_T(t)$. Equation (14.7) is the general relation for the velocity correlation time, valid under the assumption of molecular chaos. We further assume small dissipation which assures that the velocity distribution is well approximated by the Sonine polynomial expansion which is cut after the first non-vanishing coefficient a_2:

$$\tilde{f}(\vec{c},t) = \phi(c)\left[1 + a_2(t)S_2\left(c^2\right)\right]. \qquad (14.8)$$

Then τ_v may be expressed by

$$\tau_v^{-1}(t) = \frac{1}{12\sqrt{2\pi}}\tau_c^{-1}(t)$$

$$\times\int d\vec{c}_1 d\vec{c}_2\int d\vec{e}\,\Theta\left(-\vec{c}_{12}\cdot\vec{e}\right)|\vec{c}_{12}\cdot\vec{e}|\,(\vec{c}_{12}\cdot\vec{e})^2(1+\varepsilon) \qquad (14.9)$$

$$\times\left\{1 + a_2(t)\left[S_2\left(c_1^2\right) + S_2\left(c_2^2\right)\right] + a_2^2(t)S_2\left(c_1^2\right)S_2\left(c_2^2\right)\right\}$$

with[14]

---

[14] Note that $\tau_c(t)$ differs from the actual mean collision time at time t if the velocity distribution differs from the Maxwell distribution.

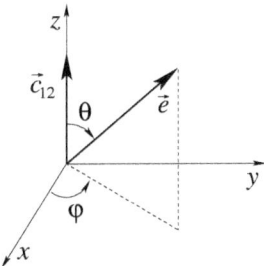

FIG. 14.1. Illustration of the integration angles φ and θ in (14.12).

$$\tau_c(t)^{-1} \equiv 4\sqrt{\pi} g_2(\sigma) \sigma^2 n \sqrt{\frac{T(t)}{m}}. \tag{14.10}$$

This expression for the velocity correlation time as a function of the second Sonine coefficient a_2 will be used below to find the self-diffusion coefficient.

14.2 Constant coefficient of restitution

14.2.1 Self-diffusion with the assumption of a Maxwell distribution

In Part II we have shown that the velocity distribution function deviates from the Maxwell distribution. Let us for the moment neglect this deviation, that is, we assume for the second Sonine coefficient $a_2 = 0$. This approximation is justified for slightly inelastic particles, $\varepsilon \lesssim 1$. In Section 14.2.2 we will consider the more general case $a_2 = a_2(\varepsilon) \neq 0$.

As previously we use the centre of mass velocity $\vec{C}$ and the relative velocity $\vec{c}_{12}$ (see (8.24)). The Jacobian of the transformation $\vec{c}_1, \vec{c}_2 \to \vec{c}_{12}, \vec{C}$ equals unity and the product $\phi(c_1)\phi(c_2)$ transforms into $\phi(c_{12})\phi(C)$. Here $\phi(C)$ is the Maxwell distribution for the centre of mass velocity, which corresponds to an effective particle of mass $2m$ and $\phi(c_{12})$ is the Maxwell distribution for the relative velocity, which corresponds to an effective particle of mass $m/2$:

$$\phi(\vec{c}_1)\phi(\vec{c}_2) \to (2\pi)^{-3/2} \exp\left(-c_{12}^2/2\right) (2/\pi)^{3/2} \exp\left(-2C^2\right) \equiv \phi(\vec{c}_{12})\phi\left(\vec{C}\right). \tag{14.11}$$

Both functions are normalized to unity (see also (8.25) and related discussion).

Choosing the axis OZ along the vector $\vec{c}_{12}$ as sketched in Fig. 14.1, equation (14.9) takes the form

$$\begin{aligned}\tau_v^{-1} &= \frac{1}{6}(1+\varepsilon) v_T g_2(\sigma) \sigma^2 n \int d\vec{C}\, \phi(C) \int d\vec{c}_{12}\, \phi(c_{12}) \\ &\quad \times \int_{\pi/2}^{\pi} \sin\theta\, d\theta\, c_{12}\, |\cos\theta|\, c_{12}^2 \cos^2\theta \int_0^{2\pi} d\varphi \\ &= \frac{1}{6}(1+\varepsilon) v_T g_2(\sigma)\sigma^2 n\, 4\pi \int_0^{\infty} dc_{12}\, c_{12}^5\, (2\pi)^{-3/2} \exp\left(-\frac{c_{12}^2}{2}\right) \frac{2\pi}{4}, \end{aligned} \tag{14.12}$$

where θ and φ are the angles of the vector $\vec{e}$ in the chosen coordinate system. To obtain the last line of (14.12) we take into account that the integration over the centre of mass velocity $\vec{C}$ yields unity (due to the normalization), the angular integration over θ and φ yields the factors $1/4$ and 2π, correspondingly. The integration over the direction of the vector $\vec{c}_{12}$ yields the factor 4π since the integrand is independent of the direction of $\vec{c}_{12}$ but it depends on the angle between $\vec{c}_{12}$ and $\vec{e}$ only. The integral over c_{12} may be easily calculated (yielding the factor 8) and we finally obtain

$$\tau_v^{-1}(t) = \frac{1}{2}(1+\varepsilon)\frac{8}{3}\sigma^2 n g_2(\sigma)\sqrt{\frac{\pi T(t)}{m}}, \qquad (14.13)$$

with the thermal velocity $v_T = \sqrt{2T/m}$. According to (14.13) the velocity correlation time for inelastic collisions is larger than for elastic collisions. This may be explained by partial suppression of backscattering of particles due to inelastic losses of their normal relative velocity. As a result, for inelastic collisions, the angle between the trajectories of colliding particles is smaller after the collision than before, while for elastic collisions both angles are identical (see Fig. 6.2 and discussion there). Therefore, the trajectories of dissipatively colliding particles are more aligned, that is, the velocity correlation time is larger than for gases of elastically colliding particles.

Using the velocity correlation function we write

$$\left\langle (\Delta r(t))^2 \right\rangle = 2\int_0^t dt_1 \frac{3T(t_1)}{m}\int_{t_1}^t dt_2 \exp\left(-\frac{|t_2-t_1|}{\tau_v(t_1)}\right). \qquad (14.14)$$

On the short time scale $t \sim \tau_c \sim \tau_v$, $T(t_1)$ and $\tau_v(t_1)$ may be considered as constants. The exponential correlation function in (14.14) decays on this time scale, thus for $t \gg \tau_c$ which is addressed when describing diffusion processes, the upper limit of the second integral may be substituted by infinity:

$$\left\langle (\Delta r(t))^2 \right\rangle = 6\int_0^t dt_1 \frac{T(t_1)}{m}\int_0^\infty dt_3 \exp\left(-\frac{t_3}{\tau_v(t_1)}\right) = 6\int_0^t \frac{T(t_1)}{m}\tau_v(t_1)\,dt_1. \qquad (14.15)$$

Comparing this equation with (12.16) we obtain the time-dependent self-diffusion coefficient (the diffusivity)

$$D(t) = \frac{T(t)}{m}\tau_v(t). \qquad (14.16)$$

Using the time dependence of temperature given by (8.64, 8.65) with $a_2 = 0$,

$$T(t) = \frac{T_0}{(1+t/\tau_0)^2}, \qquad \tau_0^{-1} = \frac{1}{6}(1-\varepsilon^2)\tau_c(0)^{-1}, \qquad (14.17)$$

with the initial mean collision time

$$\tau_c^{-1}(0) = 4\sqrt{\pi}g_2(\sigma)\sigma^2 n\sqrt{\frac{T_0}{m}}, \qquad (14.18)$$

(14.13, 14.16) yield the coefficient of self-diffusion for the case $\varepsilon = $ const.:

$$D(t) = 2D_0 \left[(1+\varepsilon)\left(1 + \frac{t}{\tau_0}\right)\right]^{-1}. \qquad (14.19)$$

The Enskog coefficient of self-diffusion for elastic particles is given by

$$D_0^{-1} = \frac{8}{3}\sigma^2 g_2(\sigma) n \sqrt{\frac{\pi m}{T_0}}. \qquad (14.20)$$

Correspondingly, with (12.16) for $t \gg \tau_0$ the mean square displacement reads

$$\left\langle (\Delta r(t))^2 \right\rangle \propto \log t \qquad (14.21)$$

14.2.2 Self-diffusion with regard to the deviations from the Maxwell distribution

Equation (14.9) for the velocity correlation time is written in form of the standard kinetic integral that was introduced in Section 8.3. This type of integrals can be evaluated by computational formula manipulation as discussed in detail in Appendix A. In linear approximation with respect to a_2 the result reads

$$\tau_v^{-1}(t) = \frac{(1+\varepsilon)}{3}\tau_c^{-1}(t)\left(1 + \frac{3}{16}a_2\right), \qquad (14.22)$$

where a_2 is given by (8.41).

Exercise 14.3 *Derive the velocity correlation time for a granular gas of particles that interact with $\varepsilon = $ const. in linear approximation with respect to a_2: first perform the manual calculations, then use Maple (see Appendix A) to get the expression for τ_v in terms of Basic Integrals. Finally, derive (14.22)!*

From (14.13, 14.16) we obtain the coefficient of self-diffusion:

$$D(t) = \frac{2D_0}{1+\varepsilon}\left(1 + \frac{3}{16}a_2\right)^{-1}\left(1 + \frac{t}{\tau_0}\right)^{-1} \qquad (14.23)$$

where D_0 is given by (14.20), while τ_0 in the linear approximation with respect to a_2 reads

$$\tau_0^{-1} = \frac{1}{6}\left(1 - \varepsilon^2\right)\left(1 + \frac{3}{16}a_2\right)\tau_c(0)^{-1}, \qquad (14.24)$$

with $\tau_c(0)$ defined in (14.18). The evolution of the mean square displacement for $t \gg \tau_0$ is still given by (14.21).

Exercise 14.4 *Find the self-diffusion coefficient up to the second order in a_2!*

14.3 Coefficient of self-diffusion for gases of viscoelastic particles

The coefficient of restitution of colliding viscoelastic particles is a function of the impact velocity as given by (3.22). If we analyse the coefficient of self-diffusion for gases of viscoelastic particles, surprisingly, we obtain a qualitatively different result than for the simplified collision model, $\varepsilon = $ const. (Brilliantov and Pöschel, 2000b, 2000c, 2001).

Again we use (14.9) and take into account that ε depends now on the impact velocity, i.e.,

$$1 + \varepsilon = 2 - C_1 \delta'(t) |\vec{c}_{12} \cdot \vec{e}|^{1/5} + \frac{3}{5} C_1^2 \delta'^2(t) |\vec{c}_{12} \cdot \vec{e}|^{2/5} + \cdots \quad (14.25)$$

with the (small) dissipative parameter $\delta'(t) = \delta[2T(t)/T_0]^{1/10}$ and δ defined by (9.14). Similarly as for the case $\varepsilon = $ const. we can apply the Maple program given below (see Appendix A) to find τ_v.

```
1  DefDimension(3);
2  DefJ();
3  Expr:=tau_c_inv/(12*sqrt(2*Pi))*(1+epsilon)*c12D0Te*c12D0Te
4       *(1+a2*(S(2,c1p)+S(2,c2p))):
5  tau_v_inv:=getJexpr(1,Expr,2);
6  coeff(tau_v_inv, dprime,0);
7  coeff(coeff(tau_v_inv, dprime,1),a2,0);
```

In linear order with respect to δ, the computation yields

$$\tau_v^{-1}(t) = \frac{2}{3}\tau_c^{-1}(t)\left[1 + \frac{3}{16}a_2 - \frac{\omega_0}{4\sqrt{2\pi}}\delta'(t)\right]. \quad (14.26)$$

We do not have explicit expressions for the time dependence of $a_2(t)$ and $T(t)$, however, for $t \gg \tau_0$ (τ_0 is the temperature relaxation time) (9.51, 9.58) provide linear approximations with respect to δ:

$$a_2(t) \simeq -\delta\, h \left(\frac{t}{\tau_0}\right)^{-1/6} \simeq -\delta\, h \left[\frac{T(t)}{T_0}\right]^{1/10} \quad (14.27)$$

$$\frac{T(t)}{T_0} \simeq \left(\frac{t}{\tau_0}\right)^{-5/3} + \delta\left(2q_1 + \frac{12}{25}h\right)\left(\frac{t}{\tau_0}\right)^{-11/6}, \quad (14.28)$$

where τ_0 and the constants $h = 0.415$ and $q_1 = 1.534$ are given by (9.29, 9.30, 9.46). We arrive at the velocity correlation time in linear approximation with respect to the dissipative constant δ:

$$\tau_v^{-1}(t) = \frac{2}{3}\tau_c^{-1}(t)\left[1 + \delta\left(\frac{T(t)}{T_0}\right)^{1/10}\left(4q_0 + \frac{3}{16}h\right)\right], \quad (14.29)$$

with the constant $q_0 = 0.173$ as introduced in (9.30). Finally, from (14.29, 14.28, 14.16) we obtain the coefficient of self-diffusion for $t \gg \tau_0$:

$$D = D_0\sqrt{\frac{T(t)}{T_0}}\left[1 - \delta\left(\frac{T(t)}{T_0}\right)^{1/10}\left(4q_0 + \frac{3h}{16}\right)\right], \quad (14.30)$$

with D_0 given by (14.20). Keeping only linear terms with respect to δ we obtain finally

$$\frac{D(t)}{D_0} \simeq \left(\frac{t}{\tau_0}\right)^{-5/6} + \delta\left(q_1 - 4q_0 - \frac{21}{400}h\right)\left(\frac{t}{\tau_0}\right)^{-1} = \left(\frac{t}{\tau_0}\right)^{-5/6} + c\delta\left(\frac{t}{\tau_0}\right)^{-1} \quad (14.31)$$

with $c = 0.82$. The first term, which is independent of δ, represents the coefficient of self diffusion for a gas of elastic particles at temperature $T(t)$. Its time dependence enters only via the time dependence of the temperature. The second term corresponds to the dissipative properties of the particle collisions. It arises from the alignment of the particle trajectories (the angle of the traces after a collision is smaller than the angle before the collision) and from the deviation of the velocity distribution from the Maxwell distribution. The latter contribution is $171/400\,h \approx 0.177$ which is about 3% of the value of c.

Correspondingly, the mean square displacement reads for $t \gg \tau_0$

$$\boxed{\left\langle (\Delta r(t))^2 \right\rangle \sim t^{1/6} + c\delta\,\log t} \quad (14.32)$$

This dependence holds true for

$$\tau_c(0)/\delta \ll t \ll \tau_c(0)/\delta^{11/5}, \quad (14.33)$$

where the first inequality follows from the condition $\tau_0 \ll t$, while the second condition corresponds to $\tau_c(t) \ll \tau_0$ which is required for the physically meaningful definition of the temperature. Indeed, since $\tau_0 \sim \tau_c(0)/\delta$:

$$\tau_c(t) = \tau_c(0)\left[\frac{T(t)}{T_0}\right]^{-1/2} \sim \tau_c(0)\left(\frac{t}{\tau_0}\right)^{5/6} \sim \tau_c(0)^{1/6}\delta^{5/6}t^{5/6}. \quad (14.34)$$

This way we obtain the condition for the upper limit of time (the second inequality in (14.33)).

Exercise 14.5 *Find the velocity correlation time τ_v up to the second order in δ and in a_2 by means of Maple!*

The laws of self-diffusion in gases of viscoelastic particles, (14.32), and gases of particles which interact with a constant coefficient of restitution, (14.21), reveal a qualitative difference: whereas the mean square displacement of viscoelastic particles grows with time due to a power law $\sim t^{1/6}$, for $\varepsilon = $ const. we obtain a much slower logarithmic growth $\sim \log t$.

Independently from the discussion in this section, the functional form of (14.32) and (14.21) may be obtained by scaling arguments: the average particle velocity scales as $\bar{v} \sim \sqrt{T}$ and, therefore, with the time dependence of temperature, as $\sim t^{-1}$ for $\varepsilon = $ const. and as $\sim t^{-5/6}$ for viscoelastic particles, respectively. The diffusivity in granular gas scales as $D \sim l^2/\tau_c$, where $l \sim \sigma^{-2} n^{-1}$ is the mean free path (which is time-independent in the homogeneous cooling state) and $\tau_c \sim l/\bar{v}$ is the mean collision time, thus, $D \sim l\bar{v} \sim \sqrt{T}$. Therefore, for $\varepsilon = $ const. we obtain for the mean square displacement $\int^t D(t) dt \sim \log t$, whereas for viscoelastic particles it scales as $\sim t^{1/6}$.

From the qualitatively different functional form of the mean square deviation as a function of time, (14.32, 14.21), we conclude that the particles of cooling gases of viscoelastic particles spread much wider than particles which interact via $\varepsilon = $ const. The more intensive mixing in cooling gases of viscoelastic particles gives rise to the expectation that clustering in such gases occurs retardedly (or may even be suppressed completely) as compared with the case $\varepsilon = $ const. We will come back to this interesting problem in Chapter 26.

14.4 Inherent time scales

Throughout this book time is defined as the laboratory time, measured in seconds. In part of the the literature, however, time is defined by the cumulative number of collisions per particle $\mathcal{N}$. The laboratory time t and the inherent time $\mathcal{N}$ are related via $d\mathcal{N} = \tau_c(t)^{-1} dt$ (van Noije et al., 1997). Integration of this relation yields for $\varepsilon = $ const. $\mathcal{N}(t) \sim \log t$ and for gases of viscoelastic particles $\mathcal{N}(t) \sim t^{1/6}$, respectively. Consequently, the mean square displacement grows with time for both cases due to

$$\left\langle (\Delta r (\mathcal{N}))^2 \right\rangle \sim \mathcal{N}. \tag{14.35}$$

Therefore, with respect to the inherent time scale $\mathcal{N}$ the dynamical behaviour of granular gases of viscoelastic particles and gases of particles which interact with $\varepsilon = $ const. are identical.

Let us give two consequences of defining the time by the cumulative number of collisions $\mathcal{N}$:

- Assume two granular particles which move on a ring, that is, a one dimensional box with periodic boundary conditions. At each collision they lose part of their relative velocity, therefore, the particles slow down gradually due to dissipative collisions. In the inherent time scale $\mathcal{N}$, however, the number of collisions per unit of time is time-independent, that is, the system appears to be energy conserving, although there occur dissipative collisions.

- Assume there occurs an inelastic collapse of three particles in a three-dimensional system. According to the definition of a collapse (see Section 4.1) the particles lose the energy of their relative motion in axial direction in *finite* time (here laboratory time is meant) by performing an *infinite* number of collisions. Since the collapse concerns only these three particles, the dynamics of the granular gas is not affected by the collapse. Insofar, for the description of the granular gas a collapse of three particles is an unimportant event. If time is measured by the number of collisions $\mathcal{N}$, however, *all* particles of the gas get frozen at their actual positions as soon as the collapse occurs.

From these examples we conclude that the definition of time by the number of collisions $\mathcal{N}$ this is not an adequate description of physical reality.

Another inherent time scale is based on the temperature of the granular gas: $\mathcal{T}^{-1} \equiv T(t)/T_0$. An important argument to use a temperature-based time has been given by Goldhirsch and Zanetti (1993). They argue that in the regime of homogeneous cooling inelastic collisions lead to a cooling rate $\sim T^{3/2}$ (for $\varepsilon = $ const.) At the same time viscous heating occurs at rate $\sim \sqrt{T}$. Starting with a homogeneous system, initially the first contribution is dominating, but when temperature decays, the second contribution takes over and the shear mode adiabatically enslaves the temperature field. (The different regimes of the evolution of a granular gas will be the subject of Chapters 24–27.) This corresponds to the nonlinear stage of the gas evolution when clustering starts. Thus, the temperature may indicate the stage of the evolution of a granular gas.

The application of $\mathcal{T}$ as an inherent time scale is also supported by numerical simulations: Brito and Ernst (1998a) have investigated the time $\mathcal{N}_c$ and $\mathcal{T}_c \equiv T(\mathcal{N}_c)/T_0$, respectively, when granular gases leave the homogeneous cooling regime and start to develop inhomogeneities (which finally grow to pronounced clusters). It was shown that while $\mathcal{N}_c$ differs by more than a factor of 3 for two different systems ($\mathcal{N}_c = 70$ for a system with $\varepsilon = 0.9$ and packing fraction $\phi = 0.245$ and $\mathcal{N}_c = 23$ for $\varepsilon = 0.6$ and $\phi = 0.05$), the values of $\mathcal{T}_c$ are very close ($\mathcal{T}_c \approx 0.0031$ for the first system and $\mathcal{T}_c \approx 0.0027$ for the second one).

Given a granular gas of viscoelastic particles and a gas of particles which interact with $\varepsilon = $ const. where both gases are of the same density and the same initial temperature T_0, the time $\mathcal{T}$ allows a comparison of their evolution. With $T(\mathcal{N})/T_0 \sim \exp(-2\gamma_0 \mathcal{N})$ for $\varepsilon = $ const. and $T(\mathcal{N})/T_0 \sim \mathcal{N}^{-10}$ for gases of viscoelastic particles, we obtain

$$\left\langle (\Delta r)^2 \right\rangle \sim \log \mathcal{T} \quad \text{for } \varepsilon = \text{const.} \quad (14.36)$$

$$\left\langle (\Delta r)^2 \right\rangle \sim \mathcal{T}^{1/10} \quad \text{for gases of viscoelastic particles.} \quad (14.37)$$

Thus, for the temperature-based inherent time scale in agreement with (14.21, 14.32) we again obtain a power-law dependence for gases of viscoelastic particles and a logarithmically slow time dependence for $\varepsilon = $ const.

14.5 Coefficient of self-diffusion beyond the adiabatic approximation

To go beyond the adiabatic approximation, (13.51, 14.1), we write

$$\langle \vec{v}_1(t') \cdot \vec{v}_1(t) \rangle = v_T(t') v_T(t) \langle \vec{c}_1(t') \cdot \vec{c}_1(t) \rangle . \tag{14.38}$$

Consider the case $\varepsilon = $ const. We measure time in units of the mean collision time:

$$d\tau = \frac{dt}{\tau_c(t)}, \qquad \tau(t) = \int_0^t \frac{dt'}{\tau_c(t')} \tag{14.39}$$

with $\tau_c(t) \propto 1/\sqrt{T(t)}$. In this new time, the dynamics of the system is stationary (as discussed on page 91), that is, the temperature and the collision frequency are independent of time. Hence, the time-correlation function of the reduced velocity, $\langle \vec{c}_1(\tau') \cdot \vec{c}_1(\tau) \rangle$, is also stationary, that is, it depends only on the difference $|\tau'-\tau|$, but not on τ' and τ themselves. Under the assumption of molecular chaos (which implies a Markov process) the velocity correlations decay exponentially (Resibois and de Leener, 1977), that is,

$$\langle \vec{c}_1(\tau') \cdot \vec{c}_1(\tau) \rangle = \langle c_1^2 \rangle \exp\left(-\frac{|\tau-\tau'|}{\hat{\tau}_v}\right) = \frac{3}{2} \exp\left(-\frac{|\tau-\tau'|}{\hat{\tau}_v}\right), \tag{14.40}$$

with $\hat{\tau}_v$ being the velocity relaxation time, measured in collision units $\tau_c(t)$, thus, it is constant. Hence, the derivative of the correlation function taken at $t' = t$ reads

$$\frac{d}{dt} \langle \vec{v}_1(t') \cdot \vec{v}_1(t) \rangle |_{t'=t} = \frac{d}{dt} \left(\frac{3}{2} v_T(t') v_T(t) \exp\left[-\frac{|\tau(t)-\tau(t')|}{\hat{\tau}_v}\right] \right)\Big|_{t'=t}$$

$$= \frac{3}{2} v_T(t) \frac{dv_T(t)}{dt} - \frac{3}{2} v_T^2(t) \hat{\tau}_v^{-1} \frac{d\tau}{dt} = \frac{3}{2} v_T^2 \left[\frac{1}{2T}\dot{T} - (\hat{\tau}_v \tau_c(t))^{-1}\right] \tag{14.41}$$

$$= -\langle v_1^2(t) \rangle \left[\frac{1}{2}\zeta(t) + \tau_v^{-1}(t)\right],$$

where we use the relations

$$\frac{dv_T}{dt} = \frac{1}{2T}\dot{T}\, v_T(t) = -\frac{1}{2}\zeta(t)v_T(t), \qquad \frac{3}{2}v_T^2(t) = \langle v_1^2(t) \rangle_t \tag{14.42}$$

and notice that the *constant* reduced relaxation time $\hat{\tau}_v$ corresponds to the *time-dependent* relaxation time $\tau_v(t) = \hat{\tau}_v \tau_c(t)$. On the other hand we, obtain for the same quantity

$$\frac{d}{dt}\langle \vec{v}_1(t')\cdot\vec{v}_1(t)\rangle|_{t'=t} = \frac{d}{dt}\int d\Gamma \rho(t')\, \vec{v}_1 \cdot e^{\mathcal{L}(t-t')}\vec{v}_1 \Big|_{t'=t} = \langle \vec{v}_1 \cdot \mathcal{L}\vec{v}_1\rangle_t \tag{14.43}$$

$$= (N-1)\left\langle \vec{v}_1 \cdot \hat{T}_{12}\vec{v}_1 \right\rangle_t .$$

The last relation in (14.43) follows from the properties of the Liouville operator (13.52, 14.2). Comparing (14.43) and (14.41), we obtain the velocity relaxation time $\tau_v(t)$:

$$\tau_v^{-1}(t) = -(N-1)\frac{\langle \vec{v}_1 \cdot \hat{T}_{12}\vec{v}_1 \rangle_t}{\langle \vec{v}_1 \cdot \vec{v}_1 \rangle_t} - \frac{1}{2}\zeta(t) = \tau_{v,\,\text{ad}}^{-1}(t) - \frac{1}{2}\zeta(t) \qquad (14.44)$$

with $\tau_{v,\,\text{ad}}(t)$ being the velocity relaxation time in adiabatic approximation (14.2). This quantity reads [see also (14.6)]:

$$\tau_{v,\,\text{ad}}^{-1}(t) = \frac{m\sigma^2 g_2(\sigma)}{12nT} \int d\vec{v}_1 d\vec{v}_2 \int d\vec{e}\,\Theta\left(-\vec{v}_{12}\cdot\vec{e}\right)|\vec{v}_{12}\cdot\vec{e}|\,f(\vec{v}_1,t)\,f(\vec{v}_2,t) \\ \times (1+\varepsilon)\,(\vec{v}_{12}\cdot\vec{e})^2\,. \qquad (14.45)$$

The relaxation time $\tau_{v,\,\text{ad}}$ has been already calculated for both cases, $\varepsilon = \text{const.}$ (14.22) and for viscoelastic particles (14.29).

To find the mean square displacement it is convenient to use the reduced velocity time correlation function (14.40) and correspondingly the reduced length

$$\vec{r} = r_0\hat{\vec{r}}, \qquad r_0 \equiv v_T(t)\,\tau_c(t)\,. \qquad (14.46)$$

The reduced mean square displacement reads

$$\left\langle (\Delta\hat{r}(\tau))^2 \right\rangle = 2\int_0^\tau d\tau_1 \int_{\tau_1}^\tau d\tau_2\,\frac{3}{2}\exp\left[-\frac{\tau_2-\tau_1}{\hat{\tau}_v}\right] \qquad (14.47)$$

and correspondingly its reduced time derivative for $\tau \gg \hat{\tau}_v$ is

$$\frac{d}{d\tau}\left\langle (\Delta\hat{r}^2) \right\rangle = 3\int_0^\tau d\tau_1\,\exp\left[-\frac{\tau-\tau_1}{\hat{\tau}_v}\right] = 3\hat{\tau}_v\left[1-\exp\left(-\tau/\hat{\tau}_v\right)\right] \simeq 3\hat{\tau}_v\,. \qquad (14.48)$$

Returning to unscaled variables, we obtain

$$\frac{d}{dt}\left\langle(\Delta r^2(t))\right\rangle = \frac{r_0^2}{\tau_c}\frac{d}{d\tau}\left\langle(\Delta\hat{r}^2(\tau))\right\rangle = \frac{r_0^2}{\tau_c}3\hat{\tau}_v = \frac{r_0^2}{\tau_c^2}3\tau_v \\ = \frac{v_T^2(t)\tau_c^2(t)}{\tau_c^2(t)}3\tau_v = 6\frac{T(t)}{m}\tau_v = 6D(t)\,. \qquad (14.49)$$

Hence $D = (T/m)\tau_v$ and therefore,

$$D(t) = \frac{T}{m}\left[\tau_{v,\,\text{ad}}^{-1}(t) - \frac{1}{2}\zeta(t)\right]^{-1}\,. \qquad (14.50)$$

This expression provides the corrections to the adiabatic approximation for D for the case of a constant coefficient of restitution, with $\tau_{v,\,\text{ad}}(t)$ and $\zeta(t)$ given respectively by (14.22) and (8.12). The same expression may be obtained by the rather different Chapman–Enskog method, which is addressed in Chapter 18.

For the case of viscoelastic particles one cannot obtain a stationary correlation function just by rescaling time as for $\varepsilon = $ const., since the coefficient of restitution, which defines the dynamics of the system, varies in time itself (see discussion on page 91). The calculation of the self-diffusion coefficient beyond the adiabatic approximation for a gas of viscoelastic particles is addressed in Chapter 22.

15

BROWNIAN MOTION IN GRANULAR GASES

The stochastic motion of large granular particles in a gas of much smaller granular particles may be described by Brownian dynamics. The velocities of the Brownian particles obey a Maxwell distribution with a temperature which differs from the temperature of the embedding granular gas.

15.1 Boltzmann equation for the velocity distribution function of Brownian particles

Consider the motion of large granular particles[15] of mass m_b and diameter σ_b embedded in a granular gas of smaller particles of mass $m \ll m_b$ and diameter $\sigma \ll \sigma_b$. We assume that the concentration of the large particles is very small, that is, we can neglect collisions between them. According to the significant difference of the masses a collision between large and small particles alter the velocity of large particles only slightly. Subjected to these uncorrelated impacts, the large particles exhibit Brownian motion. We will call them Brownian particles. In spite of the deviation of the velocity distribution of the gas particles from the Maxwell distribution, the velocities of the Brownian particles obey a Maxwell distribution. This follows from the fact that the velocity of a Brownian particle is determined by a large number of successive uncorrelated impacts. Then (with mild preconditions), from the central limit theorem follows that the sum of a large number of random variables obeys a Gaussian distribution.[16] Whereas the shape of the velocity distribution of the Brownian particles follows already from general statistical arguments, the result for their average kinetic energy is not obvious.

The Brownian motion of heavy particles in a gas of smaller granular particles was first investigated by (Brey et al., 1999b, 1999b). Here we will follow mainly their arguments.

Collisions between gas particles are characterised by the coefficient of restitution $\varepsilon = $ const., collisions between a gas particle and a Brownian particle are characterised by $\varepsilon_b = $ const. The collision rule for gas particles of identical mass

---

[15]To distinguish the properties of gas particles from the properties of Brownian particles, in this chapter all variables which refer to Brownian particles wear the subscript 'b'.

[16]As it is shown below (see equations (15.12) and (15.24)) the increment of the velocity of the Brownian particle due to a collision is almost independent of its velocity. Hence, the current velocity of a Brownian particle is a sum of independent random variables.

is given by (6.5), while the collision of a Brownian particle with a gas particle (of different mass) is described by the general collision rule (2.7). With

$$\Delta \equiv \frac{m}{m_b} \ll 1, \qquad \frac{m^{\text{eff}}}{m_b} \equiv \frac{\Delta}{1+\Delta}, \qquad \frac{m^{\text{eff}}}{m} \equiv \frac{1}{1+\Delta}, \qquad (15.1)$$

the velocities after a collision read

$$\vec{v}_b' = \vec{v}_b - \frac{(1+\varepsilon_b)\Delta}{1+\Delta}(\vec{w}\cdot\vec{e})\,\vec{e}, \qquad \vec{v}' = \vec{v} + \frac{1+\varepsilon_b}{1+\Delta}(\vec{w}\cdot\vec{e})\,\vec{e}, \qquad (15.2)$$

with

$$\vec{w} \equiv \vec{v}_b - \vec{v}, \qquad \vec{e} \equiv \frac{\vec{r}_b - \vec{r}}{|\vec{r}_b - \vec{r}|}, \qquad (15.3)$$

where $\vec{r}_b$, $\vec{r}$, $\vec{v}_b$ and $\vec{v}$ are, respectively, the positions and velocities of the Brownian particle and of the gas particle before the collision. Correspondingly, the rule of the inverse (restituting) collision is given by

$$\vec{v}_b'' = \vec{v}_b - \frac{(1+\varepsilon_b)\Delta}{\varepsilon_b(1+\Delta)}(\vec{w}\cdot\vec{e})\,\vec{e}, \qquad \vec{v}'' = \vec{v} + \frac{1+\varepsilon_b}{\varepsilon_b(1+\Delta)}(\vec{w}\cdot\vec{e})\,\vec{e}. \qquad (15.4)$$

These velocities are used in the collision integral of the Boltzmann equation for the distribution function of the ensemble of Brownian particles $f_b(\vec{r}_b, \vec{v}_b, t)$:

$$\left(\frac{\partial}{\partial t} + \vec{v}_b \cdot \vec{\nabla}_b\right) f_b(\vec{r}_b, \vec{v}_b, t) = g_2(\sigma_0)\, I(f_b, f)$$

$$= \sigma_0^2 g_2(\sigma_0) \int d\vec{v} \int d\vec{e}\, \Theta(-\vec{w}\cdot\vec{e})\, |\vec{w}\cdot\vec{e}|$$

$$\times \left[\frac{1}{\varepsilon_b^2} f_b(\vec{r}_b, \vec{v}_b'', t) f(\vec{v}'', t) - f_b(\vec{r}_b, \vec{v}_b, t) f(\vec{v}, t)\right], \qquad (15.5)$$

where $f(\vec{v}, t)$ is the velocity distribution function of the (small) granular gas particles and $g_2(\sigma_0)$ is the contact value of the pair distribution function for a Brownian particle and a gas particle (see below). The second term on the LHS accounts for the gradient of the concentration of the Brownian particles (since the distribution function f_b depends on the coordinate $\vec{r}_b$). The Boltzmann equation of the large particles in the gas of small particles (15.5) is also called *Boltzmann–Lorentz* equation. Despite possible concentration gradients of the Brownian particles, due to their small concentration we assume homogeneous cooling of the embedding granular gas of uniform density. Therefore, the results of Chapters 5–11 for the velocity distribution function of granular gases remain valid. In particular, for the case of constant coefficient of restitution the time-dependent velocity distribution function obeys a simple scaling form (see (7.1))

$$f(\vec{v}, t) = \frac{n}{v_T^3(t)} \tilde{f}(\vec{c}), \qquad \text{with} \qquad v_T(t) = \sqrt{\frac{2T(t)}{m}}, \qquad (15.6)$$

where n is the number density of the granular gas. The temperature of the gas evolves due to (8.64):

$$T(t) = \frac{T_0}{(1+t/\tau_0)^2} \tag{15.7}$$

with the characteristic time τ_0 given by (8.65). The contact value of the pair distribution function for a Brownian particle and a gas particle $g_2(\sigma_0)$ with $\sigma_0 \equiv (\sigma + \sigma_b)/2$ needs some explanation: In the Boltzmann equation (6.21) of identical particles of diameter σ the argument of the pair distribution function is σ since the collision occurs at the instant when the particle distance equals their diameter. For a mixture of $N^* = \sum_a N_a$ hard spheres of different sizes σ_a a contact of two spheres of diameters σ_a and σ_b occurs at the distance $\sigma_{ab} \equiv (\sigma_a + \sigma_b)/2$. The corresponding contact values of the pair distribution function $g_2^{ab}(\sigma_{ab})$ have been derived by Lebowitz (1964):

$$g_2^{ab}(\sigma_{ab}) = \frac{1}{1-\eta} + \frac{1}{4}\frac{\sigma_a \sigma_b}{\sigma_{ab}} \frac{\rho s}{(1-\eta)^2}. \tag{15.8}$$

with the definitions

$$\eta = \frac{\pi}{6} \sum_a \frac{N_a}{N^*} \sigma_a^3 \frac{1}{V} \qquad s = \pi \sum_a \frac{N_a}{N^*} \sigma_a^2. \tag{15.9}$$

Here $\rho \equiv N^*/V$ and V is the system volume.

For the case which is addressed here, $N^* = N + N_b$ with $N_b \ll N$,

$$\eta \simeq \eta = \frac{\pi}{6}\sigma^3 n, \qquad \rho \simeq n, \qquad s \simeq \pi\sigma^2, \qquad \sigma_0 \equiv \frac{\sigma_b + \sigma}{2} \simeq \frac{\sigma_b}{2} \tag{15.10}$$

and (15.8) turns into

$$g_2(\sigma_0) = \frac{1 + \left(\frac{3\sigma_b}{2\sigma_0} - 1\right)\eta}{(1-\eta)^2} \simeq \frac{1+2\eta}{(1-\eta)^2}. \tag{15.11}$$

15.2 Fokker–Planck equation for Brownian particles

When a heavy Brownian particle collides with a small gas particle, the velocity of the Brownian particle is changed by only a very small amount,

$$|\vec{v}_b'' - \vec{v}_b| = \left|\frac{(1+\varepsilon_b)\Delta}{\varepsilon_b(1+\Delta)}(\vec{w}\cdot\vec{e})\vec{e}\right| \propto \Delta, \tag{15.12}$$

that is, $\vec{v}_b'' \approx \vec{v}_b$. This suggests an expansion of $f_b(\vec{r}_b, \vec{v}_b'', t)$ in the collision integral in (15.5) around the velocity $\vec{v}_b$. This expansion will lead to the Fokker–Planck equation for the distribution function of the Brownian particles. The expansion of $f_b(\vec{r}_b, \vec{v}_b'', t)$ around $\vec{v}_b$ can be done by straightforward algebra.

Here we prefer to follow the elegant approach which has been used by Brey et al. (1999b): consider an arbitrary function $H(\vec{v}_b)$ and its integral

$$J(H) \equiv g_2(\sigma_0) \int d\vec{v}_b H(\vec{v}_b) I(f_b, f)$$
$$= \sigma_0^2 g_2(\sigma_0) \int d\vec{v}_b H(\vec{v}_b) \int d\vec{v} \int d\vec{e}\,\Theta(-\vec{w}\cdot\vec{e})|\vec{w}\cdot\vec{e}| \qquad (15.13)$$
$$\times \left[\frac{1}{\varepsilon_b^2} f_b(\vec{v}_b'') f(\vec{v}'') - f_b(\vec{v}_b) f(\vec{v})\right],$$

where for more compact notation we dropped the space and time dependences. Now we apply the property of the collision integral (6.25):[17]

$$J(H) = \sigma_0^2 g_2(\sigma_0) \int d\vec{v}_b \int d\vec{v} f_b(\vec{v}_b) f(\vec{v}) \int d\vec{e}\,\Theta(-\vec{w}\cdot\vec{e})|\vec{w}\cdot\vec{e}|[H(\vec{v}_b') - H(\vec{v}_b)]$$
$$= \sigma_0^2 g_2(\sigma_0) \int d\vec{v}_b \int d\vec{v} f_b(\vec{v}_b) f(\vec{v}) \int d\vec{e}\,\Theta(-\vec{w}\cdot\vec{e})|\vec{w}\cdot\vec{e}|[H(\vec{v}_b - \delta\vec{v}_b) - H(\vec{v}_b)], \qquad (15.14)$$

where with (15.2)

$$\vec{v}_b' = \vec{v}_b - \delta\vec{v}_b = \vec{v}_b - \overline{\Delta}(\vec{w}\cdot\vec{e})\vec{e}, \qquad \overline{\Delta} \equiv \frac{\Delta(1+\varepsilon_b)}{1+\Delta}. \qquad (15.15)$$

Expanding $H(\vec{v}_b - \delta\vec{v}_b)$ up to the second order of the small parameter Δ, we obtain

$$H(\vec{v}_b - \delta\vec{v}_b) - H(\vec{v}_b) = -\overline{\Delta}(\vec{w}\cdot\vec{e}) e_i \frac{\partial H}{\partial v_{bi}} + \frac{1}{2}\overline{\Delta}^2 (\vec{w}\cdot\vec{e})^2 e_i e_j \frac{\partial^2 H}{\partial v_{bi}\partial v_{bj}}, \qquad (15.16)$$

with e_i and v_{bj} ($\{i,j\} = \{x,y,z\}$) being the components of the vectors $\vec{e}$ and $\vec{v}_b$. Summation over repeated indexes is implied, for example,

$$e_i \frac{\partial}{\partial v_i} \equiv e_x \frac{\partial}{\partial v_{bx}} + e_y \frac{\partial}{\partial v_{by}} + e_z \frac{\partial}{\partial v_{bz}}. \qquad (15.17)$$

If we substitute (15.16) in (15.14) and perform integration by parts, taking into account that f_b and f vanish at the boundary of the integration domain, that is, for infinite velocities, we obtain

$$J(H) = \sigma_0^2 g_2(\sigma_0) \int d\vec{v}_b H(\vec{v}_b) \left\{\int d\vec{v} \int d\vec{e} \left[\overline{\Delta} e_i \frac{\partial}{\partial v_{bi}}(\vec{w}\cdot\vec{e})\right.\right.$$
$$\left.\left. + \frac{1}{2}\overline{\Delta}^2 e_i e_j \frac{\partial^2}{\partial v_{bi}\partial v_{bj}}(\vec{w}\cdot\vec{e})^2\right]|\vec{w}\cdot\vec{e}|\Theta(-\vec{w}\cdot\vec{e}) f_b(\vec{v}_b) f(\vec{v})\right\}. \qquad (15.18)$$

---

[17] Equation (6.25) has been derived previously for the collision of identical particles. With the same arguments it can also be derived for the case of particles of different size.

Comparing the latter equation with (15.13), we recognize that the expression in curled brackets equals the collision integral $I(f_b, f)$ since $H(\vec{v}_b)$ is an arbitrary function. Therefore,

$$I[f_b, f] = \frac{\partial}{\partial v_i}[A_i(\vec{v}_b) f_b(\vec{v}_b)] + \frac{1}{2}\frac{\partial^2}{\partial v_{bi}\partial v_{bj}}[N_{ij} f_b(\vec{v}_b)] \qquad (15.19)$$

with

$$A_i(\vec{v}_b) = \overline{\Delta}\sigma_0^2 g_2(\sigma_0) \int d\vec{v} \int d\vec{e}\, e_i\, (\vec{w}\cdot\vec{e}) |\vec{w}\cdot\vec{e}|\, \Theta(-\vec{w}\cdot\vec{e}) f(\vec{v}) \qquad (15.20)$$

and

$$N_{ij}(\vec{v}_b) = \overline{\Delta}^2 \sigma_0^2 g_2(\sigma_0) \int d\vec{v} \int d\vec{e}\, e_i e_j\, (\vec{w}\cdot\vec{e})^2 |\vec{w}\cdot\vec{e}|\, \Theta(-\vec{w}\cdot\vec{e}) f(\vec{v}). \qquad (15.21)$$

After integration over the vector $\vec{e}$, these quantities read

$$\vec{A} = \frac{\pi}{2}\overline{\Delta}\sigma_0^2 g_2(\sigma_0) \int d\vec{v} f(\vec{v})\, w\vec{w} \qquad (15.22)$$

$$N_{ij} = \frac{\pi}{12}\overline{\Delta}^2 \sigma_0^2 g_2(\sigma_0) \int d\vec{v} f(\vec{v})\, w\, (\delta_{ij} w^2 + 3 w_i w_j), \qquad (15.23)$$

where δ_{ij} is the Kronecker delta. Equation (15.19) is an example of the general Kramers–Moyal expansion (van Kampen, 1992) where terms up to the second order in $\delta \vec{v}_b$ have been retained.

Exercise 15.1 *Prove (15.22) and (15.23)!*

These expressions can be simplified further if we take into account that the thermal velocity of the heavy Brownian particles is much smaller than the thermal velocity of the gas particle

$$w = |\vec{v}_b - \vec{v}| \approx |\vec{v}|. \qquad (15.24)$$

With the assumption that the main contributions to $\vec{A}$ and N_{ij} originate from collisions with particles moving with velocities close to the thermal velocity, the integral in (15.22) can be approximated by

$$\int d\vec{v} f(\vec{v})\, w\vec{w} = \int d\vec{v} f(\vec{v}) |\vec{v}_b - \vec{v}|(\vec{v}_b - \vec{v})$$

$$\approx \int d\vec{v} f(\vec{v}) |\vec{v}| \vec{v}_b - \int d\vec{v} f(\vec{v}) |\vec{v}| \vec{v} \qquad (15.25)$$

$$= \vec{v}_b n v_T \int d\vec{c}\, \tilde{f}(\vec{c})\, c = \vec{v}_b n v_T \langle c \rangle,$$

with the definition (15.6) of the scaled velocity distribution $\tilde{f}$. For the integration we used that all components of the vectorial integral vanish since the integrand is an odd function:

$$\int d\vec{v} f(\vec{v}) |\vec{v}| v_i = 0, \qquad i = x, y, z. \tag{15.26}$$

Similarly, the integrals in (15.23) simplify as

$$\int d\vec{v} f(\vec{v}) \delta_{ij} w^3 \approx \int d\vec{v} f(\vec{v}) \delta_{ij} v^3 = \delta_{ij} v_T^3 n \int d\vec{c}\, \tilde{f}(\vec{c}) c^3 = \delta_{ij} v_T^3 n \langle c^3 \rangle \tag{15.27}$$

and

$$\int d\vec{v} f(\vec{v}) w w_i w_j \approx \int d\vec{v} f(\vec{v}) v v_i v_j = \frac{1}{3} \delta_{ij} \int d\vec{v} f(\vec{v}) (v_x^2 + v_y^2 + v_z^2) v$$
$$= \frac{1}{3} \delta_{ij} \int d\vec{v} f(\vec{v}) v^3 = \frac{1}{3} \delta_{ij} v_T^3 n \langle c^3 \rangle. \tag{15.28}$$

In this case the integrals with $i \neq j$ vanish because the integrand is an odd function. The integrals with $i = j$ are identical due to symmetry.

The approximation (15.24) is a strong simplification, in a more rigorous approach $|\vec{v}_b - \vec{v}|$ is expanded in terms of Δ which results in additional numerical factors in the integrals (15.25, 15.27, 15.28). The derivation of these factors can be found in Brey et al. (1999b). Combining (15.22, 15.23) and (15.25, 15.27, 15.28) (with these factors included) we give the solution for the coefficients $\vec{A}$ and N_{ij}:

$$\vec{A}(\vec{v}_b, t) \simeq \gamma(t) \vec{v}_b; \qquad N_{ij} \simeq 2\overline{\gamma}(t) \delta_{ij}, \tag{15.29}$$

with

$$\gamma(t) = \frac{2\pi}{3}(1+\varepsilon_b) n \sigma_0^2 g_2(\sigma_0) v_T \langle c \rangle \Delta$$
$$\overline{\gamma}(t) = \frac{\pi}{6}(1+\varepsilon_b)^2 n \sigma_0^2 g_2(\sigma_0) v_T \frac{T}{m} \langle c^3 \rangle \Delta. \tag{15.30}$$

With (15.5, 15.19, 15.29) we arrive at the Fokker–Planck equation:

$$\left(\frac{\partial}{\partial t} + \vec{v}_b \cdot \vec{\nabla}_b\right) f_b(\vec{v}_b, t) = \frac{\partial}{\partial \vec{v}_b} \cdot \left(\gamma(t) \vec{v}_b + \overline{\gamma}(t) \frac{\partial}{\partial \vec{v}_b}\right) f_b(\vec{v}_b, t) \tag{15.31}$$

For Brownian particles in molecular gases in equilibrium, the coefficients γ and $\overline{\gamma}$ are independent of time and are related via the fluctuation–dissipation relation (van Kampen, 1992)

$$\overline{\gamma} = \frac{T}{m_b} \gamma. \tag{15.32}$$

For granular gases, the coefficients γ and $\overline{\gamma}$ depend on time. They are related by

$$\overline{\gamma} = \frac{T}{m_b} \gamma \left((1+\varepsilon_b) \frac{1}{4} \frac{\langle c^3 \rangle}{\langle c \rangle}\right). \tag{15.33}$$

In the limit of elastic collisions $\varepsilon_b \to 1$ the common fluctuation–dissipation relation holds true since $\tilde{f} = \pi^{-3/2} \exp(-c^2)$, that is, $\langle c^3 \rangle / \langle c \rangle = 2$.

VELOCITY DISTRIBUTION FUNCTION FOR BROWNIAN PARTICLES

The derivation of the Fokker–Planck equation is based on the assumption that the velocity of the large Brownian particles is much smaller than the thermal velocity of the gas particles. If the granular gas cools too fast, this condition may be violated. Therefore, the condition $1 - \varepsilon \ll 1$ is necessary for the validity of the Fokker–Planck equation (Brey et al., 1999b).

15.3 Velocity distribution function for Brownian particles

The Fokker–Planck equation (15.31) for the velocity distribution function of Brownian particles is the direct counterpart of the Boltzmann equation for the velocity distribution function of the gas particles. There are different approaches for its analysis: it can be solved directly (Brey et al., 1999b), or in a more elegant way it can be mapped to the Fokker–Planck equation for elastic particles (Brey et al., 1999b) which allows for the application of the standard tools which have been developed for the analysis of the Fokker–Planck equation (e.g. Gardiner, 1983; Risken, 1996). Here we perform a simplified analysis to obtain the velocity distribution of Brownian particles in a granular gas.

We assume that the time dependence enters the distribution function only via the time dependence of the temperature, the shape of the distribution function is assumed to be time-independent. The velocity distribution function of the Brownian particles may then be represented by the scaling form (7.1):

$$f_b(\vec{v}_b, t) = \frac{n_b}{v_{Tb}^3(t)} \tilde{f}_b\left(\frac{\vec{v}_b}{v_{Tb}(t)}\right) = \frac{n_b}{v_{Tb}^3(t)} \tilde{f}_b(\vec{c}_b), \quad (15.34)$$

with $T_b(t) = \frac{1}{2} m_b v_{Tb}^2(t)$ as in (7.2). We also assume that the Brownian particles are distributed uniformly in the system, hence, the spatial derivative on the LHS of (15.31) drops out. Then with (15.34) the LHS of (15.31) reads (see also (8.2)):

$$\frac{\partial f_b}{\partial t} = -\frac{n_b}{v_{Tb}^4} \frac{dv_{Tb}}{dt}\left(3 + c_b \frac{d}{dc_b}\right)\tilde{f}_b = -\frac{n_b}{v_{Tb}^4} \frac{dv_{Tb}}{dt}\frac{\partial}{\partial \vec{c}_b} \cdot \vec{c}_b \tilde{f}_b. \quad (15.35)$$

Its RHS reads

$$\frac{1}{v_{Tb}}\frac{\partial}{\partial \vec{c}_b} \cdot \left(\gamma v_{Tb} \vec{c}_b + \overline{\gamma} \frac{1}{v_{Tb}}\frac{\partial}{\partial \vec{c}_b}\right)\frac{n_b}{v_{Tb}^3}\tilde{f}_b = \frac{n_b}{v_{Tb}^3}\gamma\frac{\partial}{\partial \vec{c}_b} \cdot \left(\vec{c}_b + a\frac{\partial}{\partial \vec{c}_b}\right)\tilde{f}_b, \quad (15.36)$$

where with (15.33)

$$a \equiv \frac{1}{v_{Tb}^2}\frac{\overline{\gamma}}{\gamma} = \frac{T}{8T_b}(1+\varepsilon_b)\frac{\langle c^3 \rangle}{\langle c \rangle}. \quad (15.37)$$

From (15.35, 15.36) follows the equation for the scaled function $\tilde{f}_b$,

$$-\frac{1}{v_{Tb}}\frac{dv_{Tb}}{dt}\frac{\partial}{\partial \vec{c}_b} \cdot \vec{c}_b \tilde{f}_b = \gamma \frac{\partial}{\partial \vec{c}_b} \cdot \left(\vec{c}_b + a\frac{\partial}{\partial \vec{c}_b}\right)\tilde{f}_b, \quad (15.38)$$

which may be recast into the form

$$-\frac{1}{\gamma(t)}\frac{1}{v_{Tb}}\frac{dv_{Tb}}{dt} = \frac{\frac{\partial}{\partial \vec{c}_b} \cdot \left(\vec{c}_b + a\frac{\partial}{\partial \vec{c}_b}\right)\tilde{f}_b}{\frac{\partial}{\partial \vec{c}_b} \cdot \vec{c}_b \tilde{f}_b(c_b)} \equiv b. \tag{15.39}$$

Let us assume $a \propto T/T_b$ does not depend on time (which will be checked a posteriori). The LHS of (15.39) depends only on time, while the RHS depends only on the rescaled velocity $\vec{c}_b$, therefore, $b = \text{const}$. Thus, we obtain from (15.39):

$$\frac{\partial}{\partial \vec{c}_b} \cdot \left[\vec{c}_b(1-b)\tilde{f}_b + a\frac{\partial}{\partial \vec{c}_b}\tilde{f}_b\right] = 0, \tag{15.40}$$

that is, the term in square brackets is a constant. From the condition $\lim_{c_b \to \infty} \tilde{f}_b(c) = 0$, we conclude that this constant is zero, therefore,

$$a\frac{d}{dc_b}\tilde{f}_b = -(1-b)c_b\tilde{f}_b. \tag{15.41}$$

The solution of this equation reads

$$\tilde{f}_b(c_b) = \mathcal{A}\exp\left(-\frac{1-b}{2a}c_b^2\right), \tag{15.42}$$

with $\mathcal{A}$ being the normalization constant. To find the constants a and b we write for the mean square velocity using (15.34, 15.42)

$$\langle v_b^2 \rangle = \frac{\int v_b^2 d\vec{v}_b \frac{n_b}{v_{Tb}^3}\mathcal{A}\exp\left(-\frac{1-b}{2a}\frac{v_b^2}{v_{Tb}^2}\right)}{\int d\vec{v}_b \frac{n_b}{v_{Tb}^3}\mathcal{A}\exp\left(-\frac{1-b}{2a}\frac{v_b^2}{v_{Tb}^2}\right)} = \frac{3v_{Tb}^2}{2}\frac{2a}{1-b}. \tag{15.43}$$

On the other hand, according to the definition of v_{Tb}, (8.15), $\langle v_b^2 \rangle = 3v_{Tb}^2/2$, that is,

$$a = \frac{1-b}{2}. \tag{15.44}$$

Hence, we conclude that the velocity distribution function of the Brownian particles is a Maxwell distribution (7.5):

$$\tilde{f}_b(c_b) = \pi^{-3/2}\exp\left(-c_b^2\right). \tag{15.45}$$

We wish to emphasize that the velocities of the Brownian particles obey a Maxwell distribution, even if the velocity distribution function of the embedding particles deviates from the Maxwell distribution.

VELOCITY DISTRIBUTION FUNCTION FOR BROWNIAN PARTICLES

From (15.42), it follows that $b < 1$, otherwise the distribution diverges at $c_b \to \infty$. With (15.39) and the relation (8.6) between v_{Tb} and T_b we obtain the constant

$$b = -\frac{1}{\gamma(t)} \frac{1}{2T_b} \frac{dT_b}{dt}. \qquad (15.46)$$

According to our assumption $a \propto T/T_b = \text{const.}$, therefore

$$\frac{1}{2T_b}\frac{dT_b}{dt} = \frac{1}{2T}\frac{dT}{dt} = -\frac{1}{2T}\left(\frac{2}{3}BT\mu_2\right) = -\gamma(t)b, \qquad (15.47)$$

where we use (8.11) for the temperature decay of the granular gas, which is not affected by the presence of the Brownian particles due to their small concentration. The value of μ_2 depends only on the coefficient of restitution, according to (8.45, 8.48), while B is a function of time:

$$B(t) = v_T(t) g_2(\sigma) \sigma_b^2 n. \qquad (15.48)$$

Thus,

$$b = \frac{B\mu_2}{3\gamma} = \frac{1}{2\pi}\frac{\mu_2}{1+\varepsilon_b}\frac{g_2(\sigma)}{g_2(\sigma_0)}\frac{\sigma^2}{\sigma_0^2}\frac{1}{\Delta \langle c \rangle}. \qquad (15.49)$$

Hence, the assumptions $b = \text{const.}$ and, thus, $a = \text{const.}$ are consistent. A more detailed analysis (Brey et al., 1999b) shows that the velocity distribution of the Brownian particles relaxes to a Maxwell distribution (and b relaxes to its stationary value given by (15.49)) within few collisions per gas particle.

For nearly elastic granular gas particles ($\varepsilon \to 1$), the velocity distribution can be approximated by a Maxwell distribution, which implies $\mu_2 = \sqrt{2\pi}\left(1 - \varepsilon^2\right)$ and $\langle c \rangle = 2/\sqrt{\pi}$. For small gas density the contact value of the correlation function is approximated by $g_2(\sigma) = g_2(\sigma_0) = 1$ and we obtain

$$b = \frac{1}{2\sqrt{2}}\frac{(1-\varepsilon^2)}{(1+\varepsilon_b)}\left(\frac{\sigma}{\sigma_0}\right)^2\frac{1}{\Delta}. \qquad (15.50)$$

Since the condition $b < 1$ is imposed and since $\Delta \to 0$, we conclude that this may be fulfilled only for nearly elastic gas with $1 - \varepsilon^2 \to 0$. Thus, the requirement of self-consistency implies two limits to be satisfied simultaneously, $\Delta \to 0$ and $\varepsilon \to 1$ (Brey et al., 1999b). For a mixture of elastic particles in equilibrium the temperature of the gas particles and the Brownian particles are identical. This is not the case for Brownian particles which are embedded in a dissipative granular gas. From (15.37) it follows that

$$T_b = \frac{T}{8a}(1+\varepsilon_b)\frac{\langle c^3 \rangle}{\langle c \rangle}, \qquad (15.51)$$

and with (15.44) one obtains the temperature of Brownian particles in a granular gas, whose particles collide almost elastically, as

$$T_b(t) = T(t)\frac{1+\varepsilon_b}{2(1-b)} \tag{15.52}$$

where we take into account $\langle c^3 \rangle / \langle c \rangle = 2$ for the Maxwell distribution.

Surprisingly, depending on the values of Δ, ε, ε_b and on the diameters of the particles σ and σ_b, the temperature of the Brownian particles may be larger or smaller than the temperature of the embedding granular gas.

15.4 Diffusion of Brownian particles

15.4.1 Langevin equation

Consider a granular gas in the homogeneous cooling state as a viscous fluid. A large Brownian particle moving in this fluid feels a force $-\gamma \vec{v}_b$, which counteracts its velocity $\vec{v}_b$, where γ is the friction coefficient. Moreover, the Brownian particle feels a stochastic force $\vec{\mathcal{F}}(t)$ due to uncorrelated random impacts of the gas particles. The Langevin equation for the motion of the Brownian particle reads

$$\frac{d\vec{v}_b}{dt} = -\gamma \vec{v}_b + \vec{\mathcal{F}}(t). \tag{15.53}$$

Since the granular gas is homogeneous and isotropic and the random impacts occur uncorrelatedly (δ-correlated), the stochastic force has the properties

$$\langle \vec{\mathcal{F}}(t) \rangle = 0, \qquad \langle \vec{\mathcal{F}}(t)\vec{\mathcal{F}}(t') \rangle = \Gamma \delta(t-t'). \tag{15.54}$$

The amplitude of the stochastic force fluctuations (its mean square) is characterised by Γ. Such processes are called *white-noise* processes. For equilibrium systems the friction coefficient γ and the coefficient Γ are related according to the fluctuation–dissipation theorem by the Einstein relation $\Gamma = 2T_b\gamma$, where T_b is measured in units of the Boltzmann constant k_B (van Kampen, 1992; Resibois and de Leener, 1977).

The important difference between the Langevin equation and common differential equations is that the variable $v(t)$ is a stochastic variable. This means that (15.53) describes an *ensemble* of trajectories where a particular realization of $\vec{v}_b(t)$ corresponds to a particular realization of the stochastic force $\vec{\mathcal{F}}(t)$. In contrast, the Fokker–Planck equation deals with the probability distribution function which is a deterministic (non-stochastic) variable. In equilibrium statistical mechanics it is shown that both approaches, based on the Langevin equation and on the Fokker–Planck equation, are equivalent (see, e.g. van Kampen, 1992) for the precise meaning of this statement). Moreover, for each Langevin equation one can derive the according Fokker–Planck equation and vice versa. Namely, the Langevin equation for a certain variable $y(t)$

$$\dot{y} = -\gamma y + L(t) \tag{15.55}$$

with Gaussian white noise[18] specified by

$$\langle L(t) \rangle = 0, \qquad \langle L(t)L(t') \rangle = \Gamma \delta(t - t') \qquad (15.56)$$

represents the same process as the following Fokker–Planck equation for the distribution function of this variable $F(y,t)$:

$$\frac{\partial F}{\partial t} = \gamma \frac{\partial}{\partial y} y F + \frac{\Gamma}{2} \frac{\partial^2}{\partial y^2} F. \qquad (15.57)$$

The process (15.55) is also called the Ornstein–Uhlenbeck process and (15.57) is the Rayleigh equation. Equations (15.55–15.57) refer to the one-dimensional case, the generalization to the multi-dimensional case is obvious.

The situation is more subtle for the case of non-equilibrium processes, when the coefficients of the Fokker–Planck equation depend on time. However, we can try to eliminate the time dependence by an appropriate choice of the variables. Again we consider the transformation

$$f_b(\vec{v}_b, t) = \frac{n_b}{v_{Tb}^3(t)} \tilde{f}_b(\vec{c}_b, t) \qquad (15.58)$$

with the scaled velocity, $\vec{c}_b = \vec{v}_b / v_{Tb}(t)$. Then

$$\frac{\partial f_b}{\partial t} = \frac{n_b}{v_{Tb}^3} \frac{\partial \tilde{f}_b}{\partial t} - \frac{n_b}{v_{Tb}^4} \frac{dv_{Tb}}{dt} \frac{\partial}{\partial \vec{c}_b} \cdot \vec{c}_b \tilde{f}_b = \frac{n_b}{v_{Tb}^3} \left(\frac{\partial \tilde{f}_b}{\partial t} + \gamma(t) b \frac{\partial}{\partial \vec{c}_b} \cdot \vec{c}_b \tilde{f}_b \right), \qquad (15.59)$$

where we use (15.39) which relates dv_{Tb}/dt and b. If we now use the RHS of the Fokker–Planck equation (15.36) written for the rescaled variables with $a = (1-b)/2$ (see (15.44)) and collect the velocity derivatives, we obtain

$$\frac{\partial \tilde{f}_b}{\partial t} = \gamma(t)(1-b) \frac{\partial}{\partial \vec{c}_b} \cdot \left(\vec{c}_b + \frac{1}{2} \frac{\partial}{\partial \vec{c}_b} \right) \tilde{f}_b. \qquad (15.60)$$

The only time-dependent coefficient in this equation is $\gamma(t)$. With the new time variable

$$dt = t_0 d\tau, \qquad \text{with} \qquad t_0(t) \equiv [\gamma(t)(1-b)]^{-1} \qquad (15.61)$$

the coefficients of the Fokker–Planck equation become time-independent. To simplify the notations we consider the Langevin and Fokker–Planck equations for one component of the velocity only, say v_{bx} (and c_{bx}), and we omit the subscript. Then, the Fokker–Planck equation with constant coefficients reads

$$\frac{\partial \tilde{f}_b}{\partial \tau} = \frac{\partial}{\partial c_b} \left(c_b + \frac{1}{2} \frac{\partial}{\partial c_b} \right) \tilde{f}_b, \qquad (15.62)$$

---

[18] $L(t)$ is called a Gaussian variable if all odd moments of $L(t)$ vanish, while all even moments may be expressed in terms of its second moment (15.56) (van Kampen, 1992).

which corresponds to the Langevin equation

$$\frac{dc_b}{d\tau} = -c_b + L(\tau),\qquad(15.63)$$

with

$$\langle L(\tau)\rangle = 0, \qquad \langle L(\tau)L(\tau')\rangle = \delta(\tau - \tau').\qquad(15.64)$$

Multiplying both sides of (15.63) by $v_{Tb}(t)$, with (15.61) and using $dv_{Tb}/dt = -b\gamma(t)v_{Tb}$ according to (15.39), we obtain the Langevin equation in the variables v_b and t:

$$\frac{dv_b(t)}{dt} = -\gamma(t)v_b(t) + \mathcal{F}(t),\qquad(15.65)$$

with

$$\langle \mathcal{F}(t)\rangle = 0,$$
$$\langle \mathcal{F}(t)\mathcal{F}(t')\rangle = 2\overline{\gamma}\delta(t - t').\qquad(15.66)$$

To obtain (15.66) we use the relation $v_{Tb}^2 = \overline{\gamma}/(a\gamma)$, which follows from (15.37), $dv_{Tb}/dt = -b\gamma(t)v_{Tb}$ from (15.39) and $a = (1-b)/2$. The above equations (15.65, 15.66) coincide with the phenomenological equations (15.53, 15.54); now, however, we have microscopic expressions for the coefficients γ and $\Gamma = 2\overline{\gamma}$, given by (15.30).

Exercise 15.2 *Prove the second line of* (15.66)!

15.4.2 *Diffusion coefficient for Brownian particles*

The diffusion coefficient can be obtained by integrating the velocity correlation function (see (12.11, 14.14)); it is more convenient however to use the reduced velocity correlation function. We multiply the Langevin equation (15.63) by $c_b(\tau')$ ($\tau' \le \tau$) and average over the ensemble:

$$\frac{d}{d\tau}\langle c_b(\tau)c_b(\tau')\rangle = -\langle c_b(\tau)c_b(\tau')\rangle,\qquad(15.67)$$

where we take into account

$$\langle c_b(\tau')L(\tau)\rangle = 0\qquad(15.68)$$

since the force $L(\tau)$ is δ-correlated.[19] Equation (15.67) has the solution

$$\langle c_b(\tau)c_b(\tau')\rangle = \langle c_b^2(\tau')\rangle \exp[-(\tau - \tau')].\qquad(15.69)$$

For the three-dimensional system using the isotropic properties of the system, we can write

---

[19] A rigorous proof of (15.68) for equilibrium systems can be found, for example, in Berne and Harp (1970).

$$\langle c_{bx}(\tau) c_{bx}(\tau')\rangle = \langle c_{by}(\tau) c_{by}(\tau')\rangle = \langle c_{bz}(\tau) c_{bz}(\tau')\rangle \quad (15.70)$$

and obtain the velocity correlation function

$$\langle \vec{c}_b(\tau) \cdot \vec{c}_b(\tau')\rangle = 3\langle c_{bx}^2(\tau')\rangle \exp\left[-(\tau-\tau')\right] = \frac{3}{2}\exp\left[-(\tau-\tau')\right]. \quad (15.71)$$

This coincides with the reduced velocity correlation function addressed in Section 14.5 (see (14.40)) with the reduced relaxation time $\hat{\tau}_v = 1$.

Similar to the calculation on page 147 we consider the mean square displacement in the reduced units,

$$\vec{r} = r_0 \hat{\vec{r}}, \qquad r_0 \equiv v_{Tb}(t) t_0(t). \quad (15.72)$$

where $t_0(t)$ has been defined in (15.61). Performing exactly the same steps which previously led us to (14.49), substituting $\tau_c(t)$ in this equation by $t_0(t)$ and taking into account that $\hat{\tau}_v = 1$, we obtain for a Brownian particle

$$\frac{d}{dt}\langle (\Delta r^2(t))\rangle = 3\frac{r_0^2}{t_0} = 3\frac{v_{Tb}^2 t_0^2}{t_0} = 6\frac{T_b(t)}{m_b(1-b)\gamma(t)} = 6D(t). \quad (15.73)$$

From (15.73) we finally obtain the diffusion coefficient

$$D(t) = \frac{T_b(t)}{m_b(1-b)\gamma(t)}. \quad (15.74)$$

Using the relation between the temperature of the Brownian particles and the temperature of the granular gas (15.52) the diffusion coefficient reads in the elastic limit $\varepsilon_b \to 1$

$$D(t) = \frac{T(t)}{\gamma_e(t)(1-b)^2} = \frac{D_e(t)}{(1-b)^2}, \quad (15.75)$$

where $\gamma_e(t)$ is the friction coefficient[20] of a gas of elastic particles ($\varepsilon, \varepsilon_b \to 1$) at the same temperature:

$$\gamma_e = \frac{8}{3}\sqrt{\pi}\sigma_0^2 g_2(\sigma_0) n v_T \Delta \quad (15.76)$$

and $D_e = T_b/\gamma_e$ is the corresponding diffusion coefficient for the elastic case.

We obtain the surprising result that the coefficient of diffusion in a dissipative granular gas is enhanced by the factor $(1-b)^{-2} > 1$ as compared with a gas of elastic hard spheres at the same temperature.

---

[20] Obviously, the friction force, which decelerates a moving Brownian particle also exists in fluids of elastic particles.

16
TWO-DIMENSIONAL GRANULAR GASES

It has been shown for molecular fluids that the velocity-time correlation function in d dimensions exhibits a power-law dependence for $t \gg 1$:

$$\langle \vec{v}(0) \cdot \vec{v}(t) \rangle \propto t^{-d/2}. \tag{16.1}$$

This function reveals a long time tail as compared with an exponentially decaying function. According to the fluctuation–dissipation theorem (Resibois and de Leener, 1977), the kinetic coefficients are expressed by time integrals of the corresponding correlation functions. For $d > 2$ these integrals converge, while for two-dimensional systems the convergence of these integrals is problematic and, hence, the existence of the kinetic coefficients is questionable (see, e.g. Resibois and de Leener, 1977 for discussion). For force-free granular fluids, however, one can expect convergence of these integrals, since there exists an additional decay of the correlation functions due to the decay of temperature. This problem, however, is not yet resolved for granular gases. With this reservation we give below the results for d-dimensional gases, including the case $d = 2$.

The mean-square displacement of tagged particles depends on time in general dimension as

$$\left\langle (\Delta r(t))^2 \right\rangle = 2d \int^t D(t')\,dt', \tag{16.2}$$

with the self-diffusion coefficient $D(t)$ which reads, for a two-dimensional gas of particles with $\varepsilon = \text{const.}$, as

$$D(t) = \frac{4D_0(t)}{(1+\varepsilon)^2 \left(1 + \frac{3}{16}a_2\right)^2}. \tag{16.3}$$

The Enskog self-diffusion coefficient $D_0(t)$ reads

$$D_0(t) = \frac{d\,\Gamma(d/2)}{4\pi^{(d-1)/2} n g_2(\sigma)\sigma^{d-1}} \sqrt{\frac{T}{m}} \tag{16.4}$$

A similar result was obtained by Brey et al. (2000) (see footnote on page 215).

The corresponding expression for a two-dimensional gas of viscoelastic particles is

$$\frac{D(t)}{D_0} = \left[1 + \frac{89}{160}\frac{\omega_0}{\sqrt{2\pi}}\delta' + \left(\frac{646337}{5120000}\frac{\omega_0^2}{\pi} - \frac{2251}{3520}\frac{\omega_1}{\sqrt{2\pi}}\right)\delta'^2 + \cdots\right]. \tag{16.5}$$

The d-dimensional motion of a granular Brownian particle has essentially the same properties as for the three-dimensional case. Namely, the temperature of the ensemble of Brownian particles $T_b(t)$ may differ from the temperature of the embedding gas $T(t)$:

$$T_b(t) = T(t)\frac{1+\varepsilon_b}{2(1-b)}, \tag{16.6}$$

with

$$b = \frac{(1-\varepsilon^2)}{2\sqrt{2}(1+\varepsilon_b)\Delta}\frac{g_2(\sigma)}{g_2(\sigma_b)}\left(\frac{\sigma}{\sigma_b}\right)^{d-1}\left(1+\frac{3}{16}a_2(\varepsilon)\right), \tag{16.7}$$

where ε is the coefficient of restitution describing the collision of two gas particles and ε_b is the coefficient for the contact of a gas particle with a Brownian particle. The velocity distribution function of Brownian particles is Maxwellian as in the three-dimensional case.

The diffusion coefficient for Brownian particles reads

$$D(t) = \frac{T(t)}{\gamma_e(t)(1-b)^2}, \tag{16.8}$$

where $\gamma_e(t)$ is the friction coefficient in the elastic limit ($\varepsilon, \varepsilon_b \to 1$) at the same temperature:

$$\gamma_e(t) = \frac{4\pi^{(d-1)/2}\sigma_0^{d-1}\Delta}{d\Gamma(d/2)}nv_T(t). \tag{16.9}$$

SUMMARY

We analyse single-particle transport and self-diffusion in granular gases in the regime of homogeneous cooling. We consider the process of self-diffusion, when tagged particles are mechanically identical with the particles of the embedding gas, as well as Brownian motion, when large Brownian particles diffuse in the gas. Both cases, $\varepsilon = $ const. and viscoelastic particles have been addressed. The take-home message is:

1. Contrary to the case of molecular gases where the coefficient of self-diffusion is time-independent, granular gases are characterized by a time-dependent coefficient.
2. The self-diffusion coefficient in granular gases is larger than the corresponding value for molecular gases of the same temperature and density. This holds true because the trajectories of granular particles are more stretched owing to smaller deflection angle at the dissipative collisions.
3. The mean-square displacement of particles in a granular gas with a simplified collision model of $\varepsilon = $ const. increases with time logarithmically slow, $\langle \Delta r^2(t) \rangle \propto \log(t)$. This is qualitatively different from the power-law time dependence of this quantity in gases of viscoelastic particles, where $\langle \Delta r^2(t) \rangle \propto t^{1/6}$.
4. The velocity distribution function of an ensemble of Brownian particles which are embedded in a granular gas is always Maxwellian, although the velocity distribution of the embedding gas may deviate noticeably from a Maxwell distribution.
5. The temperature of an ensemble of Brownian particles may significantly differ from the temperature of the embedding gas. Depending on the parameters of the system it may be larger or smaller than the temperature of the gas.
6. The diffusion coefficient of the Brownian particle in a granular gas is enhanced as compared to the corresponding system of elastic particles.

IV. Transport Processes and Kinetic Coefficients

Inhomogeneous granular gases are characterized by macroscopic fluxes of mass, momentum and energy. Transport processes can be described by continuum mechanics provided the characteristic size of the inhomogeneities is large as compared with the mean free path of the granular particles and if the characteristic time of the evolution of these structures is large as compared with the mean collision time. Assuming that macroscopic and microscopic length and time scales are well separated, we derive hydrodynamic equations for granular gases. We apply the Chapman–Enskog approach to obtain the velocity distribution function for inhomogeneous granular gases and to calculate the kinetic coefficients, such as viscosity, thermal conductivity, etc. The kinetic coefficients are found for both cases, the simplified model with constant coefficient of restitution as well as for gases of viscoelastic particles.

17

GRANULAR GAS AS A CONTINUUM: HYDRODYNAMIC EQUATIONS

We consider a slightly inhomogeneous granular gas with small spatial gradients of temperature and density. Equations for the zeroth, the first and the second moments of the velocity distribution function are derived which characterize the density, the flow velocity and the temperature of the gas. Introducing phenomenological kinetic coefficients, such as shear viscosity, thermal conductivity, etc. we obtain a closed set of hydrodynamic equations.

17.1 Macro- and microscales of inhomogeneous granular gas

So far the homogeneous cooling state of granular gases has been addressed which implies uniform temperature T, density n and lack of macroscopic flows. Now we consider the more general case of small gradients of temperature and density as well as macroscopic flows. By 'small' we mean that the characteristic lengths L of these gradients, that is, the lengths at which the change of some quantity is of the order of the quantity itself, are much larger than the mean free path l of the gas particles. For example, the condition of small temperature gradient reads

$$\vec{\nabla} T \sim \frac{T}{L}, \qquad L \gg l. \qquad (17.1)$$

An equivalent condition is assumed for the density gradient. Another condition concerns the velocity of the macroscopic flow $\vec{u}$ (the precise definition of $\vec{u}$ is given below), which is assumed to be much smaller than the thermal velocity of the granular gas:

$$u \ll \sqrt{\langle v^2 \rangle} \sim v_T \sim \sqrt{T}. \qquad (17.2)$$

Situations for which the condition (17.2) is not fulfilled correspond to supersonic dynamics of the gas, here we restrict ourselves to the regime of subsonic motion. Note that the conditions of small gradients (17.1) and small average velocity (17.2) imply the existence of two length and time scales: the macroscopic scale with the characteristic length L and the characteristic time $\mathcal{T}$ and the microscopic scale with the characteristic length l and the characteristic time τ_c (mean collision time), respectively. The most important precondition of the results of this chapter is that these time and length scales are well separated, that is,

$$l \ll L, \qquad \tau_c \ll \mathcal{T} = L/u. \qquad (17.3)$$

Although the conditions (17.3) are, in general, fulfilled for molecular gases, they are not necessarily fulfilled for granular gases (e.g. Goldhirsch, 2001 and references therein). Moreover, in a cooling granular gas, due to spontaneous formation of density inhomogeneities in the course of time, there occur transitions from subsonic behaviour to supersonic and possibly back to subsonic (Esipov and Pöschel, 1997). This interesting scenario will be discussed in Chapter 26. If the conditions (17.3) are not fulfilled the dynamics of granular gases is mesoscopic, i.e., micro- and macroscales are not well separated and the hydrodynamic description is problematic (Tan and Goldhirsch, 1998; Dufty and Brey, 1999; Kadanoff, 1999). Here we assume that all necessary requirements for the application of hydrodynamics are given. The hydrodynamic description of granular gases in the presence of gradients of density and temperature comprises mainly the following topics:

- the hydrodynamic equations which describe the evolution of the macroscopic variables $T(\vec{r},t)$, $n(\vec{r},t)$ and $\vec{u}(\vec{r},t)$,
- the kinetic coefficients of the hydrodynamic equations as functions of the microscopic properties of the granular particles,
- the velocity distribution for the case when there exist gradients of the macroscopic variables.

17.2 Hydrodynamic fields

For non-homogeneous granular gases, the velocity distribution function is a function of the position, that is, $f(\vec{r},\vec{v},t)$, for a precise definition see (6.1). The macroscopic fields of local number density $n(\vec{r},t)$, average particle velocity $\vec{u}(\vec{r},t)$, and temperature (kinetic energy of the random motion) $T(\vec{r},t)$ are then defined by means of the moments of the velocity distribution function, similar as for the homogeneous case (6.4):

$$n(\vec{r},t) \equiv \int d\vec{v} f(\vec{r},\vec{v},t), \qquad (17.4)$$

$$n(\vec{r},t)\vec{u}(\vec{r},t) \equiv \int d\vec{v}\, \vec{v}\, f(\vec{r},\vec{v},t), \qquad (17.5)$$

$$\frac{3}{2}n(\vec{r},t)T(\vec{r},t) \equiv \int d\vec{v}\, \frac{mV^2}{2} f(\vec{r},\vec{v},t). \qquad (17.6)$$

For the definition of the temperature field (17.6) we need the local velocity,

$$\vec{V}(\vec{r},t) \equiv \vec{v} - \vec{u}(\vec{r},t), \qquad (17.7)$$

that is, the velocity of a particle with respect to the average velocity $\vec{u}(\vec{r},t)$ of the close neighbours at the same time. The local velocity represents the velocity of the random motion of the particles where the systematic flow is subtracted. The Boltzmann equation for the velocity distribution function (Sec. 6.3) for the case of non-uniform density is given by

$$\left(\frac{\partial}{\partial t} + \vec{v}_1 \cdot \vec{\nabla}\right) f(\vec{r}, \vec{v}_1, t) = g_2(\sigma) I(f, f) , \quad (17.8)$$

with the collision integral

$$\begin{aligned} I(f,f) = \sigma^2 \int d\vec{v}_2 \int d\vec{e}\, \Theta(-\vec{g}\cdot\vec{e})\,|\vec{g}\cdot\vec{e}| \\ \times \left[\chi f(\vec{r}, \vec{v}_1'', t) f(\vec{r}, \vec{v}_2'', t) - f(\vec{r}, \vec{v}_1, t) f(\vec{r}, \vec{v}_2, t)\right] . \end{aligned} \quad (17.9)$$

For an inhomogeneous gas, the velocity distribution function evolves not only due to inelastic collisions, but also due to the flux of particles against the density gradient. This contribution is described by the second term on the LHS of (17.8).

17.3 Hydrodynamic equations for granular gases

From the definition of the hydrodynamic fields $n(\vec{r}, t)$, $\vec{u}(\vec{r}, t)$, $T(\vec{r}, t)$, obviously, the hydrodynamic equations are obtained by multiplying the Boltzmann equation, respectively, with 1, $\vec{v}_1$ and $\frac{1}{2} m v_1^2$ and integrating over $\vec{v}_1$. With $\psi(\vec{v})$ standing, respectively, for 1, $m\vec{v}$ and $\frac{1}{2} m v^2$ and using the property of the collision integral (6.25), the RHS of (17.8) after this transformation is

$$\begin{aligned} \int d\vec{v}_1 \psi(\vec{v}_1) I(f, f) = \frac{1}{2}\sigma^2 \int d\vec{v}_1 d\vec{v}_2 \int d\vec{e}\, \Theta(-\vec{g}\cdot\vec{e})\,|\vec{g}\cdot\vec{e}| \\ \times f(\vec{r}, \vec{v}_1, t) f(\vec{r}, \vec{v}_2, t) \Delta\left[\psi(\vec{v}_1) + \psi(\vec{v}_2)\right] , \end{aligned} \quad (17.10)$$

where

$$\Delta\left[\psi(\vec{v}_1) + \psi(\vec{v}_2)\right] \equiv \left[\psi(\vec{v}_1') + \psi(\vec{v}_2') - \psi(\vec{v}_1) - \psi(\vec{v}_2)\right] . \quad (17.11)$$

As previously, the prime denotes after-collision values. For the three cases of interest $\psi(\vec{v}) = \{1, m\vec{v}, mv^2/2\}$:

$$\begin{aligned} &\Delta(1+1) = 1 + 1 - 1 - 1 = 0 \\ &\Delta(m\vec{v}_1 + m\vec{v}_2) = m\vec{v}_1' + m\vec{v}_2' - m\vec{v}_1 - m\vec{v}_2 = 0 \\ &\Delta(mv_1^2 + mv_2^2) = mv_1'^2 + mv_2'^2 - mv_1^2 - mv_2^2 = -\frac{1}{2} m (1 - \varepsilon^2) (\vec{v}_{12} \cdot \vec{e})^2 , \end{aligned} \quad (17.12)$$

where we use the conservation of momentum and (8.28) for the change of the kinetic energy due to an inelastic collision. Thus, (17.10) vanishes for the first and the second cases and gives a non-vanishing result only for $\psi = mv^2/2$.

After multiplication by $\psi = 1$ and integration over $\vec{v}_1$ the LHS of the Boltzmann equation (17.8) turns into

$$\frac{\partial}{\partial t}\int d\vec{v}_1 f + \int d\vec{v}_1 \left(\vec{v}_1 \cdot \vec{\nabla}\right) f = \frac{\partial n}{\partial t} + \vec{\nabla} \cdot \int d\vec{v}_1 \vec{v}_1 f = \frac{\partial n}{\partial t} + \vec{\nabla} \cdot (n\vec{u}) , \quad (17.13)$$

where the definitions of n and $\vec{u}$ (17.4, 17.5) have been used. Combining the LHS and the RHS, we arrive at the first macroscopic equation:

$$\boxed{\frac{\partial n}{\partial t} + \vec{\nabla} \cdot (n\vec{u}) = 0} \qquad (17.14)$$

The same transformation may be performed for $\psi = m\vec{v}_1$. The LHS for this case (we drop the subscript and write $\vec{v}$ instead of $\vec{v}_1$) is given by

$$\frac{\partial}{\partial t}\int d\vec{v}\, m\vec{v} f + \int d\vec{v}\, m\vec{v}\left(\vec{v}\cdot\vec{\nabla}\right) f = \frac{\partial}{\partial t} mn\vec{u} + \vec{\nabla}\cdot\int m\vec{v}\vec{v} f d\vec{v}, \qquad (17.15)$$

where we use the definition of $\vec{u}$ (17.5) and take into account, that the differential operator $\vec{\nabla} \equiv \partial/\partial\vec{r}$ may be extracted from the integral.

At this point, we introduce new notations, that are widely used in the gas kinetic theory. Consider a 3×3 tensor $\hat{D}$ whose components are obtained from the components of two three-dimensional vectors $\vec{a}$ and $\vec{b}$ by

$$\mathcal{D}_{ij} \equiv \left(\vec{a}\vec{b}\right)_{ij} = a_i b_j\,. \qquad (17.16)$$

Whenever two vectors are written together without any sign in between, a 3×3 tensor or *dyad* is meant. Its trace is

$$\mathrm{Tr}(\vec{a}\vec{b}) = a_x b_x + a_y b_y + a_z b_z = \vec{a}\cdot\vec{b}\,. \qquad (17.17)$$

The product of a vector $\vec{c}$ and a dyad $\vec{a}\vec{b}$ is

$$\vec{a}\vec{b}\cdot\vec{c} = \vec{a}\left(\vec{b}\cdot\vec{c}\right), \qquad \vec{c}\cdot\vec{a}\vec{b} = (\vec{c}\cdot\vec{a})\vec{b}\,. \qquad (17.18)$$

Exercise 17.1 *Prove (17.18) for the multiplication of a vector and a dyad!*

The same notations are applied when one of the vectors in the dyad is the differential operator $\vec{\nabla}$, that is,

$$\left(\vec{\nabla}\vec{a}\right)_{ij} = \frac{\partial}{\partial x_i} a_j, \qquad \left(\vec{a}\vec{\nabla}\right)_{ij} = a_i \frac{\partial}{\partial x_j}, \qquad (17.19)$$

where $x_i = \{x,y,z\}$ for $i = \{1,2,3\}$. With these notations, the first term on the RHS of (17.15) reads

$$\frac{\partial}{\partial t} mn\vec{u} = \vec{u}\frac{\partial mn}{\partial t} + mn\frac{\partial \vec{u}}{\partial t} = mn\frac{\partial \vec{u}}{\partial t} - \vec{u}\vec{\nabla}\cdot(nm\vec{u})\,, \qquad (17.20)$$

where the balance equation (17.14) has been employed. The second term at the RHS of (17.15) (without the operator $\vec{\nabla}$) can be written using the local velocity (17.7):

$$\int m\vec{v}\vec{v} f d\vec{v} = \int m\left(\vec{V}+\vec{u}\right)\left(\vec{V}+\vec{u}\right) f(\vec{v}) d\vec{v}$$

$$= \int m\vec{V}\vec{V} f(\vec{v}) d\vec{v} + m\vec{u}\vec{u}\int f(\vec{v}) d\vec{v} + 2m\vec{u}\int \vec{V} f(\vec{v}) d\vec{v}\,. \qquad (17.21)$$

The first term in this equation is the definition of the kinetic component of the pressure tensor (see below):

$$\hat{P}(\vec{r}, t) \equiv \int m\vec{V}\vec{V} f(\vec{r}, \vec{v}, t) d\vec{v}. \qquad (17.22)$$

The second term equals $m\vec{u}\vec{u}n$, according to the definition of n and the last term vanishes due to the definition of the local velocity:

$$\int \vec{V} f(\vec{v}) d\vec{v} = \int \vec{v} f(\vec{v}) d\vec{v} - \vec{u} \int f(\vec{v}) d\vec{v} = n\vec{u} - n\vec{u} = 0. \qquad (17.23)$$

Thus, the second term of the RHS of (17.15) may be written in the form

$$\begin{aligned}\vec{\nabla} \cdot \int m\vec{v}\vec{v} f d\vec{v} &= \vec{\nabla} \cdot \hat{P} + \vec{\nabla} \cdot \vec{u}\vec{u} mn \\ &= \vec{\nabla} \cdot \hat{P} + \vec{u}\vec{\nabla} \cdot (mn\vec{u}) + \left(mn\vec{u} \cdot \vec{\nabla}\right)\vec{u}.\end{aligned} \qquad (17.24)$$

Summing up (17.20) and (17.24), we obtain the LHS of the Boltzmann equation after the transformation, whereas the RHS vanishes due to (17.12). We arrive, finally, at the hydrodynamic equation for the flow velocity $\vec{u}$:

$$\boxed{\frac{\partial \vec{u}}{\partial t} + \vec{u} \cdot \vec{\nabla} \vec{u} + (nm)^{-1} \vec{\nabla} \cdot \hat{P} = 0} \qquad (17.25)$$

We return to the discussion of the pressure tensor which was defined by (17.22). Pressure arises from the transport of momentum by two distinct mechanisms: momentum is transmitted by collision-free motion, when particles move together with their momenta $m\vec{v}_i$ from one position to another; the velocities of the particles do not change, but the particles change their positions. This kind of momentum transfer is described by the *kinetic* component of the pressure tensor. The second mechanism of momentum transfer is due to particle collisions. Here the velocities of the particles are modified, but not their positions. This mechanism is described by the *collisional* component of the pressure tensor.

For dilute gases, where collisions occur rarely, the kinetic part which is linear in the density n dominates. When density increases, the frequency of collisions increases and the collisional part (proportional to n^2) takes over. Here we assume that the density n is small so that the collision component can be neglected.

Consider a small surface element dS in the comoving frame of the local average velocity $\vec{u}$. In our coordinate system the normal of this surface element is directed along the z-axis (Fig. 17.1). Any particle which crosses the plane contributes to a transfer of momentum. Hence, during the time span dt some momentum $\Delta \vec{p}$ is transferred through dS. The contribution to its y-component originating from particles with velocities from the interval $(\vec{V}, d\vec{V})$ is given by

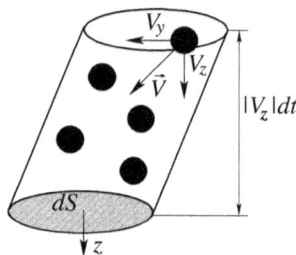

FIG. 17.1. Transfer of momentum through a small surface element dS with normal directed along the z-axis. The volume of the collision cylinder is $V_z dt dS$ and the number of particles in this collision cylinder with velocities from the interval $(\vec{V}, d\vec{V})$ is $V_z dt dS\, f(\vec{v})d\vec{v}$, where $\vec{V} = \vec{v} - \vec{u}$ is the local velocity.

the number of such particles multiplied by the y-component of the momentum carried by these particles:

$$\Delta p_y = mV_y \left[V_z dt dS\, f(\vec{v}) d\vec{v} \right] . \tag{17.26}$$

Division by dt and dS yields the zy component of the stress tensor due to the particles from the velocity interval $(\vec{V}, d\vec{V})$. By integration over $\vec{v}$ we find the zy-component of the stress tensor, which coincides with the zy-component of (17.22). It is convenient to write the diagonal part of the stress tensor explicitly, using the definition of temperature (17.6):

$$\begin{aligned} P_{ij} &= \int m \left(V_i V_j - \frac{1}{3} \delta_{ij} V^2 \right) f(\vec{r},\vec{v},t) d\vec{v} + \frac{1}{3}\delta_{ij} \int m V^2 f(\vec{r},\vec{v},t) d\vec{v} \\ &= \int D_{ij} f d\vec{v} + nT \delta_{ij} \end{aligned} \tag{17.27}$$

with

$$D_{ij} \equiv m \left(V_i V_j - \frac{1}{3} \delta_{ij} V^2 \right) . \tag{17.28}$$

The diagonal part of the stress tensor is called *hydrostatic pressure*. In the dilute limit it may be approximated by the kinetic component nT as in (17.27). For dense systems, however, the collision component is important, which characterises the (instantaneous) momentum transfer between colliding particles. The derivation of the hydrostatic pressure containing both components can be found in Hansen and McDonald (1986), Luding and Strauss (2001), Soto and Mareschal (2001). In the following it is not used, therefore, we give only the final result:

$$p = nT \left(1 + \frac{1+\varepsilon}{3} \pi n \sigma^3 g_2(\sigma) \right) . \tag{17.29}$$

The last hydrodynamic equation for the energy (temperature) corresponds to $\psi = mv^2/2$. In this case the RHS of the Boltzmann equation does not vanish after this transformation:

$$\int d\vec{v}_1 \frac{mv_1^2}{2} g_2(\sigma) I(f,f) = -\frac{3}{2} nT\zeta, \qquad (17.30)$$

which defines the cooling coefficient ζ:

$$\zeta(\vec{r},t) = -\frac{mg_2(\sigma)}{3nT} \int d\vec{v}_1 v_1^2 I(f,f)$$

$$= \frac{mg_2(\sigma)}{3nT} \frac{\sigma^2}{4} \int d\vec{v}_1 d\vec{v}_2 \int d\vec{e}\, \Theta(-\vec{v}_{12}\cdot\vec{e}) |\vec{v}_{12}\cdot\vec{e}| f(\vec{v}_1) f(\vec{v}_2) (\vec{v}_{12}\cdot\vec{e})^2 (1-\varepsilon^2)$$

$$= \frac{\pi m g_2(\sigma)\sigma^2}{24 nT} \int d\vec{v}_1 d\vec{v}_2 v_{12}^3 f(\vec{r},\vec{v}_1,t) f(\vec{r},\vec{v}_2,t)(1-\varepsilon^2). \qquad (17.31)$$

To obtain ζ as given by (17.31), the property of the collision integral (17.10), and the relation (17.12) for inelastic collisions has been used and integration over the unit vector $\vec{e}$ has been performed which yields the factor $\pi/2$.

Evaluating the LHS in exactly the same way as before, we eventually arrive at the hydrodynamic equation

$$\boxed{\frac{\partial T}{\partial t} + \vec{u}\cdot\vec{\nabla} T + \frac{2}{3n}\left(\hat{P}:\vec{\nabla}\vec{u} + \vec{\nabla}\cdot\vec{q}\right) + \zeta T = 0} \qquad (17.32)$$

where the vector $\vec{q}$ denotes the heat flux

$$\vec{q}(\vec{r},t) \equiv \int \frac{mV^2}{2} \vec{V} f(\vec{r},\vec{v},t) d\vec{v}. \qquad (17.33)$$

We have used the common short-hand notation for the product of two tensors:

$$\hat{P}:\vec{\nabla}\vec{u} \equiv \mathrm{Tr}\left(\hat{P}\cdot\vec{\nabla}\vec{u}\right) = P_{ij}\nabla_j u_i, \qquad (17.34)$$

where summation over repeating indices is performed. With (17.23) we can write the heat flux in a different form which will be useful below:

$$\vec{q} = \int \vec{S}\left(\vec{V}\right) f(\vec{r},\vec{v},t) d\vec{v} \quad \text{with} \quad \vec{S} \equiv \left(\frac{mV^2}{2} - \frac{5}{2}T\right)\vec{V}. \qquad (17.35)$$

Exercise 17.2 *Derive the hydrodynamic equation (17.32) from the Boltzmann equation (17.8) by multiplying it with $mv_1^2/2$ and integrating over $\vec{v}_1$!*

The hydrodynamic equations (17.14, 17.25, 17.32) describe the evolution of a granular gas as a continuum. The main condition for their application is the assumption of small gradients. From the hydrodynamic equations follows that this condition implies that the time derivatives are small too. Hence, the hydrodynamic fields change on a large space scale $L \gg l$ and, correspondingly, on a large time scale $T \gg \tau_c$.

To close the hydrodynamic equations for the fields n, $\vec{u}$ and T, the pressure tensor $\hat{P}$ and the heat flux $\vec{q}$ have to be expressed in terms of these fields themselves and their gradients by the constitutive relations. In linear approximation with respect to the gradients they may be written in the simplest form which satisfies the physical requirements (see discussion below)

$$P_{ij} = p\delta_{ij} - \eta \left(\nabla_i u_j + \nabla_j u_i - \frac{2}{3}\delta_{ij}\vec{\nabla} \cdot \vec{u} \right)$$
$$\vec{q} = -\kappa \vec{\nabla} T - \mu \vec{\nabla} n$$

(17.36)

where p is the (isotropic) hydrostatic pressure and η, κ and μ are phenomenological coefficients.[21] The coefficients η and κ are called respectively *shear viscosity* and *thermal conductivity*. As for molecular gases, the shear viscosity characterizes the flux of momentum due to the gradient of the flow velocity, while the thermal conductivity characterises the heat flux due to the gradient of temperature. The coefficient μ does not have an analogue in the hydrodynamics of molecular fluids. It relates the heat flux with the density gradient (Brey et al., 1998; Brey and Cubero, 2001). It appears due to the increased collision frequency (sink of energy) in regions of enhanced density.

The pressure tensor in (17.36) is written as a sum of the diagonal part, which is in zeroth order with respect to the gradients, and the traceless part, which is in first-order with respect to the velocity gradient. The former is the usual hydrostatic pressure, while the latter is written as a symmetrised combination of the velocity gradients.[22]

Keeping, in (17.36), only the linear-order terms with respect to the gradients corresponds to the *Navier–Stokes hydrodynamics*. Keeping the next order gradient terms corresponds to the *Burnett* or *super-Burnett* hydrodynamics (Chapman and Cowling, 1970; Schram, 1991). In this book we restrict ourselves to the Navier–Stokes level, that is, we will consider only linear-order gradient terms.

So far, the coefficients η, κ and μ have been introduced as phenomenological coefficients. They have to be determined from the (microscopic) properties of the gas particles.

---

[21] Relations (17.36) may be also obtained by regular methods, for example, by the Chapman–Enskog approach addressed in the next section, (e.g. Schram, 1991). In this approach the transport coefficients η, κ μ are expressed in terms of the velocity distribution function.

[22] The stress in a fluid which is contained in a uniformly rotating vessel must vanish, which is assured by this part of the stress tensor. Indeed, for angular velocity $\vec{\Omega}$ the linear velocity of a volume element at $\vec{r}$ is $\vec{u} = \vec{\Omega} \times \vec{r}$ (the rotation axis is assumed to cross the origin $\vec{r} = 0$). It can be easily checked that the non-diagonal part of $\hat{P}$ vanishes for any tensor of the form $a(\nabla_i u_j + \nabla_j u_i) + b\delta_{ij}\vec{\nabla} \cdot \vec{u}$, where a and b are arbitrary coefficients.

18

CHAPMAN–ENSKOG APPROACH FOR NON-UNIFORM GRANULAR GASES

For slightly non-uniform gases with small gradients of the hydrodynamic fields, the velocity distribution function can be written as a perturbation expansion $f = f^{(0)} + f^{(1)} + f^{(2)} + \cdots$, where $f^{(k)}$ depends on the kth order of the gradients. We obtain a set of entangled equations for $f^{(k)}$, the first of them is the Boltzmann equation for the homogeneous gas. The Chapman–Enskog scheme provides a technique for solving these equations and calculating the transport coefficients.

18.1 Basic idea of the Chapman–Enskog scheme

The Chapman–Enskog scheme is a method to derive the velocity distribution function for slightly non-uniform gases, when the field gradients are small. In Chapters 5–10 we have discussed the velocity distribution function of a homogeneous cooling granular gas, that is, in the absence of density gradients. If the gradients are small the distribution function may be written as a perturbation expansion with respect to the gradients of the fields: The term in zeroth order corresponds to the homogeneous gas, the first-order term depends linearly on the gradients, etc. A method to determine this expansion was proposed independently by D. Enskog and S. Chapman. Their method has been applied to dissipative gases by a number of authors (Lun *et al.*, 1984; Jenkins and Richman, 1985; Goldshtein and Shapiro, 1995; Brey *et al.*, 1998; Sela and Goldhirsch, 1998; Brey and Cubero, 2001). A rigorous discussion of the Chapman–Enskog approach can be found, for example, in Chapman and Cowling (1970) and Ferziger and Kaper (1972). Here we briefly sketch the application of this method to granular gases (see also Brey *et al.*, 1998; Brilliantov and Pöschel, 2003).

To simplify the notations we assume that the density is small enough to drop the factor $g_2(\sigma) \simeq 1$ in the collision integral. This factor may be easily taken into account by the replacing σ^2 by $\sigma^2 g_2(\sigma)$ in the final expressions.

The goal of the Chapman–Enskog approach is to construct a perturbation expansion for the velocity distribution function in escalating powers of the fields gradients

$$f = f^{(0)} + \lambda f^{(1)} + \lambda^2 f^{(2)} + \cdots, \qquad (18.1)$$

where each power of λ corresponds to the order of the gradient. Thus, $f^{(0)}$ denotes the unperturbed solution (without gradients), $f^{(1)}$ is the solution obtained in linear-order approximation with respect to the fields gradients, $f^{(2)}$ is the solution in second-order approximation, etc. The final solution is then obtained by taking $\lambda = 1$. As usual for perturbation approaches, we construct a system of equations where the first one contains only $f^{(0)}$, the second one contains $f^{(1)}$ and $f^{(0)}$, with $f^{(0)}$ found from the first equation, the third one contains $f^{(2)}$, $f^{(1)}$ and $f^{(0)}$, with $f^{(1)}$ and $f^{(0)}$ obtained in previous equations, etc.

Obviously, the solution in zeroth order, $f^{(0)}$, is the distribution function of the homogeneous cooling state. For a granular gas consisting of particles that collide with a constant coefficient of restitution, $f^{(0)}$ depends on time only via the time-dependent temperature. We assume that a similar property of the distribution function is justified for slightly inhomogeneous systems as well. In contrast to the homogeneous cooling state where only the temperature depends on time, now the density and the flow velocity depend on time too. Hence, we assume that the distribution function depends on time and space only via its moments: density n, average velocity $\vec{u}$ and temperature T. Then the time derivative of the distribution function reads

$$\frac{\partial f}{\partial t} = \frac{\partial f}{\partial n}\frac{\partial n}{\partial t} + \frac{\partial f}{\partial \vec{u}}\cdot\frac{\partial \vec{u}}{\partial t} + \frac{\partial f}{\partial T}\frac{\partial T}{\partial t}. \quad (18.2)$$

The time derivatives of n, $\vec{u}$ and T may be expressed using the the hydrodynamic equations (17.14, 17.25, 17.32). Substituting the expressions (17.36) for the pressure tensor and the heat flux into the hydrodynamic equations and marking each power of the gradients with the corresponding power of λ, we obtain:

$$\frac{\partial n}{\partial t} = -\lambda \vec{\nabla}(n\vec{u})$$
$$\frac{\partial \vec{u}}{\partial t} = -\lambda \left(\vec{u}\cdot\vec{\nabla}\vec{u} + \frac{1}{nm}\vec{\nabla}p\right) + \lambda^2 \frac{\eta}{nm}\left(\nabla^2\vec{u} + \frac{1}{3}\vec{\nabla}(\vec{\nabla}\cdot\vec{u})\right) \quad (18.3)$$
$$\frac{\partial T}{\partial t} = -\zeta T - \lambda\left(\vec{u}\cdot\vec{\nabla}T + \frac{2}{3n}p(\vec{\nabla}\cdot\vec{u})\right) + \lambda^2 G,$$

where

$$G = \frac{2\eta}{3n}\left[(\nabla_i u_j)(\nabla_j u_i) + (\nabla_j u_i)(\nabla_j u_i) - \frac{2}{3}\left(\vec{\nabla}\cdot\vec{u}\right)^2\right]$$
$$+ \frac{2}{3n}\left(\kappa\nabla^2 T + \mu\nabla^2 n\right). \quad (18.4)$$

Exercise 18.1 *Derive equations (18.3) and (18.4)!*

Note that (18.3) and (18.4) do not contain terms with powers of λ larger than two. This corresponds to the Navier–Stokes level of the hydrodynamic description; the next-order levels (Burnett, super-Burnett) would contain higher orders of λ. If we substitute (18.3) into (18.2), we notice that the time derivative of the distribution function is given as a series of terms with escalating powers of λ:

$$\frac{\partial}{\partial t} = \frac{\partial^{(0)}}{\partial t} + \lambda \frac{\partial^{(1)}}{\partial t} + \lambda^2 \frac{\partial^{(2)}}{\partial t} + \cdots \qquad (18.5)$$

where each order k of the time derivative, $\partial_t^{(k)} \equiv \partial^{(k)}/\partial t$, corresponds to the related order of the space gradient. This property applies to the hydrodynamic fields themselves too (see (18.3)). Physically, the expansion of the time derivatives is based on the requirement that the higher the order of the space gradient, the slower the time variation it causes. This means that the fastest variations are caused by the zeroth order gradient term, then follows the first-order term, the second-order, etc. To illustrate this statement, we compare the continuity equation (17.14) which describes the variation of the density according to the flux of particles with the hydrodynamic velocity $\vec{u}$ with (12.3) which relates the variation of the concentration with the diffusion process. While in (17.14) the density varies due to the first-order gradient of the fields, in (12.3) the second order gradient is involved. Hence, obviously the diffusion equation describes much slower changes of local concentrations (or densities) of particles, than the continuity equation.

To illustrate the precise meaning of the operator $\partial^{(k)}/\partial t$, let us consider an example,

$$\frac{\partial^{(0)}}{\partial t} F^{(2)} \equiv \frac{\partial F^{(2)}}{\partial n} \frac{\partial^{(0)} n}{\partial t} + \frac{\partial F^{(2)}}{\partial \vec{u}} \cdot \frac{\partial^{(0)} \vec{u}}{\partial t} + \frac{\partial F^{(2)}}{\partial T} \frac{\partial^{(0)} T}{\partial t}. \qquad (18.6)$$

This expression explicitly implies that the function $F^{(2)}$ depends on time *only* via the hydrodynamic fields (the upper index shows that $F^{(2)}$ depends on the second order space gradients). Hence, the time derivative is to be transmitted to the fields. For the time derivatives $\partial^{(0)}/\partial t$ of the fields, one should take only zero-order gradient terms in the RHS of the hydrodynamic equations (18.3), that is, $\partial_t^{(0)} n = 0$, $\partial_t^{(0)} \vec{u} = 0$, etc. (see also equations (18.9) below).

Now we have all ingredients to construct the perturbation expansion. Substituting the expansions (18.1) for f and (18.5) for its time-derivative into the Boltzmann equation (17.8) we obtain

$$\begin{aligned}\left(\frac{\partial}{\partial t} + \vec{v}_1 \vec{\nabla}\right) f &= \left(\frac{\partial^{(0)}}{\partial t} + \lambda \frac{\partial^{(1)}}{\partial t} + \cdots + \lambda \vec{v}_1 \cdot \vec{\nabla}\right) \left(f^{(0)} + \lambda f^{(1)} + \cdots\right) \\ &= I\left[\left(f^{(0)} + \lambda f^{(1)} + \cdots\right), \left(f^{(0)} + \lambda f^{(1)} + \cdots\right)\right],\end{aligned} \qquad (18.7)$$

where the collision integral I is given by (17.9). Collecting terms with the same power of λ, we arrive at a set of entangled equations.

18.2 Equations of zeroth order of the Chapman–Enskog expansion

Collecting the terms $\mathcal{O}\left(\lambda^0\right)$ in (18.7) we obtain the equation in zeroth order of the Chapman–Enskog scheme:

$$\frac{\partial^{(0)}}{\partial t} f^{(0)} = I\left(f^{(0)}, f^{(0)}\right). \tag{18.8}$$

Correspondingly, the hydrodynamic equations in this order are

$$\boxed{\frac{\partial^{(0)}}{\partial t} n = 0, \quad \frac{\partial^{(0)}}{\partial t} \vec{u} = 0, \quad \frac{\partial^{(0)}}{\partial t} T = -\zeta^{(0)} T} \tag{18.9}$$

where the zeroth order coefficient $\zeta^{(0)}$ may be calculated using (17.31) with the distribution function $f^{(0)}$ for the homogeneous gas. For the case of a constant coefficient of restitution, the distribution function reads

$$f^{(0)} = \frac{n}{v_T^3} \tilde{f}^{(0)} \left(\frac{\vec{V}}{v_T}\right) \quad \text{with} \quad \vec{V} \equiv \vec{v} - \vec{u}, \quad \vec{c} = \frac{\vec{V}}{v_T}, \tag{18.10}$$

where for simpler notation we drop the space and time dependence of $n(\vec{r}, t)$ and $v_T(\vec{r}, t)$ and where the characteristic velocity $v_T(\vec{r}, t) \equiv \sqrt{2T(\vec{r}, t)/m}$. The detailed form of $\tilde{f}^{(0)}$ has been derived in Chapters 8 and 9, where the homogeneous cooling state has been addressed. With (18.10) the coefficient $\zeta^{(0)}$ reads

$$\begin{aligned}
\zeta^{(0)} &= -\frac{m g_2(\sigma)}{3nT} \int d\vec{v}_1 v_1^2 I\left(f^{(0)}, f^{(0)}\right) \\
&= -\frac{m g_2(\sigma)}{3nT} v_T^5 \sigma^2 n^2 v_T^{-2} \int d\vec{c}_1 c_1^2 \tilde{I}\left(\tilde{f}^{(0)}, \tilde{f}^{(0)}\right) \\
&= \frac{2}{3} n g_2(\sigma) \sigma^2 \sqrt{\frac{2T}{m}} \mu_2,
\end{aligned} \tag{18.11}$$

where we use (8.3, 8.8) for the definitions of the dimensionless collision integral and its moment μ_2. Naturally, (18.11) coincides with (8.12) as derived previously for a granular gas in the homogeneous cooling state. For the case $\varepsilon = $ const. the moment μ_2 is a constant too, hence, (18.11) implies the following temperature and density derivatives of $\zeta^{(0)}$:

$$\frac{\partial \zeta^{(0)}}{\partial n} = \frac{1}{n} \zeta^{(0)}, \quad \frac{\partial \zeta^{(0)}}{\partial T} = \frac{1}{2T} \zeta^{(0)}. \tag{18.12}$$

18.3 First-order equations of the Chapman–Enskog expansion

Collecting the terms of the order λ^1 in (18.7) we obtain the first-order equation

$$\frac{\partial^{(0)}}{\partial t}f^{(1)} + \left(\frac{\partial^{(1)}}{\partial t} + \vec{v}_1 \cdot \vec{\nabla}\right)f^{(0)} = I\left(f^{(0)}, f^{(1)}\right) + I\left(f^{(1)}, f^{(0)}\right). \quad (18.13)$$

To obtain the RHS we use the obvious property of the collision integral

$$I\left[(h_1 + \lambda h_2 + \cdots), (h_1 + \lambda h_2 + \cdots)\right] \\ = I(h_1, h_1) + \lambda I(h_1, h_2) + \lambda I(h_2, h_1) + \lambda^2 I(h_2, h_2) + \cdots, \quad (18.14)$$

where $h_1, h_2, \ldots$ are functions of $(\vec{r}, \vec{v}, t)$.

Exercise 18.2 *Prove (18.14)!*

The corresponding first-order hydrodynamic equations are then given by

$$\frac{\partial^{(1)}}{\partial t}n = -\vec{\nabla}(n\vec{u})$$

$$\frac{\partial^{(1)}}{\partial t}\vec{u} = -\vec{u} \cdot \vec{\nabla}\vec{u} - \frac{1}{nm}\vec{\nabla}p \quad (18.15)$$

$$\frac{\partial^{(1)}}{\partial t}T = -\vec{u} \cdot \vec{\nabla}T - \frac{2}{3}T\vec{\nabla} \cdot \vec{u} - \zeta^{(1)}T$$

where we use $p = nT$. For a gas of elastic particles where $\zeta^{(1)} = 0$, the set (18.15) is called *Euler equations*. It describes an ideal (energy conserving) non-viscous fluid. The first-order coefficient $\zeta^{(1)}$ is obtained by substituting $f = f^{(0)} + \lambda f^{(1)}$ into (17.31) and keeping terms $\mathcal{O}(\lambda)$:

$$\zeta^{(1)} = 2\frac{(1-\varepsilon^2)m\pi\sigma^2 g_2(\sigma)}{24nT} \int d\vec{v}_1 d\vec{v}_2 v_{12}^3 f^{(0)}(\vec{v}_1) f^{(1)}(\vec{v}_2). \quad (18.16)$$

Again, in the notation we have dropped the dependences of the fields on $\vec{r}$ and t.

Since the distribution function $f^{(0)}$ is known, we can easily calculate the terms in (18.13) which depend only on $f^{(0)}$. For example, the time derivative of $f^{(0)}$ reads

$$\frac{\partial^{(1)} f^{(0)}}{\partial t} = \frac{\partial f^{(0)}}{\partial n}\frac{\partial^{(1)} n}{\partial t} + \frac{\partial f^{(0)}}{\partial \vec{u}} \cdot \frac{\partial^{(1)} \vec{u}}{\partial t} + \frac{\partial f^{(0)}}{\partial T}\frac{\partial^{(1)} T}{\partial t}, \quad (18.17)$$

where the time derivatives of n, $\vec{u}$ and T are given by (18.15). The prefactors are

$$\frac{\partial f^{(0)}}{\partial n} = \frac{1}{n} f^{(0)}$$

$$\frac{\partial f^{(0)}}{\partial \vec{u}} = -\frac{\partial f^{(0)}}{\partial \vec{V}} \qquad (18.18)$$

$$\frac{\partial f^{(0)}}{\partial T} = -\frac{1}{2T} \frac{\partial}{\partial \vec{V}} \cdot \left(\vec{V} f^{(0)} \right).$$

Exercise 18.3 *Find the derivatives of $f^{(0)}$ with respect to n, $\vec{u}$ and T, that is, derive (18.18)!*

Therefore,

$$\frac{\partial^{(1)} f^{(0)}}{\partial t} = -\frac{1}{n} f^{(0)} \left(\vec{u} \cdot \vec{\nabla} n + n \vec{\nabla} \cdot \vec{u} \right) + \frac{\partial f^{(0)}}{\partial V_i} \left((\vec{u} \cdot \vec{\nabla}) u_i + \frac{1}{nm} \nabla_i p \right)$$

$$- \frac{\partial f^{(0)}}{\partial T} \left(\vec{u} \cdot \vec{\nabla} T + \frac{2T}{3} \vec{\nabla} \cdot \vec{u} + \zeta^{(1)} T \right). \qquad (18.19)$$

Similarly, the other terms in (18.13) which depends only on $f^{(0)}$ can be evaluated:

$$\vec{v}_1 \cdot \vec{\nabla} f^{(0)} = \frac{1}{n} f^{(0)} \vec{v}_1 \cdot \vec{\nabla} n - \frac{\partial f^{(0)}}{\partial V_i} \left(\vec{v}_1 \cdot \vec{\nabla} \right) u_i + \frac{\partial f^{(0)}}{\partial T} \vec{v}_1 \cdot \vec{\nabla} T, \qquad (18.20)$$

where we again use (18.18) for the derivatives of $f^{(0)}$ with respect to the hydrodynamic fields. Substituting (18.19, 18.20) into the first-order equation (18.13) and using $\vec{v}_1 = \vec{V} + \vec{u}$ we arrive at the first-order equation of the Chapman–Enskog expansion:

$$\frac{\partial^{(0)} f^{(1)}}{\partial t} + J^{(1)} \left(f^{(0)}, f^{(1)} \right) - \zeta^{(1)} T \frac{\partial f^{(0)}}{\partial T}$$

$$= f^{(0)} \left(\vec{\nabla} \cdot \vec{u} - \vec{V} \cdot \vec{\nabla} \log n \right) + \frac{\partial f^{(0)}}{\partial T} \left(\frac{2}{3} T \vec{\nabla} \cdot \vec{u} - \vec{V} \cdot \vec{\nabla} T \right) \qquad (18.21)$$

$$+ \frac{\partial f^{(0)}}{\partial V_i} \left((\vec{V} \cdot \vec{\nabla}) u_i - \frac{1}{nm} \nabla_i p \right),$$

where

$$J^{(1)} \left(f^{(0)}, f^{(1)} \right) \equiv -I \left(f^{(0)}, f^{(1)} \right) - I \left(f^{(1)}, f^{(0)} \right). \qquad (18.22)$$

The RHS of this equation depends only on the known function $f^{(0)}$ and linearly on the field gradients $\vec{\nabla} n$, $\vec{\nabla} \vec{u}$, $\vec{\nabla} T$.

18.4 Solution of the first-order equation

Using the relation

$$\frac{1}{nm}\vec{\nabla}_i p = \frac{1}{nm}\vec{\nabla}_i(nT) = \frac{T}{m}\vec{\nabla}_i \log T + \frac{T}{m}\vec{\nabla}_i \log n \qquad (18.23)$$

the RHS of (18.21) yields

$$-\left(\vec{V}T\frac{\partial f^{(0)}}{\partial T} + \frac{T}{m}\frac{\partial f^{(0)}}{\partial \vec{V}}\right) \cdot \vec{\nabla} \log T - \left(\vec{V} f^{(0)} + \frac{T}{m}\frac{\partial f^{(0)}}{\partial \vec{V}}\right) \cdot \vec{\nabla} \log n$$
$$+ \left[\left(f^{(0)} + \frac{2}{3}T\frac{\partial f^{(0)}}{\partial T}\right)\delta_{ij} + \frac{\partial f^{(0)}}{\partial V_i}V_j\right]\nabla_j u_i, \quad (18.24)$$

where we use $\vec{\nabla} \cdot \vec{u} = \delta_{ij}\nabla_j u_i$ with the summation convention for the repeating indices. Thus, the first-order equation (18.21) reads

$$\boxed{\begin{aligned}\frac{\partial^{(0)} f^{(1)}}{\partial t} + J^{(1)}\left(f^{(0)}, f^{(1)}\right) - \zeta^{(1)} T \frac{\partial f^{(0)}}{\partial T} \\ = \vec{A} \cdot \vec{\nabla} \log T + \vec{B} \cdot \vec{\nabla} \log n + C_{ij} \nabla_j u_i\end{aligned}} \qquad (18.25)$$

with

$$\vec{A}(\vec{V}) = \frac{1}{2}\vec{V}\frac{\partial}{\partial \vec{V}} \cdot \left(\vec{V}f^{(0)}\right) - \frac{T}{m}\frac{\partial}{\partial \vec{V}}f^{(0)} = \vec{V}\left[\frac{T}{m}\left(\frac{mV^2}{2T} - 1\right)\frac{1}{V}\frac{\partial}{\partial V} + \frac{3}{2}\right]f^{(0)}$$

$$\vec{B}(\vec{V}) = -\vec{V}f^{(0)} - \frac{T}{m}\frac{\partial}{\partial \vec{V}}f^{(0)} = -\vec{V}\left(\frac{T}{m}\frac{1}{V}\frac{\partial}{\partial V} + 1\right)f^{(0)}$$

$$C_{ij}(\vec{V}) = \frac{\partial}{\partial V_i}\left(V_j f^{(0)}\right) - \frac{1}{3}\delta_{ij}\frac{\partial}{\partial \vec{V}} \cdot \left(\vec{V}f^{(0)}\right) = \left(V_i V_j - \frac{1}{3}\delta_{ij}V^2\right)\frac{1}{V}\frac{\partial f^{(0)}}{\partial V}.$$
$$(18.26)$$

To obtain these coefficients we use (18.18) and the isotropy of the function $f^{(0)}$, that is, $f^{(0)}(\vec{V}) = f^{(0)}(V)$ and, therefore,

$$\frac{\partial f^{(0)}}{\partial \vec{V}} = \frac{\vec{V}}{V}\frac{\partial f^{(0)}}{\partial V}, \qquad \frac{\partial f^{(0)}}{\partial V_i} = \frac{V_i}{V}\frac{\partial f^{(0)}}{\partial V}. \qquad (18.27)$$

Exercise 18.4 Derive the coefficients $\vec{A}$, $\vec{B}$ and C_{ij} in (18.26)!

The velocity distribution function $f^{(0)}$ of a homogeneous granular gas is known. Hence, the coefficients $\vec{A}$, $\vec{B}$ and C_{ij}, which are functions of the hydrodynamic fields n, $\vec{u}$ and T via $f^{(0)}$, are also known. From their definitions

(18.26) it follows that the vectorial coefficients $\vec{A}$ and $\vec{B}$ are directed along the vector $\vec{V}$, while the tensorial coefficient C_{ij} is traceless:

$$\text{Tr}\left(\hat{C}\right) = C_{ii} = (V^2 - V^2) \frac{1}{V} \frac{\partial f^{(0)}}{\partial V} = 0. \tag{18.28}$$

From the form of the RHS of the first-order equation (18.25) we expect the form of its solution

$$f^{(1)} = \vec{\alpha} \cdot \vec{\nabla} \log T + \vec{\beta} \cdot \vec{\nabla} \log n + \gamma_{ij} \nabla_j u_i, \tag{18.29}$$

which is the most general form of a scalar function which depends linearly on the vectorial gradients $\vec{\nabla} T$, $\vec{\nabla} n$ and on the tensorial gradients $\nabla_j u_i$. The coefficients $\vec{\alpha}$, $\vec{\beta}$ and γ_{ij} are functions of $\vec{V}$ and of the hydrodynamic fields n, $\vec{u}$ and T.

From the comparison of (18.29) and (18.25) we expect that the vectorial coefficients $\vec{\alpha}$ and $\vec{\beta}$ are proportional to $\vec{A}$ and $\vec{B}$ and, hence, they too are directed along $\vec{V}$ (see (18.26)). Similarly, we expect that the tensorial coefficient γ_{ij}, being proportional to C_{ij}, is also traceless (see (18.28)). Motivated by the form of (18.29) for $f^{(1)}$ we assume $\zeta^{(1)} = 0$ and check this assumption *a posteriori* (see Exercise 18.5). From the definition (18.16) of $\zeta^{(1)}$ we may draw this conclusion as well: The coefficient $\zeta^{(1)}$ is obtained by integration of the product $f^{(0)} f^{(1)}$ over $d\vec{v}_1 d\vec{v}_2 = d\vec{V}_1 d\vec{V}_2$. The function $f^{(1)}$ contains only terms proportional to $\vec{V}_2$ since $\vec{\alpha} \propto \vec{V}_2$, $\vec{\beta} \propto \vec{V}_2$ and to $\left(V_{2,i} V_{2,j} - \frac{1}{3}\delta_{ij} V_2^2\right)$ since $\gamma_{ij} \propto C_{ij}$. All these terms vanish upon integration over $\vec{V}_2$. The physical reason for this is the absence of any preferred direction in the comoving frame ($\vec{V}_2$ is the local velocity in the comoving frame).

Exercise 18.5 *Prove that the coefficient $\zeta^{(1)}$, defined by (18.16), vanishes! Use the first-order distribution function $f^{(1)}$ as given on page 201.*

We now derive equations for the coefficients $\vec{\alpha}$, $\vec{\beta}$ and γ_{ij} by substituting $f^{(1)}$ as given by (18.29) into the first-order equation (18.25) and equating the coefficients of the same gradients. To this end we need $\partial_t^{(0)} f^{(1)}$ and, therefore, the time derivatives of the coefficients $\vec{\alpha}$, $\vec{\beta}$, γ_{ij}:

$$\frac{\partial^{(0)}\vec{\alpha}}{\partial t} = \frac{\partial \vec{\alpha}}{\partial T}\frac{\partial^{(0)} T}{\partial t} + \frac{\partial \vec{\alpha}}{\partial n}\frac{\partial^{(0)} n}{\partial t} + \frac{\partial \vec{\alpha}}{\partial u_i}\frac{\partial^{(0)} u_i}{\partial t} = \frac{\partial \vec{\alpha}}{\partial T}\frac{\partial^{(0)} T}{\partial t} = -\zeta^{(0)} T \frac{\partial \vec{\alpha}}{\partial T}, \tag{18.30}$$

where we use the first-order equations (18.9). Similarly we obtain

$$\frac{\partial^{(0)}\vec{\beta}}{\partial t} = -\zeta^{(0)} T \frac{\partial \vec{\beta}}{\partial T}$$

$$\frac{\partial^{(0)}\gamma_{ij}}{\partial t} = -\zeta^{(0)} T \frac{\partial \gamma_{ij}}{\partial T}, \tag{18.31}$$

and, correspondingly, with (18.9, 18.12), the time derivatives of the gradients:

$$\frac{\partial^{(0)}}{\partial t}\vec{\nabla}\log n = \vec{\nabla}\frac{1}{n}\frac{\partial^{(0)}n}{\partial t} = 0$$

$$\frac{\partial^{(0)}}{\partial t}\nabla_j u_i = \nabla_j \frac{\partial^{(0)} u_i}{\partial t} = 0 \qquad (18.32)$$

$$\frac{\partial^{(0)}}{\partial t}\vec{\nabla}\log T = \vec{\nabla}T^{-1}\frac{\partial^{(0)}T}{\partial t} = -\vec{\nabla}\zeta^{(0)} = -\frac{\partial \zeta^{(0)}}{\partial n}\vec{\nabla}n - \frac{\partial \zeta^{(0)}}{\partial T}\vec{\nabla}T$$

$$= -\zeta^{(0)}\left(\vec{\nabla}\log n + \frac{1}{2}\vec{\nabla}\log T\right).$$

From (18.30–18.32) we obtain

$$\frac{\partial^{(0)} f^{(1)}}{\partial t} = -\left(\zeta^{(0)}T\frac{\partial \vec{\alpha}}{\partial T} + \frac{1}{2}\zeta^{(0)}\vec{\alpha}\right)\cdot\vec{\nabla}\log T$$
$$-\left(\zeta^{(0)}T\frac{\partial \vec{\beta}}{\partial T} + \zeta^{(0)}\vec{\alpha}\right)\cdot\vec{\nabla}\log n - \zeta^{(0)}T\frac{\partial \gamma_{ij}}{\partial T}\nabla_j u_i. \qquad (18.33)$$

If we insert now $\partial_t^{(0)} f^{(1)}$ into the first-order equation (18.25) and equate the coefficients of the gradients we arrive at a set of equations for the coefficients $\vec{\alpha}$, $\vec{\beta}$ and γ_{ij}:

$$-\zeta^{(0)}\left(T\frac{\partial}{\partial T} + \frac{1}{2}\right)\vec{\alpha} + J^{(1)}\left(f^{(0)},\vec{\alpha}\right) = \vec{A}$$

$$-\zeta^{(0)}T\frac{\partial \vec{\beta}}{\partial T} - \zeta^{(0)}\vec{\alpha} + J^{(1)}\left(f^{(0)},\vec{\beta}\right) = \vec{B} \qquad (18.34)$$

$$-\zeta^{(0)}T\frac{\partial \gamma_{ij}}{\partial T} + J^{(1)}\left(f^{(0)},\gamma_{ij}\right) = C_{ij},$$

where the first-order collision integral $J^{(1)}$ is defined by (18.22) and where the coefficients α_i, β_i and γ_{ij} ($i,j = 1,2,3$) are functions of the velocity $\vec{V}$.

18.5 Kinetic coefficients expressed by the velocity distribution function

The kinetic coefficients can be derived using the first-order distribution function $f^{(1)}$ which depends linearly on the gradients of the hydrodynamic fields. First we express the pressure tensor P_{ij} as a linear function of the gradient $\nabla_j u_i$. The coefficient of this gradient yields then the coefficient of viscosity η.

From the definition of the pressure tensor (17.27) and its phenomenological expression (17.36) follows

$$\int D_{ij}\left(\vec{V}\right)\left(f^{(0)} + \vec{\alpha}\cdot\vec{\nabla}\log T + \vec{\beta}\cdot\vec{\nabla}\log n + \gamma_{kl}\nabla_l u_k\right) d\vec{V}$$
$$= -\eta\left(\nabla_i u_j + \nabla_j u_i - \frac{2}{3}\delta_{ij}\vec{\nabla}\cdot\vec{u}\right), \quad (18.35)$$

where we use $f = f^{(0)} + f^{(1)}$ and the form (18.29) for $f^{(1)}$. D_{ij} is defined by (17.27). Similarly as for the calculation of $\zeta^{(1)}$ on page 182:

$$\int D_{ij}\left(\vec{V}\right) f^{(0)} d\vec{V} = 0$$
$$\int D_{ij}\left(\vec{V}\right) \vec{\alpha} d\vec{V} = \int D_{ij}\left(\vec{V}\right) \vec{\beta} d\vec{V} = 0 \quad (18.36)$$

since D_{ij} is a traceless tensor, $f^{(0)}$ depends on $\vec{V}$ isotropically and the vectors $\vec{\alpha}, \vec{\beta}$ are directed along $\vec{V}$.

Exercise 18.6 *Prove (18.36)!*

Therefore, (18.35) reduces to

$$\left(\int D_{ij}\left(\vec{V}\right) \gamma_{kl}\left(\vec{V}\right) d\vec{V}\right)\nabla_l u_k = -\eta\left(\nabla_i u_j + \nabla_j u_i - \frac{2}{3}\delta_{ij}\vec{\nabla}\cdot\vec{u}\right) \quad (18.37)$$

and, hence,

$$\int D_{ij}\left(\vec{V}\right)\gamma_{kl}\left(\vec{V}\right) d\vec{V} = -\eta\left(\delta_{li}\delta_{kj} + \delta_{lj}\delta_{ki} - \frac{2}{3}\delta_{ij}\delta_{kl}\right), \quad (18.38)$$

which can be checked by multiplying both sides of (18.38) by $\nabla_l u_k$ and summing over repeated indices. For the case $k = j$, $l = i$, the last equation changes to

$$\int D_{ij}\left(\vec{V}\right)\gamma_{ji}\left(\vec{V}\right) d\vec{V} = -\eta\left(\delta_{ii}\delta_{jj} + \delta_{ij}\delta_{ij} - \frac{2}{3}\delta_{ij}\delta_{ij}\right) = -10\eta, \quad (18.39)$$

where we use $\delta_{ii}\delta_{jj} = 9$ and $\delta_{ij}\delta_{ij} = 3$. Thus, the coefficient of viscosity is

$$\eta = -\frac{1}{10}\int D_{ij}\left(\vec{V}\right)\gamma_{ji}\left(\vec{V}\right) d\vec{V}. \quad (18.40)$$

In a similar way we derive the components of the heat flux $\vec{q}$ as defined by (17.35):

$$q_i = \int d\vec{V} S_i\left[f^{(0)} + (\alpha_j\nabla_j)\log T + (\beta_m\nabla_m)\log n + \gamma_{kl}\nabla_l u_k\right]$$
$$= -\kappa\nabla_i T - \mu\nabla_i n. \quad (18.41)$$

Again, similarly as for the calculation of $\zeta^{(1)}$ on page 182, the terms in the integrand $S_i f^{(0)}$ and $S_i \gamma_{kl}\nabla_l u_k$ vanish upon the integration. Indeed, both $S_i f^{(0)}$ ($S_i \propto V_i$) and $S_i \gamma_{kl}$ ($S_i \gamma_{kl} \propto V_i V_k V_l - \frac{1}{3}V^2 V_i \delta_{kl}$) are antisymmetric functions of the components of the velocity V_j in the comoving frame.

Exercise 18.7 Prove $\int d\vec{V} S_i f^{(0)} = 0$ and $\int d\vec{V} S_i \gamma_{kl} = 0$!

Equating the remaining terms we obtain

$$\left\{\frac{1}{T}\int d\vec{V} S_i\left(\vec{V}\right)\alpha_j\left(\vec{V}\right)\right\}\nabla_j T = -\kappa \nabla_i T$$
$$\left\{\frac{1}{n}\int d\vec{V} S_i\left(\vec{V}\right)\beta_m\left(\vec{V}\right)\right\}\nabla_m n = -\mu \nabla_i n \, . \tag{18.42}$$

Therefore

$$\frac{1}{T}\int d\vec{V} S_i\left(\vec{V}\right)\alpha_j\left(\vec{V}\right) = -\delta_{ij}\kappa$$
$$\frac{1}{n}\int d\vec{V} S_i\left(\vec{V}\right)\beta_j\left(\vec{V}\right) = -\delta_{ij}\mu \, . \tag{18.43}$$

For $i = j$ follows

$$\frac{1}{T}\int d\vec{V} S_i\left(\vec{V}\right)\alpha_i\left(\vec{V}\right) = -\delta_{ii}\kappa = -3\kappa$$
$$\frac{1}{n}\int d\vec{V} S_i\left(\vec{V}\right)\beta_i\left(\vec{V}\right) = -\delta_{ii}\mu = -3\mu \, . \tag{18.44}$$

Finally, we arrive at expressions for κ and μ:

$$\kappa = -\frac{1}{3T}\int d\vec{V} \vec{S}\left(\vec{V}\right)\cdot\vec{\alpha}\left(\vec{V}\right)$$
$$\mu = -\frac{1}{3n}\int d\vec{V} \vec{S}\left(\vec{V}\right)\cdot\vec{\beta}\left(\vec{V}\right) \, . \tag{18.45}$$

19

KINETIC COEFFICIENTS AND VELOCITY DISTRIBUTION FOR GASES OF ELASTIC PARTICLES

We apply the Chapman–Enskog approach to an inhomogeneous gas of elastic particles and derive the coefficients of viscosity and thermal conductivity as well as the velocity distribution function.

19.1 First-order Chapman–Enskog equations

To demonstrate the Chapman–Enskog formalism for the kinetic coefficients we consider first a gas of elastic particles, that is, $\varepsilon = 1$. For this case $\zeta^{(0)} = 0$ and the zeroth order velocity distribution function is the Maxwell distribution $f^{(0)}(V) = f_M(V)$. Then the coefficients $\vec{A}$, $\vec{B}$ and C_{ij}, given by (18.26), read

$$\vec{A} = -\vec{V}\left(\frac{mV^2}{2T} - \frac{5}{2}\right) f_M(V) = -\frac{1}{T}\vec{S}\left(\vec{V}\right) f_M(V)$$

$$\vec{B} = 0 \tag{19.1}$$

$$C_{ij} = -\frac{m}{T}\left(V_i V_j - \frac{1}{3}\delta_{ij} V^2\right) f_M(V) = -\frac{1}{T} D_{ij}\left(\vec{V}\right) f_M(V),$$

where $\vec{S}\left(\vec{V}\right)$ and $D_{ij}\left(\vec{V}\right)$ have been defined, respectively, by (17.35) and (17.27). With $\zeta^{(0)} = 0$ the equations (18.34) for $\vec{\alpha}$, $\vec{\beta}$ and γ_{ij} take the form

$$J^{(1)}\left(f^{(0)}, \vec{\alpha}\right) = -\frac{1}{T}\vec{S}\left(\vec{V}\right) f_M(V) \tag{19.2}$$

$$J^{(1)}\left(f^{(0)}, \vec{\beta}\right) = 0 \tag{19.3}$$

$$J^{(1)}\left(f^{(0)}, \gamma_{ij}\right) = -\frac{1}{T} D_{ij}\left(\vec{V}\right) f_M(V). \tag{19.4}$$

We conclude that $\vec{\beta}\left(\vec{V}\right) = 0$ and, of course, $\mu = 0$, since this coefficient differs from zero for inelastic particles only.

19.2 Coefficient of viscosity

The form of (19.4) suggests:[23]

---

[23] Generally, an expansion $\gamma_0 + \gamma_1 S_1^{(5/2)}(c^2) + \gamma_2 S_2^{(5/2)}(c^2) + \cdots$ in Sonine polynomials $S_k^{(5/2)}(c^2)$ is used (Chapman and Cowling, 1970; Schram, 1991); the approximation (19.5) retains only the first term.

$$\gamma_{ij}(\vec{V}) = \frac{\gamma_0}{T} D_{ij}(\vec{V}) f_M(V). \tag{19.5}$$

With this Ansatz, (19.4) for the tensor γ_{ij} transforms into an equation for the scalar coefficient γ_0. Multiplying both sides of the (19.4) by $D_{ij}(\vec{V}_1)$ and integrating over $\vec{V}_1$ yields[24]

$$\int d\vec{V}_1 D_{ij}(\vec{V}_1) J^{(1)}(f_M, \gamma_{ij}) = -\frac{1}{T} \int d\vec{V}_1 D_{ij}(\vec{V}_1) D_{ij}(\vec{V}_1) f_M(V_1). \tag{19.6}$$

With

$$\begin{aligned} D_{ij}(\vec{V}) D_{ij}(\vec{V}) &= m^2 \left(V_i V_j V_i V_j - \frac{2}{3} \delta_{ij} V^2 V_i V_j + \frac{1}{9} \delta_{ij} \delta_{ij} V^4 \right) \\ &= m^2 \left(V^4 - \frac{2}{3} V^4 + \frac{1}{3} V^4 \right) = \frac{2}{3} m^2 V^4 \end{aligned} \tag{19.7}$$

the RHS of (19.6) reads

$$-\frac{1}{T} \int d\vec{V}_1 D_{ij}(\vec{V}_1) D_{ij}(\vec{V}_1) f_M(V_1) = -\frac{2m^2}{3T} \int d\vec{V}_1 V_1^4 f_M(V_1) = -10nT. \tag{19.8}$$

Exercise 19.1 *Prove (19.8)!*

According to the definition of $J^{(1)}$ (18.22) the LHS of (19.6) reads

$$\int d\vec{V}_1 D_{ij}(\vec{V}_1) J^{(1)}(f_M, \gamma_{ij})$$
$$= -\int d\vec{V}_1 D_{ij}(\vec{V}_1) I(f_M, \gamma_{ij}) - \int d\vec{V}_1 D_{ij}(\vec{V}_1) I(\gamma_{ij}, f_M). \tag{19.9}$$

We apply the property of the collision integral (6.25)

$$\begin{aligned} \int d\vec{V}_1 D_{ij}(\vec{V}_1) I(f_M, \gamma_{ij}) &= \int d\vec{V}_1 D_{ij}(\vec{V}_1) I(\gamma_{ij}, f_M) \\ &= \frac{1}{2} \sigma^2 \int d\vec{V}_1 \int d\vec{V}_2 f_M(\vec{V}_1) \gamma_{ij}(\vec{V}_2) \int d\vec{e}\, \Theta(-\vec{V}_{12} \cdot \vec{e}) \left| \vec{V}_{12} \cdot \vec{e} \right| \\ &\quad \times \Delta \left[D_{ij}(\vec{V}_1) + D_{ij}(\vec{V}_2) \right]. \end{aligned} \tag{19.10}$$

Hence, with $\Delta \psi(\vec{v}_i) = \psi(\vec{v}_i') - \psi(\vec{v}_i)$ we obtain for the LHS of (19.6)

$$\begin{aligned} \int d\vec{V}_1 D_{ij}(\vec{V}_1) J^{(1)}(f_M, \gamma_{ij}) &= -\sigma^2 \int d\vec{V}_1 \int d\vec{V}_2 f_M(\vec{V}_1) \gamma_{ij}(\vec{V}_2) \\ &\quad \times \int d\vec{e}\, \Theta(-\vec{V}_{12} \cdot \vec{e}) \left| \vec{V}_{12} \cdot \vec{e} \right| \Delta \left[D_{ij}(\vec{V}_1) + D_{ij}(\vec{V}_2) \right]. \end{aligned} \tag{19.11}$$

---

[24] Recall that the collision integral (17.9) depends on $\vec{V}_1$, that is, we restore the subscript.

Using the dimensionless velocities $\vec{V} = v_T \vec{c}$, where $v_T = \sqrt{2T/m}$, we write the factors in the last integral as

$$D_{ij}(\vec{V}_1) = mv_T^2 \tilde{D}_{ij}(\vec{c}_1) = mv_T^2 \left(c_{1i}c_{1j} - \frac{1}{3}\delta_{ij}c_1^2 \right)$$
$$D_{ij}(\vec{V}_2) = mv_T^2 \tilde{D}_{ij}(\vec{c}_2) = mv_T^2 \left(c_{2i}c_{2j} - \frac{1}{3}\delta_{ij}c_2^2 \right) \quad (19.12)$$

and

$$\gamma_{ij}(\vec{V}_2) = \frac{\gamma_0}{T} \left(\frac{n}{v_T^3} \right) mv_T^2 \tilde{D}_{ij}(\vec{c}_2) \phi(c_2), \quad (19.13)$$

where $\phi(c)$ is the scaled Maxwell distribution, and recast (19.11) into the form

$$\int d\vec{V}_1 D_{ij}(\vec{V}_1) J^{(1)}(f_M, \gamma_{ij})$$
$$= -4\gamma_0 v_T n^2 T\sigma^2 \int d\vec{c}_1 \int d\vec{c}_2 \int d\vec{e}\, \Theta(-\vec{c}_{12}\cdot\vec{e}) |\vec{c}_{12}\cdot\vec{e}| \phi(c_1)\phi(c_2)$$
$$\times \tilde{D}_{ij}(\vec{c}_2) \Delta \left[\tilde{D}_{ij}(\vec{c}_1) + \tilde{D}_{ij}(\vec{c}_2) \right] \quad (19.14)$$
$$= -4\gamma_0 v_T n^2 T\sigma^2 \Omega_\eta^{\text{el}}.$$

Equation (19.14) defines the coefficient Ω_η^{el} which is a pure number. This coefficient has the structure of the kinetic integral (8.42) as introduced in Section 8.3.1. The coefficient Ω_η^{el} refers to the viscosity as indicated by the subscript η. The superscript 'el' indicates that elastic collisions are addressed. Equating the LHS of (19.6), given by (19.14), to its RHS, given by (19.8) we obtain

$$-4\gamma_0 v_T n^2 T\sigma^2 \Omega_\eta^{\text{el}} = -10nT, \quad \text{that is,} \quad \gamma_0^{-1} = \frac{2}{5} v_T n\sigma^2 \Omega_\eta^{\text{el}}. \quad (19.15)$$

With (19.8) and (19.15) the viscosity coefficient reads

$$\eta = -\frac{1}{10} \int d\vec{V}_1 D_{ij}(\vec{V}_1) \gamma_{ij}(\vec{V}_1) = -\frac{\gamma_0}{10} \frac{1}{T} \int d\vec{V}_1 D_{ij}(\vec{V}_1) D_{ij}(\vec{V}_1) f_M(V_1)$$
$$= -\gamma_0 nT = -\frac{5}{2\sigma^2} \sqrt{\frac{mT}{2}} \frac{1}{\Omega_\eta^{\text{el}}}. \quad (19.16)$$

To compute the numerical coefficient Ω_η^{el} we use (19.12) to express the term $\tilde{D}_{ij}(\vec{c}_2)\Delta \left[\tilde{D}_{ij}(\vec{c}_1) + \tilde{D}_{ij}(\vec{c}_2) \right]$ by

$$\left(c_{2i}c_{2j} - \frac{1}{3}c_2^2\delta_{ij} \right) \left[c'_{1i}c'_{1j} + c'_{2i}c'_{2j} - c_{1i}c_{1j} - c_{2i}c_{2j} \right.$$
$$\left. - \frac{1}{3}\left(c_1'^2 + c_2'^2 - c_1^2 - c_2^2 \right)\delta_{ij} \right]. \quad (19.17)$$

With the summation convention, for example, $c_{2i}c_{2j}c'_{1i}c'_{1j} = (\vec{c}'_1 \cdot \vec{c}_2)^2$, $c_{1i}c_{1j}\delta_{ij} = c_1^2$, etc. after some algebra we obtain

$$\tilde{D}_{ij}(\vec{c}_2) \Delta \left[\tilde{D}_{ij}(\vec{c}_1) + \tilde{D}_{ij}(\vec{c}_2) \right] = (\vec{c}'_1 \cdot \vec{c}_2)^2 + (\vec{c}'_2 \cdot \vec{c}_2)^2 - (\vec{c}_1 \cdot \vec{c}_2)^2 - (\vec{c}_2 \cdot \vec{c}_2)^2$$
$$- \frac{1}{3}c_2^2 \left(c'^2_1 + c'^2_2 - c_1^2 - c_2^2 \right).$$
(19.18)

The definition of Ω_η^{el} (19.14) then reads

$$\Omega_\eta^{\text{el}} = \int d\vec{c}_1 \int d\vec{c}_2 \int d\vec{e}\, \Theta(-\vec{c}_{12} \cdot \vec{e}) \, |\vec{c}_{12} \cdot \vec{e}| \, \phi(c_1)\phi(c_2)$$
$$\times \left[(\vec{c}'_1 \cdot \vec{c}_2)^2 + (\vec{c}'_2 \cdot \vec{c}_2)^2 - (\vec{c}_1 \cdot \vec{c}_2)^2 - (\vec{c}_2 \cdot \vec{c}_2)^2 - \frac{1}{3}c_2^2 \Delta \left(c_1^2 + c_2^2 \right) \right].$$
(19.19)

Later in this chapter, when dissipative gases will be addressed, similar integrals will occur. Therefore, we introduce a slightly more general notation:

$$\Omega_\eta [\varphi_1(c_1), \varphi_2(c_2)] \equiv \int d\vec{c}_1 \int d\vec{c}_2 \int d\vec{e}\, \Theta(-\vec{c}_{12} \cdot \vec{e}) \, |\vec{c}_{12} \cdot \vec{e}| \, \varphi_1(c_1)\varphi_2(c_2)$$
$$\times \left[(\vec{c}'_1 \cdot \vec{c}_2)^2 + (\vec{c}'_2 \cdot \vec{c}_2)^2 - (\vec{c}_1 \cdot \vec{c}_2)^2 - (\vec{c}_2 \cdot \vec{c}_2)^2 - \frac{1}{3}c_2^2 \Delta \left(c_1^2 + c_2^2 \right) \right], \quad (19.20)$$

which implies
$$\Omega_\eta^{\text{el}} = \Omega_\eta [\phi(c_1), \phi(c_2)]. \tag{19.21}$$

At this point we can apply computational formula manipulation to evaluate the kinetic integral (see below), however, since the kinetic integral appears here for the first time for the transport coefficients, we wish to show the details of the calculation.

The pre-collision velocities $\vec{c}_1, \vec{c}_2$ as well as after-collision velocities $\vec{c}'_1, \vec{c}'_2$ can be expressed in terms of the centre of mass velocity $\vec{C}$ and the relative velocity $\vec{c}_{12}$ before the collision using (8.24, 8.27):

$$\vec{c}_1 = \vec{C} + \frac{1}{2}\vec{c}_{12}, \qquad \vec{c}_2 = \vec{C} - \frac{1}{2}\vec{c}_{12}$$
$$\vec{c}'_1 = \vec{C} + \frac{1}{2}\vec{c}_{12} - (\vec{c}_{12} \cdot \vec{e})\vec{e} \qquad \vec{c}'_2 = \vec{C} - \frac{1}{2}\vec{c}_{12} + (\vec{c}_{12} \cdot \vec{e})\vec{e}.$$
(19.22)

Obviously, $\Delta \left(c_1^2 + c_2^2 \right) = 0$ due to energy conservation for elastic collisions. The factor in square brackets in (19.19) then reads

$$2 \left[(\vec{C} \cdot \vec{e})^2 (\vec{c}_{12} \cdot \vec{e})^2 - (\vec{C} \cdot \vec{e})(\vec{c}_{12} \cdot \vec{e})^3 + \frac{1}{4}(\vec{c}_{12} \cdot \vec{e})^4 + \frac{1}{2}(\vec{C} \cdot \vec{c}_{12})(\vec{c}_{12} \cdot \vec{e})^2 \right.$$
$$\left. -(\vec{C} \cdot \vec{c}_{12})(\vec{C} \cdot \vec{e})(\vec{c}_{12} \cdot \vec{e}) + \frac{1}{2}c_{12}^2(\vec{C} \cdot \vec{e})(\vec{c}_{12} \cdot \vec{e}) - \frac{1}{4}c_{12}^2(\vec{c}_{12} \cdot \vec{e})^2 \right]. \quad (19.23)$$

Substituting (19.23) into (19.19), a sum of integrals is obtained whose integrands contain the factors C^k, c_{12}^l, $(\vec{C}\cdot\vec{c}_{12})^m$, $(\vec{C}\cdot\vec{e})^n$, etc. These integrals have the structure of the basic integrals (8.30), introduced in Section 8.3.1 (see Appendix A). With the definition of the basic integrals, (8.31–8.33), and with the properties of the Γ-function $\Gamma(x+1) = x\Gamma(x)$ and $\Gamma(1/2) = \sqrt{\pi}$, we obtain

$$\Omega_\eta^{\text{el}} = 2\left(J_{000220} - J_{000130} + \frac{1}{4}J_{000040} - J_{001110} + \frac{1}{2}J_{001020}\right.$$
$$\left. + \frac{1}{2}J_{020110} - \frac{1}{4}J_{020020}\right) \quad (19.24)$$
$$= 2\sqrt{2\pi}\left(1 - 0 + \frac{1}{4}\cdot 16 - 1 + \frac{1}{2}\cdot 0 + \frac{1}{2}\cdot 0 - \frac{1}{4}\cdot 24\right)$$
$$= -4\sqrt{2\pi}\,.$$

Substituting (19.24) into (19.16) we arrive at the viscosity coefficient,

$$\eta = \frac{5}{16\sigma^2}\sqrt{\frac{mT}{\pi}}\,, \quad (19.25)$$

which is called *Enskog viscosity*. Excluded-volume effects may be taken into account by substituting $\sigma^2 \to \sigma^2 g_2(\sigma)$. The final expression for η is then

$$\boxed{\eta = \frac{5}{16\sigma^2 g_2(\sigma)}\sqrt{\frac{mT}{\pi}}} \quad (19.26)$$

The kinetic integral Ω_η^{el} may be also computed by applying Maple (for details see page 76 and Appendix A). According to the notations introduced in these sections, the core of the kinetic integral reads

$$\text{Expr}\,(\vec{c}_1', \vec{c}_2'\,\vec{c}_1, \vec{c}_2)$$
$$= \left[(\vec{c}_1'\cdot\vec{c}_2)^2 + (\vec{c}_2'\cdot\vec{c}_2)^2 - (\vec{c}_1\cdot\vec{c}_2)^2 - (\vec{c}_2\cdot\vec{c}_2)^2 - \frac{1}{3}c_2^2\Delta\left(c_1^2 + c_2^2\right)\right]. \quad (19.27)$$

Correspondingly, the body of the Maple program reads

```
1  DefDimension(3);
2  DefJ();
3  Expr :=(-1/DIM*c2pDOTc2p*Delta2+c1aDOTc2p*c1aDOTc2p+c2aDOTc2p*c2aDOTc2p
4         -c2pDOTc2p*c2pDOTc2p-c1pDOTc2p*c1pDOTc2p):
5  OmegaETA :=getJexpr(0,Expr,0);
```

For $\varepsilon = 1$ the program yields $\Omega_\eta^{\text{el}} = -4\sqrt{2\pi}$, which coincides with the result of the manual evaluation.

19.3 Coefficient of thermal conductivity

The vector $\vec{S}\left(\vec{V}\right)$, introduced in (17.35), may be expressed by

$$-\frac{1}{T}\vec{S}\left(\vec{V}\right) = \vec{V} S_1^{(3/2)}(c^2), \tag{19.28}$$

since $c^2 = V^2/v_T^2 = mV^2/2T$ and $S_1^{(3/2)}(x) = 5/2 - x$ according to the definition of the Sonine polynomials (7.14). Then with (19.1)

$$\vec{A} = \vec{V} S_1^{(3/2)}(c^2) f_M(V), \tag{19.29}$$

the equation for the coefficient $\vec{\alpha}\left(\vec{V}\right)$ reads (see (19.1, 19.2)):

$$J^{(1)}(f_M, \vec{\alpha}) = \vec{V}_1 S_1^{(3/2)}(c_1^2) f_M(V_1). \tag{19.30}$$

This suggests the solution for $\vec{\alpha}$ in the form[25]

$$\vec{\alpha} = \alpha_1 \vec{V} S_1^{(3/2)}(c^2) f_M(V) = -\frac{\alpha_1}{T} \vec{S}\left(\vec{V}\right) f_M(V), \tag{19.31}$$

which reduces (19.30) to an equation for the *scalar* coefficient α_1. We multiply (scalar product) both sides of (19.30) by $\vec{V}_1 S_1^{(3/2)}(c_1^2)$, integrate over $\vec{V}_1$ and obtain with (19.28)

$$-\frac{1}{T}\int d\vec{V}_1 \vec{S}\left(\vec{V}_1\right) \cdot J^{(1)}(f_M, \vec{\alpha}) = \int d\vec{V}_1 \vec{V}_1^2 \left(S_1^{(3/2)}(c_1^2)\right)^2 f_M(V_1). \tag{19.32}$$

Again we evaluate the LHS and the RHS of (19.32) separately. With the Maxwell distribution

$$f_M(V_1) = \left(\frac{n}{v_T^3}\right) \phi(c_1) = \left(\frac{n}{v_T^3}\right) \pi^{-3/2} \exp\left(-c_1^2\right) \tag{19.33}$$

and the properties of the Sonine polynomials (7.15), the RHS becomes

$$\int d\vec{V}_1 \vec{V}_1^2 \left(S_1^{(3/2)}(c_1^2)\right)^2 f_M(V_1)$$
$$= \left(\frac{n}{v_T^3}\right) v_T^3 v_T^2 4\pi \, \pi^{-3/2} \int_0^\infty c_1^{2\cdot\frac{3}{2}+1} \exp\left(-c_1^2\right) \left(S_1^{(3/2)}(c_1^2)\right)^2 dc_1 \tag{19.34}$$
$$= \frac{2}{\sqrt{\pi}} n v_T^2 \frac{\left(\frac{3}{2}+1\right)!}{1!} = \frac{2}{\sqrt{\pi}} n v_T^2 \Gamma\left(\frac{7}{2}\right) = \frac{15}{4} n v_T^2.$$

With the definition of $J^{(1)}$ (18.22) and the basic property of the collision integral (6.25), the LHS of (19.32) reads

---
[25] Only the first term of the expansion $\alpha = \alpha_1 S_1^{(3/2)}(c^2) + \alpha_2 S_2^{(3/2)}(c^2) + \cdots$ (Chapman and Cowling, 1970; Schram, 1991) is used, see also Exercise 19.2.

$$-\frac{1}{T}\int d\vec{V}_1 \vec{S}\left(\vec{V}_1\right) \cdot J^{(1)}\left(f_M, \vec{\alpha}\right)$$
$$=\frac{1}{T}\sigma^2 \int d\vec{V}_1 \int d\vec{V}_2 f_M(\vec{V}_1)\vec{\alpha}(\vec{V}_2) \cdot \int d\vec{e}\,\Theta(-\vec{V}_{12}\cdot\vec{e})\left|\vec{V}_{12}\cdot\vec{e}\right| \quad (19.35)$$
$$\times \Delta\left[\vec{S}\left(\vec{V}_1\right) + \vec{S}\left(\vec{V}_2\right)\right].$$

We switch to the dimensionless variables $\vec{\tilde{S}}$ and $\vec{c}$:

$$\vec{S}\left(\vec{V}\right) = v_T T \vec{\tilde{S}}\left(\vec{c}\right) = v_T T \left(c^2 - \frac{5}{2}\right)\vec{c},$$
$$\vec{\alpha}\left(\vec{V}\right) = -\alpha_1 v_T \left(\frac{n}{v_T^3}\right)\phi(c)\vec{\tilde{S}}\left(\vec{c}\right) \quad (19.36)$$

and recast the LHS of (19.32) into the form

$$-\frac{1}{T}\int d\vec{V}_1 \vec{S}\left(\vec{V}_1\right) \cdot J^{(1)}\left(f_M, \vec{\alpha}\right)$$
$$= -\alpha_1 n^2 \sigma^2 v_T^3 \int d\vec{c}_1 \int d\vec{c}_2 \int d\vec{e}\,\Theta(-\vec{c}_{12}\cdot\vec{e})\left|\vec{c}_{12}\cdot\vec{e}\right|\phi(c_1)\phi(c_2) \quad (19.37)$$
$$\times \vec{\tilde{S}}(\vec{c}_2) \cdot \Delta\left[\vec{\tilde{S}}(\vec{c}_1) + \vec{\tilde{S}}(\vec{c}_2)\right]$$
$$= -\alpha_1 n^2 \sigma^2 v_T^3 \Omega_\kappa^{\mathrm{el}},$$

where $\Omega_\kappa^{\mathrm{el}}$ is a pure number. Equating the RHS of (19.32) given by (19.34) with its LHS given by (19.37), we obtain

$$-\alpha_1 n^2 \sigma^2 v_T^3 \Omega_\kappa^{\mathrm{el}} = \frac{15}{4}nv_T^2, \quad \text{that is,} \quad \alpha_1^{-1} = -\frac{4}{15}n\sigma^2 v_T \Omega_\kappa^{\mathrm{el}}, \quad (19.38)$$

and eventually obtain for the thermal conductivity coefficient:

$$\kappa = -\frac{1}{3T}\int d\vec{V}_1 \vec{S}\left(\vec{V}_1\right) \cdot \vec{\alpha}\left(\vec{V}_1\right) = \frac{\alpha_1}{3}\int d\vec{V}_1 \vec{V}_1^2 \left(S_1^{(3/2)}(c_1^2)\right)^2 f_M(V_1)$$
$$= \frac{\alpha_1}{3}\frac{15}{4}nv_T^2 = -\frac{75}{16\sigma^2}\sqrt{\frac{2T}{m}}\frac{1}{\Omega_\kappa^{\mathrm{el}}} \quad (19.39)$$

where (18.45, 19.31, 19.34) have been employed. To obtain the numerical coefficient $\Omega_\kappa^{\mathrm{el}}$, we simplify

$$\Delta\left[\vec{\tilde{S}}(\vec{c}_1) + \vec{\tilde{S}}(\vec{c}_2)\right] = \Delta\left[\vec{c}_1\left(c_1^2 - \frac{5}{2}\right) + \vec{c}_2\left(c_2^2 - \frac{5}{2}\right)\right]$$
$$= \Delta\left(\vec{c}_1 c_1^2 + \vec{c}_2 c_2^2\right) - \frac{5}{2}\Delta\left(\vec{c}_1 + \vec{c}_2\right) = \Delta\left(\vec{c}_1 c_1^2 + \vec{c}_2 c_2^2\right), \quad (19.40)$$

where $\Delta(\vec{c}_1 + \vec{c}_2) = 0$ due to the conservation of momentum. Therefore,

$$\vec{\vec{S}}(\vec{c}_2) \cdot \Delta\left[\vec{\vec{S}}(\vec{c}_1) + \vec{\vec{S}}(\vec{c}_2)\right]$$
$$= \left(c_2^2 - \frac{5}{2}\right)\left[(\vec{c}_1' \cdot \vec{c}_2)(c_1')^2 + (\vec{c}_2' \cdot \vec{c}_2)(c_2')^2 - (\vec{c}_1 \cdot \vec{c}_2)c_1^2 - (\vec{c}_2 \cdot \vec{c}_2)c_2^2\right]. \quad (19.41)$$

Again, we introduce a more general notation

$$\Omega_\kappa[\varphi_1(c_1), \varphi_2(c_2)] \equiv \int d\vec{c}_1 \int d\vec{c}_2 \int d\vec{e}\, \Theta(-\vec{c}_{12} \cdot \vec{e})|\vec{c}_{12} \cdot \vec{e}|\varphi_1(c_1)\varphi_2(c_2)$$
$$\times \left(c_2^2 - \frac{5}{2}\right)\left[(\vec{c}_1' \cdot \vec{c}_2)(c_1')^2 + (\vec{c}_2' \cdot \vec{c}_2)(c_2')^2 - (\vec{c}_1 \cdot \vec{c}_2)c_1^2 - (\vec{c}_2 \cdot \vec{c}_2)c_2^2\right], \quad (19.42)$$

hence,

$$\Omega_\kappa^{el} = \Omega_\kappa[\phi(c_1), \phi(c_2)]. \quad (19.43)$$

The coefficient Ω_κ^{el} has the form of the kinetic integral with the core

$$\text{Expr}(\vec{c}_1', \vec{c}_2'\, \vec{c}_1, \vec{c}_2) =$$
$$= \left(c_2^2 - \frac{5}{2}\right)\left[(\vec{c}_1' \cdot \vec{c}_2)(c_1')^2 + (\vec{c}_2' \cdot \vec{c}_2)(c_2')^2 - (\vec{c}_1 \cdot \vec{c}_2)c_1^2 - (\vec{c}_2 \cdot \vec{c}_2)c_2^2\right], \quad (19.44)$$

which may be evaluated using the Maple program as described in Appendix A. The corresponding program is

```
1 DefDimension(3);
2 DefJ();
3 Expr :=(c2pDOTc2p-(DIM+2)/2)*(c1aDOTc2p*c1aDOTc1a+c2aDOTc2p*c2aDOTc2a
4          -c2pDOTc2p*c2pDOTc2p-c1pDOTc2p*c1pDOTc1p):
5 OmegaKAPPA:=getJexpr(0,Expr,0);
```

With $\Omega_\kappa^{el} = -4\sqrt{2\pi}$ obtained by the Maple program, we finally arrive at the *Enskog thermal conductivity*:

$$\kappa = \frac{75}{64\sigma^2}\sqrt{\frac{T}{\pi m}}. \quad (19.45)$$

Taking into account the finite-volume effects by substituting $\sigma^2 \to \sigma^2 g_2(\sigma)$ we obtain the corresponding generalization of (19.45):

$$\boxed{\kappa = \frac{75}{64\sigma^2 g_2(\sigma)}\sqrt{\frac{T}{\pi m}}} \quad (19.46)$$

Exercise 19.2 *In general, the coefficient $\vec{\alpha}$ is expressed as an expansion,*

$$\vec{\alpha}\left(\vec{V}\right) = \vec{V}\left[\alpha_0 + \alpha_1 S_1^{(3/2)}(c^2) + \alpha_2 S_2^{(3/2)}(c^2) + \cdots\right] f_M(V) \tag{19.47}$$

with unknown scalar coefficients α_k. In this series the coefficient α_0 is always trivial, that is, $\alpha_0 = 0$. Explain why!

Exercise 19.3 *Compute the numerical coefficient $\Omega_\kappa^{\text{el}}$ manually and confirm the result using Maple!*

19.4 Velocity distribution function of an inhomogeneous gas of elastic particles

Equations (19.16, 19.39) relate the coefficients α_1 and γ_0 with the kinetic coefficients κ and η:

$$\alpha_1 = \frac{2m}{5nT}\kappa, \qquad \gamma_0 = -\frac{1}{nT}\eta. \tag{19.48}$$

These coefficients determine the vectorial coefficient $\vec{\alpha}$ and the tensorial coefficient γ_{ij} in the lowest-order approximation (see (19.31, 19.5)) and, hence, $f^{(1)}$ via (18.29). The function $f^{(1)}$ is the first-order correction (linearly in the field gradients) to the solution in zeroth order, $f^{(0)}$, which for a gas of elastic particles is the Maxwell distribution, $f^{(0)} = f_M(V)$. With (18.29, 19.31, 19.5, 19.48) the velocity distribution function for a non-uniform gas of elastic particles is given by

$$\boxed{f(V) = f_M(V)\left(1 - \frac{2m\kappa}{5nT^3}\vec{S}(\vec{V}) \cdot \vec{\nabla} T - \frac{\eta}{nT^2} D_{ij}(\vec{V})\nabla_j u_i\right)} \tag{19.49}$$

Equation (19.49) describes the velocity distribution function up to the linear order with respect to the field gradients. To obtain the next-order terms, one can follow the same scheme as has been explained in this section.

20

KINETIC COEFFICIENTS FOR GRANULAR GASES OF SIMPLIFIED PARTICLES (ε =CONST)

The Chapman–Enskog approach for the transport coefficients is generalized for the case of granular gases whose particles collide inelastically. Under the assumption of a constant coefficient of restitution we derive the kinetic coefficients and the velocity distribution function.

20.1 Viscosity coefficient

For the derivation of the viscosity coefficient we follow the same scheme as in the preceding chapter for gases of elastic particles. The main differences from the previous case are the non-vanishing cooling rate ζ and the non-Maxwellian zeroth-order velocity distribution function $f^{(0)}$. The distribution function for the homogeneous cooling state was derived in Chapter 8:

$$f^{(0)}(V) = f_M(V) \left[1 + a_2 \left(\frac{1}{2}c^4 - \frac{5}{2}c^2 + \frac{15}{8} \right) \right] = \left(\frac{n}{v_T^3} \right) \left[1 + a_2 S_2(c^2) \right] \phi(c), \tag{20.1}$$

where $\vec{c} \equiv \vec{V}/v_T$ with $v_T = \sqrt{2T/m}$, $f_M(V)$ is the Maxwell distribution (19.33), a_2 is the first non-vanishing coefficient of the Sonine polynomial expansion. The second Sonine polynomial is given by (7.7).

The viscosity coefficient η may be expressed via the coefficient γ_{ij} of the first-order function $f^{(1)}$ (see (18.40)). In its turn, this coefficient is a solution of (18.34):

$$-\zeta^{(0)} T \frac{\partial \gamma_{ij}}{\partial T} + J^{(1)}\left(f^{(0)}, \gamma_{ij}\right) = C_{ij} \tag{20.2}$$

with the (known) tensor C_{ij}. According to the definition (18.26) of C_{ij}, we need the derivative $\partial f^{(0)}/\partial V$ and apply the relations

$$\frac{1}{V}\frac{\partial f_M}{\partial V} = -\frac{m}{T} f_M, \qquad \frac{1}{V}\frac{\partial c}{\partial V} = \frac{m}{T}\frac{1}{2c}. \tag{20.3}$$

Exercise 20.1 *Prove (20.3)!*

Using (20.3) we write

$$\begin{aligned}\frac{1}{V}\frac{\partial f^{(0)}}{\partial V} &= -\frac{m}{T}\left[1+a_2 S_2(c^2)\right] f_M + \frac{m}{T}\frac{1}{2c} a_2 \left(2c^3 - 5c\right) f_M \\ &= -\frac{m}{T}\left[1+a_2 S_2^{(3/2)}(c^2)\right] f_M,\end{aligned} \tag{20.4}$$

195

where $S_2^{(3/2)}(c^2) = c^4/2 - 7c^2/2 + 35/8$ according to the definition of the Sonine polynomials (7.14).[26] Substituting (20.4) into the definition of C_{ij} (18.26), we obtain

$$C_{ij} = -\frac{1}{T} D_{ij}(\vec{V}) \left[1 + a_2 S_2^{(3/2)}(c^2)\right] f_M(V), \quad (20.5)$$

which resembles the expression (19.1) for elastic particles. As in the case of elastic particles we assume the simplest form of the coefficient γ_{ij}, which follows from the structure of the equation for this coefficient, that is,

$$\gamma_{ij}(\vec{V}) = \frac{\gamma_0}{T} D_{ij}(\vec{V}) f_M(V). \quad (20.6)$$

As previously, we multiply (20.2) by $D_{ij}(\vec{V}_1)$ and integrate over $\vec{V}_1$. Using (18.40), which relates γ_{ij} and η, we obtain

$$-10\zeta^{(0)} T \frac{\partial \eta}{\partial T} = \int d\vec{V}_1 D_{ij}(\vec{V}_1) J^{(1)}\left(f^{(0)}, \gamma_{ij}\right) - \int d\vec{V}_1 D_{ij}(\vec{V}_1) C_{ij}(\vec{V}_1). \quad (20.7)$$

The first term in the RHS differs from the corresponding term for the case of elastic particles only with respect to the substitution of the function $f_M(V)$ by $f^{(0)}$ (recall that $f^{(0)} = f_M(V)$ for elastic particles). Therefore, we can process this term in the same way as for elastic particles:

$$\int d\vec{V}_1 D_{ij}(\vec{V}_1) J^{(1)}\left(f^{(0)}, \gamma_{ij}\right) = -4\gamma_0 v_T n^2 T \sigma^2 \Omega_\eta^\varepsilon = 4\eta n \sigma^2 \sqrt{\frac{2T}{m}} \Omega_\eta^\varepsilon, \quad (20.8)$$

where we substitute $\gamma_0 = -\eta/(nT)$ due to (19.48), which is valid as long as the approximation (19.5) is used. With the notations as introduced in the previous section (see (19.20)) the coefficient Ω_η^ε reads

$$\Omega_\eta^\varepsilon = \Omega_\eta \left[\tilde{f}^{(0)}(c_1), \phi(c_2)\right] = \Omega_\eta \left[(1 + a_2 S_2(c_1^2)) \phi(c_1), \phi(c_2)\right]. \quad (20.9)$$

The second term of the RHS of (20.7) is evaluated in the same way as for elastic particles and yields $10nT$.

Exercise 20.2 *Prove that the second term in the RHS of (20.7) equals $10nT$!*

Thus, the equation for the viscosity coefficient, (20.7), becomes

$$-\zeta^{(0)} T \frac{\partial \eta}{\partial T} - \frac{2}{5} \eta n \sigma^2 \sqrt{\frac{2T}{m}} \Omega_\eta^\varepsilon = nT. \quad (20.10)$$

The classical result by Jenkins and Richman (1985) is obtained with the approximation $\zeta^{(0)} = 0$. Then this equation may be solved and yields

$$\eta = -\frac{5}{2\sigma^2} \sqrt{\frac{mT}{2}} \frac{1}{\Omega_\eta^\varepsilon}, \quad (20.11)$$

which resembles the equivalent expression (19.16) for a gas of elastic particles.

The coefficient Ω_η^ε is determined using Maple:

---

[26] In Chapter 7 we have used Sonine polynomials with the upper index $1/2$. In this book, for simplicity, of the notation we omit the upper index if it is $1/2$.

```
1  DefDimension(3);
2  DefJ();
3  Expr :=(-1/DIM*c2pDOTc2p*Delta2+c1aDOTc2p*c1aDOTc2p+c2aDOTc2p*c2aDOTc2p
4          -c2pDOTc2p*c2pDOTc2p-c1pDOTc2p*c1pDOTc2p)*(1+a2*S(2,c1p)):
5  OmegaETA:=getJexpr(0,Expr,0);
```

with the result

$$\Omega_\eta^\varepsilon = -\sqrt{2\pi}\,(1+\varepsilon)(3-\varepsilon)\left(1-\frac{a_2}{32}\right). \tag{20.12}$$

Further, we approximate $a_2 = 0$, that is, we assume a Maxwell distribution, substitute Ω_η^ε from (20.12) into (20.11), take excluded volume effects into account by $\sigma^2 \to \sigma^2 g_2(\sigma)$ and obtain the classical result (Jenkins and Richman, 1985):

$$\eta = \frac{5}{4(1+\varepsilon)(3-\varepsilon)\sigma^2 g_2(\sigma)}\sqrt{\frac{mT}{\pi}}. \tag{20.13}$$

To go beyond this approximation we investigate the scaling of the terms of (20.10). Obviously, the second term on the LHS scales $\propto \eta\sqrt{T}$. The first term also scales $\propto \eta\sqrt{T}$, since $\zeta^{(0)} \propto \sqrt{T}$, according to (18.11) and, hence, $\zeta^{(0)} T \partial/\partial T \propto \sqrt{T}$. The RHS of (20.10) scales $\propto T$. Thus, we conclude $\eta \propto \sqrt{T}$, which implies

$$T\frac{\partial \eta}{\partial T} = \frac{1}{2}\eta. \tag{20.14}$$

Equation (20.10) for η turns then into

$$\left(\frac{1}{2}\zeta^{(0)} + \frac{2}{5}n\sigma^2 g_2(\sigma)\sqrt{\frac{2T}{m}}\Omega_\eta^\varepsilon\right)\eta = -nT. \tag{20.15}$$

Using (18.11) for the cooling coefficient $\zeta^{(0)}$ we obtain the viscosity coefficient for a granular gas with $\varepsilon = $ const. as

$$\eta = -\frac{1}{\sigma^2 g_2(\sigma)}\sqrt{\frac{mT}{2}}\frac{1}{\frac{1}{3}\mu_2 + \frac{2}{5}\Omega_\eta^\varepsilon}. \tag{20.16}$$

The moment of the collision integral μ_2 is given by (8.37) in linear approximation with respect to the second Sonine coefficient a_2:

$$\mu_2 = \sqrt{2\pi}\,(1-\varepsilon^2)\left(1+\frac{3}{16}a_2\right). \tag{20.17}$$

Using (20.12) for Ω_η^ε we obtain the viscosity coefficient in linear approximation with respect to a_2:

$$\eta = \frac{15}{2(1+\varepsilon)(13-\varepsilon)\sigma^2 g_2(\sigma)}\sqrt{\frac{mT}{\pi}}\left(1+\frac{3}{8}\frac{(4-3\varepsilon)}{(13-\varepsilon)}a_2\right) \tag{20.18}$$

We wish to remind that applying the Ansatz (20.6) we neglect the dependence of the coefficient γ_{ij} and, thus, of $f^{(1)}$ on a_2, that is, we neglect the terms

$\mathcal{O}(a_2 \nabla_j u_i)$, which are of second order with respect to the small values a_2 and $\nabla_j u_i$.

Exercise 20.3 *The relation $\gamma_0 = -\eta/(nT)$ has been obtained for gases of elastic particles. For which preconditions is it also valid for granular gases with $\varepsilon =$ const.?*

Exercise 20.4 *The viscosity coefficient (20.18) has been obtained in linear approximation with respect to a_2. How does one derive the viscosity coefficient in second order? Is it enough just to use (8.45) for μ_2, which is correct up to $\mathcal{O}\left(a_2^2\right)$?*

20.2 Thermal conductivity coefficient κ and transport coefficient μ

To compute the transport coefficients κ and μ the vectorial coefficients $\vec{\alpha}$ and $\vec{\beta}$ are required (see (18.45)). These coefficients are the solutions of (18.34), where the RHS contains (known) vectorial functions $\vec{A}$ and $\vec{B}$, defined by (18.26). They may be straightforwardly evaluated, for example,

$$\vec{B} = -\vec{V}\left\{-\frac{T}{m}\frac{m}{T}\left[1 + a_2 S_2(c^2) + a_2 S_1^{(3/2)}(c^2)\right] + \left[1 + a_2 S_2(c^2)\right]\right\} f_M(V)$$
$$= a_2 \vec{V} S_1^{(3/2)}(c^2) f_M(V) \,, \tag{20.19}$$

where the derivative of the distribution function (20.4) was used. The other vectorial coefficient reads

$$\vec{A} = \vec{V} S_1^{(3/2)}(c^2)\left[1 + a_2\left(S_2^{(3/2)}(c^2) - 3/2\right)\right] f_M(V) \,. \tag{20.20}$$

Exercise 20.5 *Prove (20.20) for $\vec{A}$ by means of (18.26) and (20.4)!*

Consider now (18.34) for $\vec{\alpha}$,

$$-\zeta^{(0)} T \frac{\partial \vec{\alpha}}{\partial T} - \frac{1}{2}\zeta^{(0)}\vec{\alpha} + J^{(1)}\left(f^{(0)}, \vec{\alpha}\right) = \vec{A}. \tag{20.21}$$

The structure of the RHS of (20.21) given by (20.20) suggests the same form of the vectorial coefficient $\vec{\alpha}$ as for the gas of elastic particles,

$$\vec{\alpha} = \alpha_1 \vec{V} S_1^{(3/2)}\left(c^2\right) f_M(V) = -\frac{\alpha_1}{T}\vec{S}\left(\vec{V}\right) f_M(V), \tag{20.22}$$

which implies that in the distribution function we neglect terms $\mathcal{O}(a_2 \nabla T)$. Following the same scheme as for the gas of elastic particles, we multiply both sides of (20.21) by $\vec{S}(\vec{V}_1)/T$ and integrate over $\vec{V}_1$. Then, with the definition of κ (18.45), we obtain

$$3\zeta^{(0)}\frac{\partial}{\partial T}\kappa T + \frac{3}{2}\zeta^{(0)}\kappa = -\frac{1}{T}\int d\vec{V}_1 \vec{S}\left(\vec{V}_1\right) \cdot J^{(1)}\left(f^{(0)}, \vec{\alpha}\right)$$
$$+ \frac{1}{T}\int d\vec{V}_1 \vec{S}\left(\vec{V}_1\right) \cdot \vec{A}\left(\vec{V}_1\right) \,. \tag{20.23}$$

The first term of the RHS is identical to the analogous expression for the case of elastic particles (19.37), except for $f_M(V) \to f^{(0)}$. Hence, it is evaluated in the same way as before and reads

$$-\frac{1}{T}\int d\vec{V_1}\vec{S}\left(\vec{V_1}\right) J^{(1)}\left(f^{(0)}, \vec{a}\right) = -\alpha_1 n^2 \sigma^2 v_T^3 \Omega_\kappa^\varepsilon = -\frac{4}{5}\kappa n \sigma^2 \sqrt{\frac{2T}{m}} \Omega_\kappa^\varepsilon. \quad (20.24)$$

We have used $\alpha_1 = 2m\kappa/5nT$ (see (19.48)) and according to our notation convention (19.42) the numerical coefficient

$$\Omega_\kappa^\varepsilon = \Omega_\kappa^\varepsilon\left[\left(1 + a_2 S_2(c_1^2)\right)\phi(c_1), \phi(c_2)\right]. \quad (20.25)$$

has been defined. The second term in the RHS of (20.23) is evaluated again in the same way as for the elastic case, yielding

$$\frac{1}{T}\int d\vec{V_1}\vec{S}\left(\vec{V_1}\right)\cdot\vec{A}\left(\vec{V_1}\right) = -\frac{15}{4}nv_T^2\left(1 + 2a_2\right). \quad (20.26)$$

Exercise 20.6 *Prove (20.26)!*

With these expressions for the terms of the RHS of (20.23), we recast this equation for κ into the form

$$3\zeta^{(0)}\frac{\partial}{\partial T}\kappa T + \frac{3}{2}\zeta^{(0)}\kappa + \frac{4}{5}\kappa n\sigma^2\sqrt{\frac{2T}{m}}\Omega_\kappa^\varepsilon = -\frac{15}{2}\frac{nT}{m}\left(1 + 2a_2\right). \quad (20.27)$$

With the approximation $\zeta^{(0)} = 0$ and $a_2 = 0$:

$$\kappa = -\frac{75}{16\sigma^2}\sqrt{\frac{2T}{m}}\frac{1}{\Omega_\kappa^\varepsilon}. \quad (20.28)$$

We use Maple to compute the numerical coefficient $\Omega_\kappa^\varepsilon$:

```
1  DefDimension(3);
2  DefJ();
3  Expr:=(c2pDOTc2p-(DIM+2)/2)*(c1aDOTc2p*c1aDOTc1a+c2aDOTc2p*c2aDOTc2a
4          -c2pDOTc2p*c2pDOTc2p-c1pDOTc2p*c1pDOTc1p)*(1+a2*S(2,c1p)):
5  OmegaKAPPA:=getJexpr(0,Expr,0);
```

and obtain

$$\Omega_\kappa^\varepsilon = -\sqrt{2\pi}(1+\varepsilon)\left(\frac{49 - 33\varepsilon}{8} + \frac{19 - 3\varepsilon}{256}a_2\right). \quad (20.29)$$

Again, the classical result by Jenkins and Richman (1985) for the coefficient of the thermal conductivity is reproduced if we approximate $\zeta^{(0)} = 0$ and $a_2 = 0$ and take into account excluded volume effects by $\sigma^2 \to \sigma^2 g_2(\sigma)$:

$$\kappa = \frac{75}{2(1+\varepsilon)(49 - 33\varepsilon)\sigma^2 g_2(\sigma)}\sqrt{\frac{T}{\pi m}}. \quad (20.30)$$

The same dimension analysis as performed for the coefficient of viscosity on page 197 applied to (20.27), shows for the thermal conductivity $\kappa \propto \sqrt{T}$. This implies

$$\frac{\partial}{\partial T} \kappa T = \kappa \frac{\partial}{\partial T} T + T \frac{\partial}{\partial T} \kappa = \kappa + \frac{1}{2}\kappa = \frac{3}{2}\kappa. \qquad (20.31)$$

Substituting (20.31) and (18.11) for $\zeta^{(0)}$ into (20.27) yields,

$$\kappa = -\frac{15}{8\sigma^2 g_2(\sigma)} \sqrt{\frac{T}{2m}} \frac{1 + 2a_2}{\mu_2 + \frac{1}{5}\Omega_\kappa^\varepsilon}. \qquad (20.32)$$

With $\Omega_\kappa^\varepsilon$ and μ_2 as given by (20.29) and (20.17), respectively, we obtain the coefficient of thermal conductivity in linear approximation with respect to a_2:

$$\kappa = \frac{75}{2(1+\varepsilon)(9+7\varepsilon)\sigma^2 g_2(\sigma)} \sqrt{\frac{T}{\pi m}} \left(1 + \frac{1}{32} \frac{797 + 211\varepsilon}{9 + 7\varepsilon} a_2\right) \qquad (20.33)$$

Exercise 20.7 *Derive the complete dependence $\kappa(a_2)$, neglecting the terms of order $\mathcal{O}(a_2 \nabla T)$!*

Equation (18.34) for the coefficient $\vec{\beta}$ reads

$$-\zeta^{(0)} T \frac{\partial \vec{\beta}}{\partial T} - \zeta^{(0)} \vec{\alpha} + J^{(1)}\left(f^{(0)}, \vec{\beta}\right) = \vec{B}, \qquad (20.34)$$

with the coefficient $\vec{B}$ given by (20.19). This equation is very similar to (20.21) for the coefficient $\vec{\alpha}$. Thus $\vec{\beta}$ has the similar structure as $\vec{\alpha}$, i.e. (see (20.22)),

$$\vec{\beta} = -\frac{\beta_1}{T} \vec{S}(\vec{V}) f_M(V) = \frac{\beta_1}{\alpha_1} \vec{\alpha}. \qquad (20.35)$$

According to (18.45) and (19.48) (which relates κ and α_1), this implies

$$\mu = \frac{T}{n}\left(\frac{\beta_1}{\alpha_1}\right)\kappa, \quad \text{that is,} \quad \mu = \frac{5T^2}{2m}\beta_1. \qquad (20.36)$$

The calculation of the coefficient μ is perfectly analogous to the coefficient κ.

Exercise 20.8 *Calculate the coefficient μ, following the same procedure as for the coefficient of thermal conductivity!*

In linear approximation with respect to a_2 the result is

$$\mu = \frac{750(1-\varepsilon)}{(1+\varepsilon)(9+7\varepsilon)(19-3\varepsilon)n\sigma^2 g_2(\sigma)} \sqrt{\frac{T^3}{\pi m}}$$
$$\times \left(1 + \frac{50201 - 30971\varepsilon - 7253\varepsilon^2 + 4407\varepsilon^3}{80(1-\varepsilon)(19-3\varepsilon)(9+7\varepsilon)} a_2\right) \qquad (20.37)$$

Equations (20.18), (20.33) and (20.37) describe the set of transport coefficients of a granular gas of particles that interact by a constant coefficient of restitution in linear approximation with respect to the second Sonine coefficient. These expressions can be used to derive the velocity distribution function for an inhomogeneous granular gas.

20.3 Velocity distribution function

In the same way as for a gas of elastic particles we express the coefficients α_1, β_1, γ_0 in terms of the transport coefficients η, κ, μ by (19.48, 20.36):

$$\alpha_1 = \frac{2m}{5nT}\kappa, \qquad \beta_1 = \frac{2m}{5T^2}\mu, \qquad \gamma_0 = -\frac{1}{nT}\eta. \qquad (20.38)$$

Then with (18.29, 19.5, 19.31, 20.35) we obtain the velocity distribution function:

$$f(V) = f^{(0)}(V) + f^{(1)}(\vec{V})$$

$$f^{(0)}(V) = \left[1 + a_2 S_2\left(\frac{mV^2}{2T}\right)\right] f_M(V) \qquad (20.39)$$

$$f^{(1)}(\vec{V}) = -\frac{1}{nT^2}\left[\frac{2m}{5T}\vec{S}(\vec{V})\cdot\left(\kappa\vec{\nabla}T + \mu\vec{\nabla}n\right) + \eta D_{ij}(\vec{V})\nabla_j u_i\right] f_M(V)$$

where

$$S_2(x) \equiv S_2^{(1/2)}(x) = \frac{1}{2}x^2 - \frac{5}{2}x + \frac{15}{8} \qquad (20.40)$$

is the second Sonine polynomial and the coefficients a_2, κ, μ and η are given, respectively, by (8.41), (20.33), (20.37) and (20.18). Relation (20.39) gives the velocity distribution function in linear approximation with respect to the gradients of the hydrodynamic fields and with respect to the second Sonine coefficient a_2. To go beyond this approximations one can keep the next-order terms and follow the same scheme as above.

Exercise 20.9 *Check the normalization of the distribution function $f(\vec{V})$ given by (20.39)!*

21
KINETIC COEFFICIENTS FOR GRANULAR GASES OF VISCOELASTIC PARTICLES

The kinetic coefficients for gases of viscoelastic particles reveal a much more complicated time and temperature dependence than for the case $\varepsilon =$ const. In particular, for such gases the transport coefficients and the cooling rate have different temperature dependences.

21.1 Chapman–Enskog approach for gases of viscoelastic particles

The Chapman–Enskog method, which has been applied in the preceding chapters to gases of elastic particles and to gases of particles that interact by a constant coefficient of restitution, was based on an important precondition: it was assumed that the velocity distribution function depends on time and space via its first three moments only, that is, on density (zeroth moment), average velocity (first moment) and temperature (second moment). Thus, the evolution of the velocity distribution function follows the evolution of these three fields, which obey the hydrodynamic equations. For the case of a constant coefficient of restitution the shape of the velocity distribution function does not depend on time.

The situation changes when we consider granular gases of viscoelastic particles: in Chapter 9 we have shown that the Sonine coefficients that describe the shape of the velocity distribution function depend explicitly on time. They characterize the moments of the velocity distribution function: a_2 characterizes the fourth moment, a_3 the sixth moment, etc. (see (7.19)). Hence, the first three moments are not sufficient to describe the evolution of $f(v,t)$ for a granular gas of viscoelastic particles.

The kinetic coefficients in the hydrodynamic equations are determined, in their turn, by the distribution function. Therefore, the conventional set of hydrodynamic equations for n, $\vec{u}$ and T is incomplete for these systems and should be supplemented by hydrodynamic equations for the higher-order moments.

A method to construct hydrodynamic equations for the moments of the velocity distribution function of escalating order has been developed by Grad (1960). Initially it was applied for gases of elastic particles, but later also for granular gases with $\varepsilon =$ const. (Jenkins and Richman, 1985; Sela and Goldhirsch, 1998; Ramírez and Cordero, 1999; Ramírez et al., 2000). Grad's method is, therefore, suited to describe an inhomogeneous granular gas of viscoelastic particles. This technically rather complicated method is, however, beyond the scope of this book and we address the interested reader to the original papers.

From the analysis in Chapter 9 for a granular gas of viscoelastic particles in the homogeneous cooling state it follows that the evolution of temperature $T(t)$ and the evolution of the Sonine coefficients $a_i(t)$ are coupled. For the inhomogeneous case there must be an equivalent coupling of the corresponding hydrodynamic equations, which complicates the problem drastically. To simplify the analysis, as in previous chapters we assume that the dissipation is small enough to neglect all Sonine coefficients except for a_2. Moreover, we assume that the characteristic times of the hydrodynamic fields are well separated from the microscopic characteristic time.

The evolution of the coefficient $a_2(t)$, starting with $a_2(0) = 0$ (Maxwell distribution) passes two different regimes: the first takes place on the collision time scale, while the second on the time scale of the temperature evolution (for a detailed discussion see Chapter 9). Since the characteristic times of the hydrodynamic fields exceed significantly the microscopic mean collision time, we can neglect the first relaxation regime of a_2 and assume an adiabatic approximation for this coefficient. The adiabatic approximation implies that the value of a_2 is determined only by the current local temperature. With these preconditions there is a relation (8.20) between the moments of the collision integral, μ_2 and μ_4 and the second Sonine coefficient a_2:

$$5\mu_2(1 + a_2) - \mu_4 = 0, \tag{21.1}$$

where the moments μ_2 and μ_4 are functions of temperature and of a_2, according to (9.20, 9.21). The adiabatic condition (21.1) is obtained from (9.27) for $a_2(t)$ omitting the time derivative da_2/dt, which describes the fast relaxation of a_2 to its adiabatic value.[27] Solving (21.1) with (9.13, 9.23, 9.24), we obtain a_2 as a function of temperature:

$$\begin{aligned}a_2 &= a_{21}\delta' + a_{22}\delta'^2 + \cdots \\ a_{21} &= -\frac{3\omega_0}{20\sqrt{2\pi}} \approx -0.388 \\ a_{22} &= \frac{12063}{640000}\frac{\omega_0^2}{\pi} + \frac{27}{40}\frac{\omega_1}{\sqrt{2\pi}} \approx 2.752,\end{aligned} \tag{21.2}$$

where $\delta' \equiv \delta(2T/T_0)^{1/10}$.

Thus, in adiabatic approximation we can still apply the Chapman–Enskog approach since, as previously, the velocity distribution function depends on time only via its first three moments. Its dependence on temperature, however, becomes more complicated, because according to (21.2) the Sonine coefficient a_2 depends now on temperature.

---

[27] More rigorously, (21.1) is valid if $a_2(t)$ decays as a power law, $a_2 \propto t^{-\nu}$ with $\nu < 1$. Then for large t one can neglect da_2/dt as compared with a_2. For the homogeneous cooling state this condition holds true for the second stage of the evolution of a_2.

For a gas of viscoelastic particles a_2 is a function of temperature, given (in adiabatic approximation) by (21.2). The temperature derivative of δ' is given by

$$T\frac{\partial \delta'}{\partial T} = \frac{1}{10}\delta', \qquad T\frac{\partial \delta'^2}{\partial T} = \frac{1}{5}\delta'^2 \qquad (21.3)$$

and, thus, the temperature derivative of the second Sonine coefficient reads

$$T\frac{\partial a_2}{\partial T} = \frac{1}{10}a_{21}\delta' + \frac{1}{5}a_{22}\delta'^2. \qquad (21.4)$$

The calculation of the transport coefficients follows the same procedure as for the case $\varepsilon = \text{const}$. The only difference is that the temperature derivatives contain now an additional term due to the dependence of a_2 on T. In particular, we obtain for the cooling rate in zeroth order, $\zeta^{(0)}$, given by (18.11):

$$\zeta^{(0)} = \frac{2}{3}n\sigma^2 g_2(\sigma)\sqrt{\frac{2T}{m}}\mu_2 = \frac{2}{3}n\sigma^2 g_2(\sigma)\sqrt{\frac{2T}{m}}\left(\omega_0\delta' - \omega_2\delta'^2 + \cdots\right), \qquad (21.5)$$

where we express μ_2 in terms of a_2 and δ' (9.20) and subsequently substitute a_2 by the adiabatic approximation (21.2). The coefficient ω_2 in (21.5) is obtained by collecting terms of the order $\mathcal{O}\left(\delta'^2\right)$ and using (9.23) for the numerical coefficients $\mathcal{A}_i$ in (9.20) for μ_2:

$$\omega_2 = \omega_1 + \frac{9}{500}\frac{\omega_0^2\sqrt{2\pi}}{\pi}. \qquad (21.6)$$

Hence, the derivatives of the cooling rate now read (instead of (18.12)):

$$\frac{\partial \zeta^{(0)}}{\partial n} = \frac{1}{n}\zeta^{(0)}$$

$$T\frac{\partial \zeta^{(0)}}{\partial T} = \frac{1}{2}\zeta^{(0)} + \frac{2}{3}v_T n\sigma^2 g_2(\sigma)\left(\frac{\omega_0}{10}\delta' + \frac{\omega_2}{5}\delta'^2 + \cdots\right). \qquad (21.7)$$

The equation for the first-order solution has the same form as (18.25) with the coefficients

$$\vec{A}(\vec{V}) = -\vec{V}T\frac{\partial}{\partial T}f^{(0)} - \frac{T}{m}\frac{\partial}{\partial \vec{V}}f^{(0)}$$

$$\vec{B}(\vec{V}) = -\vec{V}f^{(0)} - \frac{T}{m}\frac{\partial}{\partial \vec{V}}f^{(0)} \qquad (21.8)$$

$$C_{ij}(\vec{V}) = \frac{\partial}{\partial V_i}\left(V_j f^{(0)}\right) + \frac{2T}{3}\delta_{ij}\frac{\partial}{\partial T}f^{(0)}.$$

Again, we are looking for the first-order solution $f^{(1)}$ in the form (18.29) and with the same reasoning as previously we derive equations for the coefficients $\vec{\alpha}$, $\vec{\beta}$ and γ_{ij}. For $\vec{\beta}$ and γ_{ij} these equations are identical with (18.34), while the equation for $\vec{\alpha}$ is slightly different:

$$-T\frac{\partial}{\partial T}\zeta^{(0)}\vec{\alpha} + J^{(1)}\left(f^{(0)}, \vec{\alpha}\right) = \vec{A}. \qquad (21.9)$$

21.2 Viscosity coefficient

The scheme for evaluating the kinetic coefficients for granular gases of viscoelastic particles is identical with the previous case $\varepsilon =$ const. Therefore, we discuss their derivation only briefly and stress the differences with respect to the case $\varepsilon =$ const. For gases of viscoelastic particles, the coefficient

$$C_{ij}(\vec{V}) = \left(V_i V_j - \frac{1}{3}\delta_{ij} V^2\right) \frac{1}{V}\frac{\partial f^{(0)}}{\partial V} + \frac{2}{3}\delta_{ij} S_2\left(c^2\right) f_M(V) T \frac{\partial a_2}{\partial T} \qquad (21.10)$$

differs from the corresponding coefficient (18.26) for a gas of particles with $\varepsilon =$ const. by the second term. Within the same level of approximation as for the case $\varepsilon =$ const. again we assume that γ_{ij} may be found in the form (19.5) and again we arrive at (20.7). The first term in the RHS of this equation is identical to the equivalent term in (20.8)) for $\varepsilon =$ const., that is,

$$\int d\vec{V}_1 D_{ij}\left(\vec{V}_1\right) J^{(1)}\left(f^{(0)}, \gamma_{ij}\right) = 4\eta\, n\sigma^2 \sqrt{\frac{2T}{m}} \Omega_\eta^{\text{vis}}, \qquad (21.11)$$

where the coefficient Ω_η^{vis} is to be calculated using a Maple program where the collision rules for viscoelastic particles are used:

```
DefDimension(3);
DefJ();
Expr:=(-1/DIM*c2pDOTc2p*Delta2+c1aDOTc2p*c1aDOTc2p+c2aDOTc2p*c2aDOTc2p
       -c2pDOTc2p*c2pDOTc2p-c1pDOTc2p*c1pDOTc2p)*(1+a2*S(2,c1p)):
OmegaETA:=getJexpr(1,Expr,2);
```

The result reads

$$\Omega_\eta^{\text{vis}} = \Omega_\eta\left[(1+a_2 S_2(c_1^2))\,\phi(c_1), \phi(c_2)\right] = -\left(w_0 + \delta' w_1 - \delta'^2 w_2\right), \qquad (21.12)$$

with

$$w_0 = 4\sqrt{2\pi}\left(1 - \frac{1}{32} a_2\right)$$

$$w_1 = w_0\left(\frac{1}{15} - \frac{1}{500} a_2\right) \qquad (21.13)$$

$$w_2 = w_1\left(\frac{97}{165} - \frac{679}{44000} a_2\right).$$

As for the second term of the RHS of (20.7), we notice that the coefficients C_{ij} for the case of viscoelastic particles and for particles with $\varepsilon =$ const. have the same off-diagonal part. Only that part of C_{ij} contributes to this term (recall that D_{ij} is a traceless tensor, i.e. $D_{ij}\delta_{ij} = 0$), therefore, we conclude that it coincides with the previous result:

$$-\int d\vec{V} D_{ij}\left(\vec{V}\right) C_{ij}\left(\vec{V}\right) = \frac{2}{3T}\int d\vec{V}\, m^2 V^4 f^{(0)}\left(\vec{V}\right) = 10 nT. \qquad (21.14)$$

By means of (21.5, 21.11, 21.12, 21.14) equation (20.7) for the viscosity coefficient may be written as

$$\left(w_0\delta' - w_2\delta'^2\right) T \frac{\partial \eta}{\partial T} = \frac{3}{5}\left(w_0 + w_1\delta' - w_2\delta'^2\right)\eta - \frac{3}{2\sigma^2 g_2(\sigma)}\sqrt{\frac{mT}{2}}, \quad (21.15)$$

where we replace $\sigma^2 \to \sigma^2 g_2(\sigma)$ to take excluded-volume effects into account. We seek the solution as an expansion in terms of δ':

$$\eta = \eta_0 \left(1 + \delta'\tilde{\eta}_1 + \delta'^2 \tilde{\eta}_2 + \cdots\right). \quad (21.16)$$

The solution in zeroth order

$$\eta_0 = \frac{5}{16\sigma^2 g_2(\sigma)} \sqrt{\frac{mT}{\pi}} \quad (21.17)$$

is the viscosity coefficient of a gas of elastic particles, while the coefficients $\tilde{\eta}_1$ and $\tilde{\eta}_2$ account for the dissipative properties of viscoelastic particles. With (21.3) the temperature derivative of the viscosity coefficient reads

$$T \frac{\partial \eta}{\partial T} = \eta_0 \left(\frac{1}{2} + \frac{3}{5}\delta'\tilde{\eta}_1 + \frac{7}{10}\delta'^2 \tilde{\eta}_2 + \cdots\right). \quad (21.18)$$

Substituting (21.16, 21.18) into (21.15) and collecting terms of the same order of δ' we obtain equations for $\tilde{\eta}_1$, $\tilde{\eta}_2$, etc:

$$\tilde{\eta}_1 = \frac{359}{3840} \frac{\sqrt{2\pi}}{\pi} w_0 \approx 0.483$$

$$\tilde{\eta}_2 = \frac{41881}{2304000} \frac{w_0^2}{\pi} - \frac{567}{28160} \frac{w_1 \sqrt{2\pi}}{\pi} \approx 0.0942, \quad (21.19)$$

where the second Sonine coefficient a_2 has been expressed as an expansion in δ' too, according to (21.2). Thus, we arrive at the viscosity coefficient for a granular gas of viscoelastic particles:

$$\boxed{\begin{array}{l} \eta = \eta_0 \left(1 + \delta'\tilde{\eta}_1 + \delta'^2 \tilde{\eta}_2 + \cdots\right), \quad \delta'(t) = \delta \left(\frac{2T(t)}{T_0}\right)^{1/10} \\[6pt] \eta_0 = \dfrac{5}{16\sigma^2 g_2(\sigma)} \sqrt{\dfrac{mT}{\pi}} \\[6pt] \tilde{\eta}_1 \approx 0.483 \qquad\qquad\qquad\qquad \tilde{\eta}_2 \approx 0.0942 \end{array}} \quad (21.20)$$

Note that contrary to the case $\varepsilon = \text{const.}$, where $\eta \propto \sqrt{T}$, for a gas of viscoelastic particles there appears an additional temperature dependence due to the time-dependent coefficient δ'.

The evaluation of the viscosity coefficient has been done using Maple:

```
1   DefDimension(3):
2   DefJ():
3   Expr:=Delta2*(-1/2*(1+a2*(S(2,c1p)+S(2,c2p))+a2^2*S(2,c1p)*S(2,c2p))):
4   mu2:=getJexpr(1,Expr,2):
5   mu2:=convert(mu2,polynom):
6   Expr:=Delta4*(-1/2*(1+a2*(S(2,c1p)+S(2,c2p))+a2^2*S(2,c1p)*S(2,c2p))):
7   mu4:=getJexpr(1,Expr,2):
8   tmpA2:={solve(mu4=mu2/DIM*4*DIM/4*(DIM+2)*(1+a2),a2)}:
9   a2:=tmpA2[1]:
10  a2:=convert(a2,polynom):
11  xi0:=2*sigma^(DIM-1)/DIM*n*mu2*sqrt(2*T/m):
12  ExprETA:=(-1/DIM*c2pDOTc2p*Delta2+c1aDOTc2p*c1aDOTc2p+c2aDOTc2p*c2aDOTc2p
13      -c2pDOTc2p*c2pDOTc2p-c1pDOTc2p*c1pDOTc2p)*(1+a2*S(2,c1p)):
14  OmegaETA:=getJexpr(1,ExprETA,2):
15  OmegaETA:=convert(OmegaETA,polynom):
16  EqETA:=xi0*eta0*(1/2+3/5*dprime*eta1+7/10*dprime^2*eta2)+4/(DIM-1)/(DIM+2)
17      *eta0*(1+dprime*eta1+dprime^2*eta2)*n*sigma^(DIM-1)*sqrt(2*T/m)
18      *OmegaETA+n*T:
19  EqETA:=taylor(EqETA,dprime,3):
20  eta0:=solve(coeff(EqETA,dprime,0),eta0);
21  eta1:=solve(coeff(EqETA,dprime,1),eta1);
22  eta2:=solve(coeff(EqETA,dprime,2),eta2);
23  omega1:=9.28569:
24  omega0:=6.48562:
25  evalf(eta1);
26  evalf(eta2);
27  etaVIS:=eta0*(1+dprime*eta1+dprime^2*eta2);
28  plot(etaVIS/eta0,dprime=0..0.5,color=black,linestyle=[1]);
```

Note that on line 9 of the program the root for $a_2 \propto \delta'$ is to be chosen.

21.3 Coefficients κ and μ

The coefficient $\vec{B}$ is identical to its analogue for $\varepsilon = \text{const.}$ (20.19), while $\vec{A}$ differs from the previous expression (20.20) by an additional term

$$-\vec{V}\frac{\partial f^{(0)}}{\partial a_2}T\frac{\partial a_2}{\partial T} = -\vec{V}S_2\left(c^2\right)f_M(V)\left(\frac{1}{10}a_{21}\delta' + \frac{1}{5}a_{22}\delta'^2\right). \quad (21.21)$$

With the same approximations as for the case $\varepsilon = \text{const.}$ we choose $\vec{\alpha}$ in the form (19.31). Again multiplying (21.9) for $\vec{\alpha}$ by $\vec{S}(\vec{V}_1)/T$, integrating over $\vec{V}_1$ and using (18.45) for κ, we obtain

$$3\frac{\partial \zeta^{(0)}}{\partial T}\kappa T = -\frac{1}{T}\int d\vec{V}_1 \vec{S}\left(\vec{V}_1\right) J^{(1)}\left(f^{(0)},\vec{\alpha}\right) + \frac{1}{T}\int d\vec{V}_1 \vec{S}\left(\vec{V}_1\right) \vec{A}\left(\vec{V}_1\right). \quad (21.22)$$

The first term on the RHS has the same form as (20.24) (for $\varepsilon = \text{const.}$), where the numerical coefficient Ω_κ is calculated with the collision rules for viscoelastic particles:

$$-\frac{1}{T}\int d\vec{V}_1 \vec{S}\left(\vec{V}_1\right) J^{(1)}\left(f^{(0)}, \vec{\alpha}\right)$$
$$= -\frac{4}{5}\kappa n\sigma^2 \sqrt{\frac{2T}{m}} \Omega_\kappa^{\rm vis} = \frac{4}{5}\kappa n\sigma^2 \sqrt{\frac{2T}{m}} \left(u_0 + \delta' u_1 + \delta'^2 u_2\right), \quad (21.23)$$

where

$$u_0 = 4\sqrt{2\pi}\left(1 + \frac{1}{32}a_2\right)$$
$$u_1 = \omega_0 \left(\frac{17}{5} - \frac{9}{500}a_2\right) \quad (21.24)$$
$$u_2 = \omega_1 \left(\frac{1817}{440} - \frac{1113}{352000}a_2\right).$$

are obtained by the Maple program

```
1  DefDimension(3);
2  DefJ();
3  Expr:=(c2pDOTc2p-(DIM+2)/2)*(c1aDOTc2p*c1aDOTc1a+c2aDOTc2p*c2aDOTc2a
4           -c2pDOTc2p*c2pDOTc2p-c1pDOTc2p*c1pDOTc1p)*(1+a2*S(2,c1p)):
5  OmegaKAPPA:=getJexpr(1,Expr,2);
```

The second term on the RHS of (21.22) has two components, the first coincides with (20.26) (for $\varepsilon = \text{const.}$); the second component is attributed to the additional term (21.21) of the coefficient $\vec{A}(\vec{V})$. It reads

$$\frac{1}{T}\int d\vec{V}_1 \vec{S}\left(\vec{V}_1\right) \cdot \left(-\vec{V}\frac{\partial f^{(0)}}{\partial a_2} T \frac{\partial a_2}{\partial T}\right) = -\frac{15}{4} n v_T^2 T \frac{\partial a_2}{\partial T}. \quad (21.25)$$

Therefore, for the second term on the RHS of (21.22) we obtain

$$\frac{1}{T}\int d\vec{V}_1 \vec{S}\left(\vec{V}_1\right) \cdot \vec{A}\left(\vec{V}_1\right) = -\frac{15}{2}\frac{nT}{m}\left(1 + \frac{21}{10}a_{21}\delta' + \frac{11}{5}a_{22}\delta'^2\right). \quad (21.26)$$

With these expressions (and replacing $\sigma^2 \to \sigma^2 g_2(\sigma)$) we recast (21.22) for κ into the form

$$T\frac{\partial \kappa}{\partial T} T^{3/2} \left(\omega_0 \delta' - \omega_2 \delta'^2 + \cdots\right) = \frac{2}{5}\kappa T^{3/2} \left(u_0 + \delta' u_1 - \delta'^2 u_2 + \cdots\right)$$
$$- \frac{15}{4}\frac{T^{3/2}}{\sigma^2 g_2(\sigma)} \sqrt{\frac{T}{2m}} \left(1 + \frac{21}{10}a_{21}\delta' + \frac{11}{5}a_{22}\delta'^2\right). \quad (21.27)$$

We solve this equation with the Ansatz

$$\kappa = \kappa_0 \left(1 + \delta' \tilde{\kappa}_1 + \delta'^2 \tilde{\kappa}_2 + \cdots\right), \quad (21.28)$$

where

$$\kappa_0 = \frac{75}{64\sigma^2 g_2(\sigma)} \sqrt{\frac{T}{\pi m}} \quad (21.29)$$

is the thermal conductivity for a gas of elastic particles. Substituting (21.28) into (21.27) and equating the terms of the same orders of δ', we obtain the coefficients:

$$\tilde{\kappa}_1 = \frac{487}{6400}\frac{\sqrt{2\pi}\omega_0}{\pi} \approx 0.393$$
$$\tilde{\kappa}_2 = \frac{1}{\pi}\left(\frac{2872113}{51200000}\omega_0^2 + \frac{78939}{140800}\sqrt{2\pi}\omega_1\right) \approx 4.904\,.$$
(21.30)

Hence, the coefficient of thermal conductivity for a granular gas of viscoelastic particles reads in adiabatic approximation

$$\kappa = \kappa_0\left(1 + \delta'\tilde{\kappa}_1 + \delta'^2\tilde{\kappa}_2 + \cdots\right), \qquad \delta'(t) = \delta\left(\frac{2T(t)}{T_0}\right)^{1/10}$$

$$\kappa_0 = \frac{75}{64\sigma^2 g_2(\sigma)}\sqrt{\frac{T}{\pi m}}$$

$$\tilde{\kappa}_1 \approx 0.393 \qquad\qquad\qquad \tilde{\kappa}_2 \approx 4.904$$
(21.31)

Similar to the viscosity coefficient, the coefficient of thermal conductivity of a granular gas of viscoelastic particles reveals an additional temperature dependence as compared to the case $\varepsilon = \mathrm{const}$. The Maple program for the derivation of the coefficient of thermal conductivity is given below.

```
1  DefDimension(3):
2  DefJ():
3  ExprKAPPA:=(c2pD0Tc2p-(DIM+2)/2)*(c1aD0Tc2p*c1aD0Tc1a+c2aD0Tc2p*c2aD0Tc2a
4     -c2pD0Tc2p*c2pD0Tc2p-c1pD0Tc2p*c1pD0Tc1p)*(1+a2*S(2,c1p)):
5  OmegaKAPPA:=getJexpr(1,ExprKAPPA,2):
6  OmegaKAPPA:=convert(OmegaKAPPA,polynom):
7  Expr:=Delta2*(-1/2*(1+a2*(S(2,c1p)+S(2,c2p))+a2^2*S(2,c1p)*S(2,c2p))):
8  mu2:=getJexpr(1,Expr,2):
9  Expr:=Delta4*(-1/2*(1+a2*(S(2,c1p)+S(2,c2p))+a2^2*S(2,c1p)*S(2,c2p))):
10 mu4:=getJexpr(1,Expr,2):
11 mu2:=convert(mu2,polynom):
12 tmpA2:={solve(mu4=mu2/DIM*4*DIM/4*(DIM+2)*(1+a2),a2)}:
13 a2:=tmpA2[1]:
14 a2:=convert(a2,polynom):
15 mu2:=convert(mu2,polynom):
16 xi0:=2*sigma^(DIM-1)/DIM*n*mu2*sqrt(2*T/m):
17 xi01:=coeff(xi0,dprime,1)/(2/DIM*sqrt(2*T/m)*sigma^(DIM-1)*n):
18 xi02:=-coeff(xi0,dprime,2)/(2/DIM*sqrt(2*T/m)*sigma^(DIM-1)*n):
19 Tdxi0dT:=2/DIM*sqrt(2*T/m)*sigma^(DIM-1)*n*(3/5*xi01*dprime
20    -7/10*xi02*dprime^2)
21 a21:=coeff(a2,dprime,1):
22 a22:=coeff(a2,dprime,2):
23 Tda2dT:=1/10*a21*dprime+1/5*a22*dprime^2:
24 OmegaKAPPA:=convert(taylor(OmegaKAPPA,dprime,3),polynom):
25 EqKAPPA:=DIM*(kappa*Tdxi0dT+xi0*TdkappadT+xi0*kappa)+4/(DIM+2)*kappa*n
```

```
26      *sigma^(DIM-1)*sqrt(2*T/m)*OmegaKAPPA+(DIM+2)*DIM/4*n*2*T/m
27      *(1+2*a2+Tda2dT):
28   kappa:=kappa0*(1+dprime*kappa1+dprime^2*kappa2):
29   TdkappadT:=kappa0*(1/2+3/5*dprime*kappa1+7/10*dprime^2*kappa2):
30   EqKAPPA:=taylor(EqKAPPA,dprime,3):
31   kappa0:=solve(coeff(EqKAPPA,dprime,0),kappa0);
32   kappa1:=solve(coeff(EqKAPPA,dprime,1),kappa1);
33   kappa2:=solve(coeff(EqKAPPA,dprime,2),kappa2);
34   omega1:=9.28569:
35   omega0:=6.48562:
36   evalf(kappa1);
37   evalf(kappa2);
38   plot(kappa/kappa0,dprime=0..0.1,color=black,linestyle=[1]);
```

The evaluation of the coefficient μ may be performed in the same way as κ. The only difference is that the expansion of μ in terms of the dissipative parameter δ' lacks the zeroth-order term, since μ vanishes in the elastic limit. Leaving the computational details for Exercise 21.1 we present here the final result:

$$\mu = \frac{\kappa_0 T}{n} \left(\delta' \tilde{\mu}_1 + \delta'^2 \tilde{\mu}_2 + \cdots \right), \qquad \delta'(t) = \delta \left(\frac{2T(t)}{T_0} \right)^{1/10}$$

$$\kappa_0 = \frac{75}{64\sigma^2 g_2(\sigma)} \sqrt{\frac{T}{\pi m}}$$

$$\tilde{\mu}_1 = \frac{19}{80} \frac{\omega_0 \sqrt{2\pi}}{\pi} \approx 1.229$$

$$\tilde{\mu}_2 = \frac{1}{\pi} \left(\frac{58813}{640000} \omega_0^2 - \frac{1}{40} \sqrt{2\pi} \omega_1 \right) \approx 1.415$$

(21.32)

The factor $\kappa_0 T/n$ in the coefficient μ relates the two origins of heat flux. If $\vec{\nabla} T/T$ is of the same order as $\vec{\nabla} n/n$, the effect of the density gradient on the heat flux is $\mathcal{O}(\delta)$ times smaller than the effect of the temperature gradient. The influence of the density gradient on the heat flux is only noticeable if the temperature gradient is very small, while the density gradient is significant.

The velocity distribution function for the inhomogeneous granular gas of viscoelastic particles is still given by (20.39). This expression remains valid as long as (20.38) is fulfilled, which relates the kinetic coefficients η, κ and μ with the coefficients α_1, β_1 and γ_0. However, although the functional form of the first-order velocity distribution function $f^{(1)}(\vec{V})$ for $\varepsilon = \text{const.}$ coincides with that for a gas of viscoelastic particles, their time dependences are different due to an additional time dependence of the kinetic coefficients in the latter system.

Exercise 21.1 *Calculate the coefficient μ for a granular gas of viscoelastic particles! Follow the same procedure as for the derivation of the coefficient of thermal conductivity!*

22

CHAPMAN–ENSKOG METHOD FOR THE SELF-DIFFUSION COEFFICIENT

The coefficient of self-diffusion has been derived previously using the velocity-time correlation function. Now we derive this coefficient using the Chapman–Enskog scheme.

22.1 Constant coefficient of restitution

Although we have already derived the self-diffusion coefficient previously using the Liouville-operator technique and the velocity-time correlation function, it is instructive to show how this coefficient may be obtained within the Chapman–Enskog approach (Brey et al., 2000). We consider the diffusion of tagged particles (tracers) which by their mechanical properties do not differ from the embedding gas particles. They are sparsely immersed in a homogeneous cooling granular gas. We assume that the particles interact via a constant coefficient of restitution. The velocity distribution function of tracers $f_s(\vec{r}, \vec{v}, t)$ (the index 's' stands for self-diffusion) obeys then the Boltzmann equation:

$$\left(\frac{\partial}{\partial t} + \vec{v} \cdot \vec{\nabla}\right) f_s(\vec{v}_1, t) = g_2(\sigma) \sigma^2 \int d\vec{v}_2 \int d\vec{e}\, \Theta(-\vec{v}_{12} \cdot \vec{e}) |\vec{v}_{12} \cdot \vec{e}| \\ \times \left\{\frac{1}{\varepsilon^2} f_s(\vec{v}_1'', t) f(\vec{v}_2'', t) - f_s(\vec{v}_1, t) f(\vec{v}_2, t)\right\}, \quad (22.1)$$

where $f(\vec{v}, t)$ is the velocity distribution function of the gas particles. Assume further the number density of the tagged particles,

$$n_s(\vec{r}, t) = \int d\vec{v} f_s(\vec{r}, \vec{v}, t), \quad (22.2)$$

is not homogeneous, and hence, the motion of the tracers is governed by the diffusion equation

$$\frac{\partial}{\partial t} n_s(\vec{r}, t) + \vec{\nabla} \cdot \vec{J}_s(\vec{r}, t) = 0, \quad (22.3)$$

with $\vec{J}_s(\vec{r}, t)$ being the diffusion flux:

$$\vec{J}_s(\vec{r}, t) = \int d\vec{v}\, \vec{v} f_s(\vec{r}, \vec{v}, t). \quad (22.4)$$

For small gradients, it is proportional to the concentration gradient of the tagged particles,

$$\vec{J}_s(\vec{r}, t) = -D \vec{\nabla} n_s(\vec{r}, t). \quad (22.5)$$

Here D is the self-diffusion coefficient.

As usual for the Chapman–Enskog scheme, we employ the gradient expansion $f_s(\vec{r}, \vec{v}, t) = f_s^{(0)}(\vec{r}, \vec{v}, t) + \lambda f_s^{(1)}(\vec{r}, \vec{v}, t) + \cdots$, where $f_s^{(0)}(\vec{r}, \vec{v}, t)$ is the distribution function of the uniform system. Since the tagged particles are mechanically identical to the rest of the particles, $f_s^{(0)}(\vec{r}, \vec{v}, t)$ is just proportional to the distribution function of the embedding gas:

$$f_s^{(0)}(\vec{r}, t) = \frac{n_s(\vec{r}, t)}{n} f(\vec{v}, t), \qquad (22.6)$$

where $n \gg n_s$ is the number density of the gas. The Chapman–Enskog equation in zeroth order for $f_s^{(0)}(\vec{r}, t)$ yields the velocity distribution function for the homogeneous cooling state and the first-order equation reads, [see (18.7)]:

$$\frac{\partial^{(0)} f_s^{(1)}}{\partial t} + \frac{\partial^{(1)} f_s^{(0)}}{\partial t} + \vec{v} \cdot \vec{\nabla} f_s^{(0)} = g_2(\sigma) I\left(f, f_s^{(1)}\right). \qquad (22.7)$$

The time derivatives of escalating order reflect the order of the spatial gradient just as previously (see page 177). For the problem of tracer diffusion instead of the hydrodynamic equations the following equations are relevant:

$$\begin{aligned}
\frac{\partial n_s}{\partial t} &= \lambda^2 D \nabla^2 n_s = \left(\frac{\partial^{(0)}}{\partial t} + \lambda \frac{\partial^{(1)}}{\partial t} + \lambda^2 \frac{\partial^{(2)}}{\partial t} + \cdots\right) n_s \\
\frac{\partial T}{\partial t} &= -\zeta T = \left(\frac{\partial^{(0)}}{\partial t} + \lambda \frac{\partial^{(1)}}{\partial t} + \lambda^2 \frac{\partial^{(2)}}{\partial t} + \cdots\right) T,
\end{aligned} \qquad (22.8)$$

where ζ is the cooling coefficient for the homogeneous gas. We evaluate

$$\frac{\partial^{(0)} f_s^{(1)}}{\partial t} = \frac{\partial^{(0)} n_s}{\partial t} \frac{\partial f_s^{(1)}}{\partial n_s} + \frac{\partial^{(0)} T}{\partial t} \frac{\partial f_s^{(1)}}{\partial T} = -\zeta T \frac{\partial f_s^{(1)}}{\partial T}, \qquad (22.9)$$

where we take into account

$$\frac{\partial^{(0)} n_s}{\partial t} = 0, \qquad \frac{\partial^{(0)} T}{\partial t} = -\zeta T \qquad (22.10)$$

according to (22.8). Similarly, we obtain

$$\frac{\partial^{(1)} f_s^{(0)}}{\partial t} = \frac{\partial^{(1)} n_s}{\partial t} \frac{\partial f_s^{(0)}}{\partial n_s} + \frac{\partial^{(1)} T}{\partial t} \frac{\partial f_s^{(0)}}{\partial T} = 0, \qquad (22.11)$$

where we again apply (22.8). With

$$\vec{\nabla} f_s^{(0)} = \vec{\nabla} n_s \frac{\partial f_s^{(0)}}{\partial n_s} = \frac{f_s^{(0)}}{n_s} \vec{\nabla} n_s, \qquad (22.12)$$

and using (22.6) we arrive at an equation for $f_s^{(1)}$:

$$\zeta T \frac{\partial f_s^{(1)}}{\partial T} + g_2(\sigma) I\left(f, f_s^{(1)}\right) = \frac{1}{n} \left(\vec{v} \cdot \vec{\nabla} n_s\right) f, \quad (22.13)$$

The form of the LHS of the last equation shows that its solution is proportional to the gradient $\vec{\nabla} n_s$ and may be found in the form

$$f_s^{(1)} = \vec{G}(\vec{v}) \cdot \vec{\nabla} n_s(\vec{r}, t). \quad (22.14)$$

Substituting this form of $f_s^{(1)}$ into (22.13) leads to an equation for $\vec{G}(\vec{v})$:

$$\zeta T \frac{\partial}{\partial T} \vec{G}(\vec{v}) + g_2(\sigma) I\left(f, \vec{G}\right) = \frac{1}{n} \vec{v} f. \quad (22.15)$$

We can now express the diffusion flux in terms of the function $\vec{G}(\vec{v})$ by

$$\begin{aligned} J_i &= \int d\vec{v}\, v_i f_s^{(0)}(|\vec{v}|) + \int d\vec{v}\, v_i G_k(\vec{v}) \nabla_k n_s \\ &= \int d\vec{v}\, v_i G_k(\vec{v}) \nabla_k n_s = -D \nabla_i n_s, \end{aligned} \quad (22.16)$$

since the integral with the function $f_s^{(0)}(|\vec{v}|)$ vanishes (its integrand is an odd function). From this relation follows

$$\int d\vec{v}\, v_i G_k = -D \delta_{ik}. \quad (22.17)$$

For $i = k$ and summing over the indices we obtain

$$\int d\vec{v}\, v_i G_i = \int d\vec{v}\, \vec{v} \cdot \vec{G} = -D \delta_{ii} = -3D \quad (22.18)$$

and finally the diffusion coefficient

$$D = -\frac{1}{3} \int d\vec{v}\, \vec{v} \cdot \vec{G}. \quad (22.19)$$

We multiply (22.15) by $\vec{v}/3$ and integrate over $\vec{v}$ to obtain

$$\zeta T \frac{\partial}{\partial T} \frac{1}{3} \int d\vec{v}\, \vec{v} \cdot \vec{G}(\vec{v}) + \frac{g_2(\sigma)}{3} \int d\vec{v}\, \vec{v} \cdot I\left(f, \vec{G}\right) = \frac{2}{3nm} \int d\vec{v}\, \frac{mv^2}{2} f(v). \quad (22.20)$$

Obviously, the first term on the LHS of this equation equals $-\zeta T \partial_T D$, while the first term on the RHS is T/m. The structure of (22.15) for $\vec{G}(\vec{v})$ suggests the Ansatz

$$\vec{G}(\vec{v}) \propto \vec{v} f(v) = b_0 \vec{v} f(v). \quad (22.21)$$

with the unknown constant b_0 to be determined.

Substituting this Ansatz into (22.19) we relate the diffusion coefficient to the constant b_0:

$$D = -\frac{b_0}{3} \int d\vec{v}\, \vec{v} \cdot \vec{v} f(v) = -\frac{2b_0}{3m} \int d\vec{v}\, \frac{mv^2}{2} f(v) = -b_0 \frac{Tn}{m}, \qquad (22.22)$$

where we use the definition of temperature (7.3); the last relation implies $b_0 = -mD/nT$. Hence,

$$\int d\vec{v}_1 \vec{v}_1 I\left(f, \vec{G}\right) =$$
$$= \sigma^2 b_0 \int d\vec{v}_1 d\vec{v}_2 \int d\vec{e}\, \Theta(-\vec{v}_{12} \cdot \vec{e})\, |\vec{v}_{12} \cdot \vec{e}|\, f(\vec{v}_1) f(\vec{v}_2)\, \vec{v}_1 \cdot (\vec{v}_1' - \vec{v}_1)$$
$$= \sigma^2 b_0 \int d\vec{v}_1 d\vec{v}_2 \int d\vec{e}\, \Theta(-\vec{v}_{12} \cdot \vec{e})\, |\vec{v}_{12} \cdot \vec{e}|\, f(\vec{v}_1) f(\vec{v}_2)\, \vec{v}_2 \cdot (\vec{v}_2' - \vec{v}_2),$$
$$(22.23)$$

where we use the main property of the collision integral (6.25) and the symmetry of the above expressions with respect to the exchange of the particles (changing indices $1 \Leftrightarrow 2$ and simultaneously changing $\vec{e} \to -\vec{e}$ does not alter the expressions). If we also take into account

$$\vec{v}_2' - \vec{v}_2 = -(\vec{v}_1' - \vec{v}_1) = \frac{1+\varepsilon}{2} (\vec{v}_{12} \cdot \vec{e})\, \vec{e}, \qquad (22.24)$$

according to the collision law (6.5), for the second term on the LHS of (22.20) we obtain

$$\frac{g_2(\sigma)}{3} \int d\vec{v}_1 \vec{v}_1 I\left(f, \vec{G}\right)$$
$$= \frac{g_2(\sigma)}{3} \frac{b_0}{2} \sigma^2 \int d\vec{v}_1 d\vec{v}_2 \int d\vec{e}\, \Theta(-\vec{v}_{12} \cdot \vec{e})\, |\vec{v}_{12} \cdot \vec{e}|\, f(\vec{v}_1) f(\vec{v}_2)$$
$$\times (\vec{v}_1 - \vec{v}_2) \cdot (\vec{v}_1' - \vec{v}_1) \qquad (22.25)$$
$$= D\frac{1}{12}\frac{m}{nT}\sigma^2 g_2(\sigma) \int d\vec{v}_1 d\vec{v}_2 \int d\vec{e}\, \Theta(-\vec{v}_{12} \cdot \vec{e})\, |\vec{v}_{12} \cdot \vec{e}|\, f(\vec{v}_1) f(\vec{v}_2)$$
$$\times (1+\varepsilon)(\vec{v}_{12} \cdot \vec{e})^2 = D\tau_{v,\,\text{ad}}^{-1},$$

with $\tau_{v,\,\text{ad}}$ given by (14.45). Thus we recast (22.20) into the form

$$-\zeta T \frac{\partial D}{\partial T} + D\tau_{v,\,\text{ad}}^{-1} = \frac{T}{m}. \qquad (22.26)$$

Using the same reasoning for the temperature dependence of the viscosity coefficient (see discussion before (20.14) on page 197) we conclude $D \propto \sqrt{T}$, hence $T\partial_T D = D/2$. This leads to the solution of (22.26):

$$D(t) = \frac{T}{m} \left[\tau_{v,\,\text{ad}}^{-1}(t) - \frac{1}{2}\zeta(t) \right]^{-1}, \qquad (22.27)$$

which coincides with (14.50), obtained in previous chapter using the time-correlation functions.

If we substitute in the last equation (14.22) for $\tau_{v,\,\text{ad}}^{-1}(t)$ and (8.12) for $\zeta(t)$ with the expression (8.37) for the moment μ_2 we finally arrive at the self-diffusion coefficient as a function of the second Sonine coefficient a_2:

$$D(t) = \frac{4D_0(t)}{(1+\varepsilon)^2 \left[1 + (3/32)\, a_2\right]^2}, \qquad (22.28)$$

where $D_0(t)$ is the Enskog value given by (14.20).[28] The corresponding Maple program is given below:

```
1  DefDimension(3);
2  DefJ();
3  Expr:=(1+epsilon)*c12D0Te*c12D0Te*(1+a2*(S(2,c1p)+S(2,c2p))
4         +a2^2*S(2,c1p)*S(2,c2p)):
5  hattauV:= getJexpr(0,Expr,0);
6  tauVinv:=1/(2*DIM)*sqrt(2*T/m)*n*sigma^(DIM-1)*g2*hattauV;
7  Expr:=Delta2*(-1/2*(1+a2*(S(2,c1p)+S(2,c2p))+a2^2*S(2,c1p)*S(2,c2p))):
8  mu2:=getJexpr(0,Expr,0);
9  xi0:=2*sigma^(DIM-1)*g2/DIM*n*mu2*sqrt(2*T/m);
10 Dif:=(T/m)/(tauVinv-1/2*xi0);
11 DEnsk:=DIM*GAMMA(DIM/2)*sqrt(T/m)/(4*Pi^((DIM-1)/2)*n*g2*sigma^(DIM-1));
12 simplify(Dif/DEnsk);
```

22.2 Viscoelastic particles

For granular gases of viscoelastic particles, the shape of the velocity distribution depends on time via the time-dependent coefficients of the Sonine polynomials expansion. Therefore, (22.9) becomes

$$\frac{\partial^{(0)} f_s^{(1)}}{\partial t} = \frac{\partial^{(0)} n_s}{\partial t} \frac{\partial f_s^{(1)}}{\partial n_s} + \frac{\partial^{(0)} T}{\partial t} \frac{\partial f_s^{(1)}}{\partial T} + \frac{\partial^{(0)} a_2}{\partial t} \frac{\partial f_s^{(1)}}{\partial a_2}$$

$$= -\zeta T \frac{\partial f_s^{(1)}}{\partial T} + \dot{a}_2 \frac{\partial f_s^{(1)}}{\partial a_2}, \qquad (22.29)$$

where only the second Sonine coefficient $a_2(t)$ is taken into account. If we again choose $f_s^{(1)}(\vec{v})$ in the form (22.14), we obtain an equation for $\vec{G}(\vec{v})$ in the form of

---

[28] The result (22.28) differs slightly from that obtained by Brey et al. (2000). There, a more restrictive Ansatz $\vec{G}(\vec{v}) \propto \vec{v} f_M(v)$ was used, instead of $\vec{G}(\vec{v}) \propto \vec{v} f(v)$ used in (22.21). Moreover, in Brey et al. (2000) only linear terms with respect to a_2 have been retained.

(22.15), but with the additional term $\dot{a}_2 \partial \vec{G}/\partial a_2$. Multiplying this equation by $\vec{v}/3$ and integrating over $\vec{v}$ we arrive at

$$\zeta T \frac{\partial}{\partial T} \frac{1}{3} \int d\vec{v}\, \vec{v} \cdot \vec{G}(\vec{v}) - \frac{1}{3}\dot{a}_2 \int d\vec{v}\, \vec{v} \cdot \frac{\partial \vec{G}}{\partial a_2} + \frac{g_2(\sigma)}{3} \int d\vec{v}\, \vec{v} \cdot I\left(f, \vec{G}\right)$$
$$= \frac{2}{3nm} \int d\vec{v}\, \frac{mv^2}{2} f(v). \quad (22.30)$$

We substitute $\vec{G}(\vec{v})$ by means of (22.21) with $f(v) = f_M(v)\left[1 + a_2 S_2(c^2)\right]$. All terms in the last equation, except for the second term on the LHS, have been already evaluated in the previous section. The remaining term reads

$$\frac{\dot{a}_2}{3} \int d\vec{v}\, \vec{v} \cdot \frac{d\vec{G}}{da_2} = \frac{\dot{a}_2}{3} b_0 \int d\vec{v}\, \vec{v} \cdot \vec{v} f_M\, S_2(c^2)$$
$$= \frac{1}{3}\dot{a}_2 b_0 n v_T^2 \int d\vec{c}\, \phi(c) c^2 S_2(c^2) = \frac{\dot{a}_2}{3} b_0 n v_T^2 \nu_{22} = 0, \quad (22.31)$$

where we use the definition of ν_{kp} (7.12) and $\nu_{22} = 0$ (7.13). We arrive at the same equation (22.26) for the diffusion coefficient as for the case $\varepsilon = $ const.

$$-\zeta T \frac{\partial D}{\partial T} + D\tau_{v,\,\mathrm{ad}}^{-1} = \frac{T}{m}, \quad (22.32)$$

where, however, $\tau_{v,\,\mathrm{ad}}$ is given by (14.29).

If we had taken into account higher order Sonine coefficients a_k ($k > 2$) we would have obtained the same equation. In general (22.30) contains a sum of terms with factors $\dot{a}_k\, \vec{v} \cdot \partial \vec{G}/\partial a_k$. It may be shown that all these terms vanish when integrating over $\vec{v}$.

Exercise 22.1 *Prove that in (22.30) all terms that contain factors $\dot{a}_k\, \vec{v} \cdot \partial \vec{G}/\partial a_k$ vanish upon integration over $\vec{v}$ for $k \geq 2$!*

Equation (22.32) is solved by the same approach as used for the other kinetic coefficients. We write this coefficient as a perturbation expansion in terms of the small parameter δ':

$$D(t) = D_0 \left(1 + \delta' \tilde{D}_1 + \delta'^2 \tilde{D}_2 + \cdots\right) \quad (22.33)$$

and substitute it into (22.32) together with similar expansions for $\zeta(t)$ and $\tau_{v,\,\mathrm{ad}}^{-1}(t)$. The resulting equation can be solved perturbatively for each order of δ'. If we use the adiabatic approximation (21.2) for a_2 the calculations is performed analogously as for coefficients of viscosity and thermoconductivity. Therefore, we give only the final result:

$$\frac{D(t)}{D_0} = \left[1 + \frac{169}{320} \frac{\omega_0}{\sqrt{2\pi}} \delta' + \left(\frac{2867777}{20480000} \frac{\omega_0^2}{\pi} - \frac{3611}{7040} \frac{\omega_1}{\sqrt{2\pi}}\right) \delta'^2 + \cdots\right] \quad (22.34)$$

with D_0 being the Enskog coefficient of self-diffusion (14.20). The corresponding Maple program for the calculation of the self-diffusion coefficient is

```
1   DefDimension(3):
2   DefJ():
3   Expr:=(1+epsilon)*c12D0Te*c12D0Te*(1+a2*( S(2,c1p)+ S(2,c2p) )
4       +a2^2*S(2,c1p)*S(2,c2p) ):
5   hattauV:= getJexpr(1,Expr,2):
6   tauVinv:=1/(2*DIM)*sqrt(2*T/m)*n*sigma^(DIM-1)*g2*hattauV:
7   tauVinv:=convert(taylor(tauVinv,dprime,3),polynom):
8   Expr:=Delta2*(-1/2*(1+a2*(S(2,c1p)+S(2,c2p))+a2^2*S(2,c1p)*S(2,c2p))):
9   mu2:=getJexpr(1,Expr,2):
10  Expr:=Delta4*(-1/2*(1+a2*(S(2,c1p)+S(2,c2p))+a2^2*S(2,c1p)*S(2,c2p))):
11  mu4:=getJexpr(1,Expr,2):
12  tmpA2:={solve(mu4=mu2/DIM*4*DIM/4*(DIM+2)*(1+a2),a2)}:
13  a2:=tmpA2[1]:
14  a2:=convert(a2,polynom):
15  mu2:=convert(mu2,polynom):
16  xi0:=2*sigma^(DIM-1)*g2/DIM*n*mu2*sqrt(2*T/m):
17  tauVinv:=convert(taylor(tauVinv,dprime,3),polynom):
18  EqDiff:=-xi0*D0*(1/2+3/5*dprime*D1+7/10*dprime^2*D2)
19      +D0*(1+dprime*D1+dprime^2*D2)*tauVinv-T/m:
20  EqDiff:=taylor(EqDiff,dprime,3):
21  D0:=solve(coeff(EqDiff,dprime,0),D0);
22  assume(T>0,m>0);
23  DEnsk:=DIM*GAMMA(DIM/2)*sqrt(T/m)/(4*Pi^((DIM-1)/2)*n*g2*sigma^(DIM-1));
24  D1:=solve(coeff(EqDiff,dprime,1),D1);
25  D2:=solve(coeff(EqDiff,dprime,2),D2);
26  omega1:=9.28569:
27  omega0:=6.48562:
28  evalf(D1);
29  evalf(D2);
30  DVIS:=D0*(1+dprime*D1+dprime^2*D2);
31  plot(DVIS/D0,dprime=0..0.5,color=black,linestyle=[1]);
```

23

TWO-DIMENSIONAL GRANULAR GASES

23.1 Constant coefficient of restitution

In general dimension d the hydrodynamic equations read (Brey and Cubero, 2001)

$$\frac{\partial n}{\partial t} + \vec{\nabla} \cdot (n\vec{u}) = 0$$

$$\frac{\partial \vec{u}}{\partial t} + \vec{u} \cdot \vec{\nabla}\vec{u} + (nm)^{-1} \vec{\nabla} \cdot \hat{P} = 0 \qquad (23.1)$$

$$\frac{\partial T}{\partial t} + \vec{u} \cdot \vec{\nabla} T + \frac{2}{dn}\left(P_{ij}\nabla_j u_i + \vec{\nabla}\cdot\vec{q}\right) + \zeta T = 0.$$

The pressure tensor and the heat flux are

$$P_{ij} = p\delta_{ij} - \eta\left(\nabla_i u_j + \nabla_j u_i - \frac{2}{d}\delta_{ij}\vec{\nabla}\cdot\vec{u}\right)$$

$$\vec{q} = -\kappa\vec{\nabla}T - \mu\vec{\nabla}n. \qquad (23.2)$$

For dilute gases the hydrostatic pressure is $p = nT$, for higher density it reads

$$p = nT\left(1 + \frac{1+\varepsilon}{2}\frac{\pi^{d/2}}{d\Gamma(d/2)}n\sigma^d g_2(\sigma)\right). \qquad (23.3)$$

The expressions $g_2(\sigma)$ for $d = 2$ is given by (11.12). For gases whose particles interact via $\varepsilon = $ const. the transport coefficients η, κ, μ and the cooling rate ζ read (Brey and Cubero, 2001)

$$\eta(\varepsilon) = \frac{\eta_0}{\nu_1^*(\varepsilon) - \frac{1}{2}\zeta^*(\varepsilon)} = \eta_0 \eta^*(\varepsilon)$$

$$\kappa(\varepsilon) = \frac{\kappa_0\left(1 + 2a_2\right)}{\nu_2^*(\varepsilon) - \frac{2d}{d-1}\zeta^*(\varepsilon)} = \kappa_0 \kappa^*(\varepsilon)$$

$$\mu(\varepsilon) = \frac{2\kappa_0 T\zeta^*(\varepsilon)}{n}\frac{\kappa(\varepsilon)/\kappa_0 + \frac{(d-1)a_2}{d\zeta^*(\varepsilon)}}{\frac{2(d-1)}{d}\nu_2^*(\varepsilon) - 3\zeta^*(\varepsilon)} = \frac{\kappa_0 T}{n}\mu^*(\varepsilon) \qquad (23.4)$$

$$\zeta^{(0)} = \frac{nT}{\eta_0}\zeta^*(\varepsilon).$$

Here η_0 and κ_0 are the corresponding coefficients for the gas of elastic particles:

$$\eta_0 = \frac{2+d}{8\pi^{d/2}} \Gamma\left(\frac{d}{2}\right) \sqrt{\pi m T} \left(\sigma^{(d-1)} g_2(\sigma)\right)^{-1}$$

$$\kappa_0 = \frac{d(d+2)^2}{16(d-1)\pi^{d/2}} \Gamma\left(\frac{d}{2}\right) \sqrt{\frac{\pi T}{m}} \left(\sigma^{(d-1)} g_2(\sigma)\right)^{-1}, \tag{23.5}$$

while the dimensionless coefficients $\nu_1^*(\varepsilon)$, $\nu_2^*(\varepsilon)$ and $\zeta^*(\varepsilon)$ read

$$\zeta^*(\varepsilon) = \frac{2+d}{4d}\left(1-\varepsilon^2\right)\left(1+\frac{3a_2}{16}\right)$$

$$\nu_1^*(\varepsilon) = \frac{(3-3\varepsilon+2d)(1+\varepsilon)}{4d}\left(1-\frac{a_2}{32}\right) \tag{23.6}$$

$$\nu_2^*(\varepsilon) = \frac{1+\varepsilon}{d-1}\left(\frac{d-1}{2}+\frac{3(d+8)(1-\varepsilon)}{16}+\frac{4+5d-3(4-d)\varepsilon}{512}a_2\right)$$

and the second Sonine coefficient a_2 is

$$a_2 = \frac{16(1-\varepsilon)(1-2\varepsilon^2)}{9+24d+(8d-41)\varepsilon+30\varepsilon^2(1-\varepsilon)}. \tag{23.7}$$

23.2 Granular gases of viscoelastic particles

The hydrodynamic equations have the same form (23.1) as for the case $\varepsilon = \text{const}$. The kinetic coefficients, however, differ significantly. For a two-dimensional granular gas of viscoelastic particles they read

$$\eta = \eta_0 \left(1 + \delta' \tilde{\eta}_1 + \delta'^2 \tilde{\eta}_2 + \cdots\right), \tag{23.8}$$

where $\delta'(t) = \delta\left[2T(t)/T_0\right]^{1/10}$ and

$$\eta_0 = \frac{1}{2\sigma g_2(\sigma)}\sqrt{\frac{mT}{\pi}}$$

$$\tilde{\eta}_1 = \frac{29}{640}\frac{\sqrt{2\pi}}{\pi}\omega_0 \approx 0.234 \tag{23.9}$$

$$\tilde{\eta}_2 = \frac{111}{160000}\frac{\omega_0^2}{\pi} + \frac{569}{14080}\frac{\omega_1\sqrt{2\pi}}{\pi} \approx 0.308.$$

Similarly,

$$\kappa = \kappa_0 \left(1 + \delta' \tilde{\kappa}_1 + \delta'^2 \tilde{\kappa}_2 + \cdots\right) \tag{23.10}$$

with

$$\kappa_0 = \frac{2}{\sigma g_2(\sigma)}\sqrt{\frac{T}{\pi m}}$$

$$\tilde{\kappa}_1 = -\frac{433}{3200}\frac{\sqrt{2\pi}\omega_0}{\pi} \approx -0.700 \tag{23.11}$$

$$\tilde{\kappa}_2 = \frac{1}{\pi}\left(\frac{1749573}{12800000}\omega_0^2 + \frac{95619}{70400}\sqrt{2\pi}\omega_1\right) \approx 11.89$$

and
$$\mu = \frac{\kappa_0 T}{n}\left(\delta'\tilde{\mu}_1 + \delta'^2 \tilde{\mu}_2 + \cdots\right)$$
$$\tilde{\mu}_1 = \frac{7}{20}\frac{\omega_0 \sqrt{2\pi}}{\pi} \approx 1.811 \qquad (23.12)$$
$$\tilde{\mu}_2 = \frac{1}{\pi}\left(-\frac{7411}{80000}\omega_0^2 + \frac{7}{40}\sqrt{2\pi}\omega_1\right) \approx 0.0562.$$

The cooling rate $\zeta^{(0)}$ reads for a gas of viscoelastic particles
$$\zeta^{(0)} = \sqrt{\frac{2T}{m}}n\sigma g_2(\sigma)\left(\frac{1}{2}\omega_0\delta' - \tilde{\omega}_2\delta'^2 + \cdots\right) \qquad (23.13)$$
$$\tilde{\omega}_2 = \frac{1}{2}\omega_1 + \frac{9\sqrt{2\pi}}{500\pi}\omega_0^2 \approx 5.246.$$

The numerical constants $\omega_0 \approx 6.485$ and $\omega_1 \approx 9.285$ have been defined in (9.22).

Remark on the consistency of the hydrodynamic equations

The complete hydrodynamic description on the Navier–Stokes level includes the second-order gradients of the hydrodynamic fields. Hence, for consistency, one has to take into account the second-order gradient terms in the coefficient ζ which describes the cooling rate too. This quantity reads (Brey et al., 1998; Brilliantov and Pöschel, 2003)
$$\zeta^{(2)} = \zeta_1 \vec{\nabla}^2 T + \zeta_2 \vec{\nabla}^2 n + \cdots, \qquad (23.14)$$
where we omit terms of second order in the hydrodynamic fields.

To find $\zeta^{(2)}$ the second-order distribution function $f^{(2)}(\vec{V})$ is needed. Calculations may be performed within the Chapman–Enskog scheme for $\varepsilon = $ const. (Brey et al., 1998) and for a gas of viscoelastic particles (Brilliantov and Pöschel, 2003). However, they are rather lengthy and shall not be given here. Fortunately, the analysis performed by Brey et al., 1998, 1999a shows that for small enough inelasticity the second-order correction for the cooling rate $\zeta^{(2)}$ can be neglected in the hydrodynamic equations which coincides with the conclusion by Sela and Goldhirsch (1998).

For the sake of simplicity we consider the kinetic coefficients of inhomogeneous dissipative gases for small packing fraction. The corresponding analysis for dense granular gases has been performed by Garzo and Dufty (1999).

The results of the hydrodynamic theory for granular gases with $\varepsilon = $ const. and with the kinetic coefficients derived by the Chapman–Enskog approach have been checked by direct Monte Carlo simulations of the Boltzmann equation (Brey and Ruiz-Montero, 1999; Brey et al., 1999a) and a good agreement has been found even for relatively large dissipation.

SUMMARY

Granular gases may be described by hydrodynamic equations for the fields of density $n(\vec{r}, t)$, flux velocity $u(\vec{r}, t)$ and temperature $T(\vec{r}, t)$, provided the gradients of temperature and density are small and the velocity of macroscopic flows is significantly smaller than the thermal velocity. The transport coefficients are functions of the temperature, the density and microscopic parameters. The take-home message is:

1. The hydrodynamic equations of granular gases read

$$\frac{\partial n}{\partial t} + \vec{\nabla} \cdot (n\vec{u}) = 0$$

$$\frac{\partial \vec{u}}{\partial t} + \vec{u} \cdot \vec{\nabla} \vec{u} + (nm)^{-1} \vec{\nabla} \cdot \hat{P} = 0 \qquad (23.15)$$

$$\frac{\partial T}{\partial t} + \vec{u} \cdot \vec{\nabla} T + \frac{2}{3n} \left(P_{ij} \nabla_j u_i + \vec{\nabla} \cdot \vec{q} \right) + \zeta T = 0,$$

   where $\hat{P}$ denotes the stress tensor, $\vec{q}$ is the vector of heat flux and ζ is the cooling coefficient.

2. The structure of the hydrodynamic equations for granular gases equals those for gases of elastic particles, except for the sink term ζT which occurs for gases of dissipative particles only. This term causes very different behaviour of granular gases as compared with gases of elastic particles.

3. The stress tensor and the heat flux are expressed using phenomenological transport coefficients in linear approximation with respect to the field gradients:

$$P_{ij} = p\delta_{ij} - \eta \left(\nabla_i u_j + \nabla_j u_i - \frac{2}{3} \delta_{ij} \vec{\nabla} \cdot \vec{u} \right) \qquad (23.16)$$

$$\vec{q} = -\kappa \vec{\nabla} T - \mu \vec{\nabla} n,$$

   with the hydrostatic pressure $p = nT$ in the dilute limit. The viscosity coefficient η characterizes the force that appears when the granular gas is sheared. The coefficient of thermal conductivity κ and the coefficient μ characterize the heat flux due to temperature and density gradients. Equations (23.15) and (23.16) contain the field gradients only in linear order (Navier–Stokes level).

4. Applying the Chapman–Enskog method, the transport coefficients η, κ, μ and the cooling rate ζ are obtained in terms of temperature, density and microscopic parameters, such as particle size, mass and coefficient of

restitution. For dilute gases, the coefficients η and κ do not depend on density:

$$\eta = \frac{1}{\sigma^2}\sqrt{mT}\, f_\eta \qquad \kappa = \frac{1}{\sigma^2}\sqrt{\frac{T}{m}}\, f_\kappa$$

$$\frac{\mu}{T} = \frac{1}{n\sigma^2}\sqrt{\frac{T}{m}}\, f_\mu \qquad \zeta = n\sigma^2\sqrt{\frac{T}{m}}\, f_\zeta\,.$$

(23.17)

For a granular gas of particles that collide with a constant coefficient of restitution, the coefficients f_η, f_κ, f_μ and f_ζ are also constants, that is, they are functions of ε. Thus, for $\varepsilon = $ const. all coefficients η, κ, μ/T and ζ have the same temperature dependence. For a granular gas of viscoelastic particles, the coefficients f_η, f_κ, f_μ and f_ζ are dependent on temperature. The transport coefficients and the cooling rate ζ have different temperature dependences.

V. Structure Formation

The most striking phenomenon in the dynamics of force-free granular gases is the spontaneous formation of spatial density inhomogeneities. First reported in pioneering papers by McNamara (1993) and by Goldhirsch and Zanetti (1993) it was this effect which initiated the scientific interest in granular gases.

On cooling force-free granular gases, structure formation occurs due to a linear instability of long-wave modes. For gases of particles that collide with a constant coefficient of restitution these structures arise and persist forever. For granular gases of viscoelastic particles, however, the conditions of cluster formation are fulfilled only in a certain range of temperature. Such gases develop structures provided the initial temperature is large enough. As temperature decreases, these structures dissolve, that is, cluster formation is only a transient phenomenon. Nonlinear mechanisms of structure formation (mode enslaving) cannot prevent the clusters from dissolving.

24

INSTABILITY OF THE HOMOGENEOUS COOLING STATE

The homogeneous cooling state of a granular gas is inherently unstable. The formation of vortices and clusters may be explained qualitatively by simple arguments. For a more careful mathematical analysis, we derive the set of linearized hydrodynamic equations for granular gases.

24.1 Arguments for the instability of the homogeneous cooling state

A gas of elastic particles, that is a molecular gas, whose particles are homogeneously distributed, remains homogeneous in the absence of external forces. The most striking property of granular gases, which distinguishes them qualitatively from molecular gases, is their ability to develop a variety of spatio-temporal structures, even in the absence of any external influences.

Once homogeneously initialized at a certain temperature T_0, a granular gas cools due to inelastic collisions of its particles. At early times of its evolution, the gas stays homogeneous and isotropic, that is, there are neither macroscopic inhomogeneities nor flows. This stage of the evolution is called homogeneous cooling state. It was addressed in the preceding chapters. At later times of the gas evolution, vortex formation is observed. While the gas is still spatially homogeneous, there are well pronounced correlations of the velocities.

At yet later stages of its evolution, the gas becomes spatially inhomogeneous and pronounced density inhomogeneities develop. Figure 24.1 shows the formation of clusters during the evolution of a granular gas.

Spatial correlations of the particle velocities may be explained by a simple argument: when particles collide inelastically, the normal component of the relative velocity after the collision is ε times smaller than before while the tangential velocity stays unaffected. Consequently the outgoing angle of the particle traces is smaller than the incoming angle (see Fig. 6.2). This leads to an aligned motion of the particles, that is, to correlations of the velocities of adjacent particles.

The formation of clusters can be understood as well: assume fluctuations of the density in an otherwise homogeneous granular gas. In denser regions, the particles collide more frequently than in more dilute regions. Therefore, dense regions of the system cool down faster than dilute regions, hence, the local pressure in these regions decays with respect the pressure in dilute regions. The resulting pressure gradient causes a flux of particles into denser region, that is, the flux leads to further increase of the density. From these argument is follows that an initially small fluctuation of the density is enhanced, which leads to the

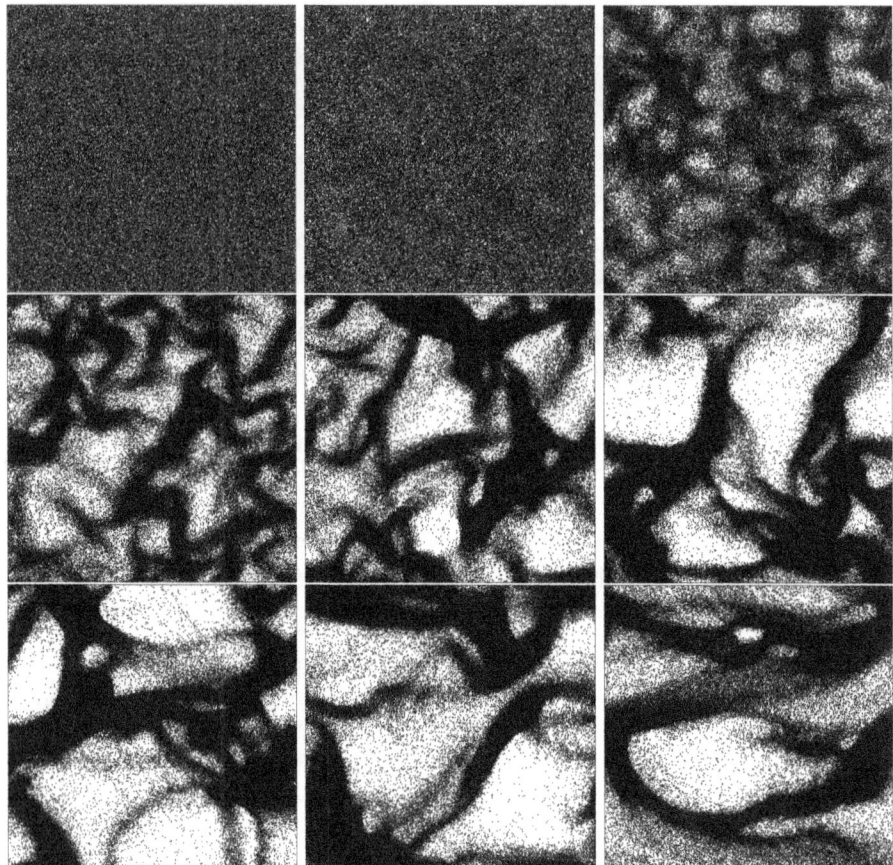

FIG. 24.1. Cluster formation in a cooling granular gas. The coefficient of restitution is $\varepsilon = 0.95$. The figures show a gas of $N = 10^5$ particles after 0, 100, 200, 300, 500, 1000, 2000, 4000 and 6000 collisions per particles (top left to bottom right). For the details of the algorithm see Appendix B.

formation of clusters. The homogeneous cooling state of a force-free granular gas is, therefore, unstable.

The scenario of cluster formation described refers to a granular gas whose particles collide with a constant coefficient of restitution. For gases of viscoelastic particles the coefficient of restitution is a function of the impact velocity: the smaller the relative velocities, that is, temperature, the more elastically the particles collide. Therefore, the particles in dense (cool) regions dissipate on average less energy per collision than particles in dilute regions. Hence, *a priori* it is not obvious whether after long time the cluster regime is stable or whether the system returns to the homogenous cooling state. We will come back to this question in Chapter 26.

24.2 Linearized hydrodynamic equations

The quantitative description of structure formation may be done in the framework of hydrodynamics, which describes the granular gas as a continuous medium.

The hydrodynamic equations are rather complicated nonlinear partial differential equations. Their full analysis is a very difficult problem which cannot be solved analytically. For this reason we will linearize the hydrodynamic equations and then Fourier transform them to obtain linear ordinary differential equations which can be investigated by the tools of standard stability analysis. The hydrodynamic fields can be written in the form

$$T(\vec{r},t) = T_h(t) + \delta T(\vec{r},t)$$
$$n(\vec{r},t) = n_h + \delta n(\vec{r},t), \quad (24.1)$$

with the constant number density $n_h = N/V$ and the spatially uniform, but time-dependent temperature $T_h(t)$ (the index 'h' stands for 'homogeneous'). The deviations of the hydrodynamic fields from their values in the homogeneous cooling state n_h, $T_h(t)$ are assumed to be small. We also assume that the flux velocity is significantly smaller than the thermal velocity, $|\vec{u}(\vec{r},t)| \ll v_T(t) \equiv \sqrt{2T_h(t)/m}$. These assumptions imply

$$\theta \equiv \frac{\delta T}{T_h} \ll 1, \qquad \rho \equiv \frac{\delta n}{n_h} \ll 1, \qquad \vec{w} \equiv \frac{\vec{u}}{v_T}; \quad |\vec{w}| \ll 1. \quad (24.2)$$

For the dimensionless fields $\theta(\vec{r},t)$, $\rho(\vec{r},t)$, $\vec{w}(\vec{r},t)$ we derive linear partial differential equations, neglecting higher-order terms with respect to these fields. First, we simplify the hydrodynamic equations (17.14, 17.25, 17.32) for the case of weak inhomogeneity.

If the transport coefficients are space-independent, equations (17.25), (17.32) together with (17.36) may be written in the form

$$\frac{\partial n}{\partial t} + \vec{\nabla} \cdot (n\vec{u}) = 0 \quad (24.3)$$

$$\frac{\partial \vec{u}}{\partial t} + \vec{u} \cdot \vec{\nabla}\vec{u} = -\frac{1}{nm}\vec{\nabla}p + \frac{\eta}{nm}\left[\nabla^2 \vec{u} + \frac{1}{3}\vec{\nabla}\left(\vec{\nabla} \cdot \vec{u}\right)\right] \quad (24.4)$$

$$\frac{\partial T}{\partial t} + \vec{u} \cdot \vec{\nabla}T = -\zeta T + \frac{2}{3n}\left(\kappa \nabla^2 T + \mu \nabla^2 n\right) - \frac{2}{3n}p\left(\vec{\nabla} \cdot \vec{u}\right) \quad (24.5)$$
$$+ \frac{2}{3n}\eta\left[(\nabla_i u_j)(\nabla_j u_i) + (\nabla_j u_i)(\nabla_j u_i) - \frac{2}{3}\left(\vec{\nabla} \cdot \vec{u}\right)^2\right],$$

where $p = nT$ is the hydrostatic pressure.

Exercise 24.1 *Derive (24.3)–(24.5)!*

With the continuity equation (24.3) in the form

$$\frac{\partial}{\partial t}(n_h + \delta n) + \vec{\nabla} \cdot (n_h + \delta n)\vec{u} \simeq \frac{\partial}{\partial t}\delta n + n_h \vec{\nabla} \cdot \vec{u} = 0 \quad (24.6)$$

and the definitions (24.2) we obtain the corresponding linearized equation

$$\frac{\partial \rho}{\partial t} + v_T \vec{\nabla} \cdot \vec{w} = 0. \qquad (24.7)$$

To linearize (24.4) for the flow velocity $\vec{u}$, we transform the time derivative of $\vec{u}$ by

$$\frac{\partial \vec{u}}{\partial t} = \frac{\partial}{\partial t}(v_T \vec{w}) = v_T \frac{\partial \vec{w}}{\partial t} + \vec{w} \frac{\partial v_T}{\partial t} = v_T \frac{\partial \vec{w}}{\partial t} - \frac{1}{2}\zeta_h^{(0)} v_T \vec{w}, \qquad (24.8)$$

where we take into account (see (8.12))

$$\frac{\partial v_T}{\partial t} = \frac{1}{2} v_T \frac{\dot{T}_h}{T_h} \quad \text{and} \quad \dot{T}_h = -\zeta_h^{(0)} T_h. \qquad (24.9)$$

The coefficient of the cooling rate ζ is approximated by its lowest-order value in the fields gradients, that is, $\zeta = \zeta_h^{(0)}$. Linearizing the pressure gradient term

$$\frac{1}{nm}\vec{\nabla}p = \frac{1}{m}(n_h + \delta n)^{-1} \vec{\nabla}(n_h + \delta n)(T_h + \delta T) = \frac{1}{m}T_h \vec{\nabla}\rho + \frac{1}{m}T_h \vec{\nabla}\theta \qquad (24.10)$$

and neglecting the nonlinear term $\vec{u} \cdot \vec{\nabla}\vec{u}$ in (24.4), we arrive at the linearized equation for the reduced flow velocity $\vec{w} \equiv \vec{u}/v_T$:

$$v_T \frac{\partial \vec{w}}{\partial t} - \frac{1}{2} v_T \zeta_h^{(0)} \vec{w} + \frac{T_h}{m}\left(\vec{\nabla}\rho + \vec{\nabla}\theta\right) = \frac{v_T \eta}{n_h m}\left[\nabla^2 \vec{w} + \frac{1}{3}\vec{\nabla}(\vec{\nabla} \cdot \vec{w})\right]. \qquad (24.11)$$

Finally, to linearize (24.5) we write

$$\zeta^{(0)}T = \left[\zeta_h^{(0)} + \delta T \left(\frac{\partial \zeta^{(0)}}{\partial T}\right)_h + \delta n \left(\frac{\partial \zeta^{(0)}}{\partial n}\right)_h\right](T_h + \delta T)$$

$$= (1+\theta)\zeta_h^{(0)} T_h + \theta T_h \left(T \frac{\partial \zeta^{(0)}}{\partial T}\right)_h + \rho T_h \left(n \frac{\partial \zeta^{(0)}}{\partial n}\right)_h \qquad (24.12)$$

and

$$\frac{\partial T}{\partial t} = \frac{\partial T_h}{\partial t}(1+\theta) = -(1+\theta)\zeta_h^{(0)} T_h + T_h \frac{\partial \theta}{\partial t}, \qquad (24.13)$$

where according to the evolution of temperature in the homogeneous cooling state

$$\frac{\partial T_h}{\partial t} = -\zeta_h^{(0)} T_h. \qquad (24.14)$$

Substituting (24.12) and (24.13) into (24.5) and neglecting the nonlinear terms, we obtain the linearized form

$$\frac{\partial \theta}{\partial t} + \theta \left(T \frac{\partial \zeta^{(0)}}{\partial T}\right)_h + \rho \left(n \frac{\partial \zeta^{(0)}}{\partial n}\right)_h$$

$$= \frac{2}{3}\left(\frac{\kappa}{n_h}\nabla^2 \theta + \frac{\mu}{T_h}\nabla^2 \rho\right) - \frac{2v_T}{3}(\vec{\nabla} \cdot \vec{w}). \qquad (24.15)$$

For the following analysis it is convenient to formulate the linearized hydrodynamic equations using dimensionless time and length. Let τ_* be the characteristic time, which will be specified later, and $l_* \equiv v_T \tau_*$ the characteristic length. Then with $t = \tau_* \tau$ and $\vec{r} = l_* \hat{\vec{r}}$ the derivatives read

$$\frac{\partial}{\partial t} = \frac{1}{\tau_*} \frac{\partial}{\partial \tau}, \qquad \vec{\nabla} = \frac{1}{l_*} \hat{\vec{\nabla}} = \frac{1}{v_T \tau_*} \hat{\vec{\nabla}}. \qquad (24.16)$$

With these definitions, we obtain the linearized hydrodynamic equations for the dimensionless fields in dimensionless time and space variables:

$$\frac{\partial \rho}{\partial \tau} + \hat{\vec{\nabla}} \vec{w} = 0$$

$$\frac{\partial \vec{w}}{\partial \tau} - \frac{1}{2} \tau_* \zeta^{(0)} \vec{w} = -\frac{1}{2} \left(\hat{\vec{\nabla}} \rho + \hat{\vec{\nabla}} \theta \right) + \frac{\eta}{2 n_h T_h \tau_*} \left[\hat{\nabla}^2 \vec{w} + \frac{1}{3} \hat{\vec{\nabla}} \left(\hat{\vec{\nabla}} \cdot \vec{w} \right) \right]$$

$$\frac{\partial \theta}{\partial \tau} + \theta \tau_* \left(T \frac{\partial \zeta^{(0)}}{\partial T} \right)_h + \rho \tau_* \left(n \frac{\partial \zeta^{(0)}}{\partial n} \right)_h$$

$$= \frac{2}{3 v_T^2 \tau_*} \left(\frac{\kappa}{n_h} \hat{\nabla}^2 \theta + \frac{\mu}{T_h} \hat{\nabla}^2 \rho \right) - \frac{2}{3} \left(\hat{\vec{\nabla}} \cdot \vec{w} \right).$$

(24.17)

This set of equations may be simplified further by appropriate choice of the unit of time τ_*. In Chapters 25 and 26 we will investigate these equations separately for the case $\varepsilon = \text{const.}$ and for granular gases of viscoelastic particles.

25

STRUCTURE FORMATION FOR $\varepsilon =$ CONST.

The hydrodynamic modes of the linearized hydrodynamic equations are unstable for long wavelengths. The instability of the transverse modes leads to the formation of vortices, while the instability of the heat modes causes clustering.

25.1 Linearized hydrodynamic equations for $\varepsilon =$ const.

For a granular gas of particles that interact by a constant coefficient of restitution all coefficients of the hydrodynamic equations have the same dependence on temperature. From (20.18, 19.45) we obtain

$$\eta = \eta_0 \eta^*(\varepsilon) \qquad \eta_0 = \frac{5}{16\sigma^2 g_2(\sigma)} \sqrt{\frac{mT}{\pi}} \propto \sqrt{T}$$

$$\kappa = \kappa_0 \kappa^*(\varepsilon) \qquad \kappa_0 = \frac{75}{64\sigma^2 g_2(\sigma)} \sqrt{\frac{T}{\pi m}} \propto \sqrt{T} \qquad (25.1)$$

and similarly from (20.37) and (18.11) with (8.37) for μ_2 follows

$$\frac{\mu}{T} = \frac{\kappa_0}{n} \mu^*(\varepsilon) \propto \sqrt{T} \qquad (25.2)$$

$$\zeta^{(0)} = \frac{nT}{\eta_0} \zeta^*(\varepsilon) \propto \sqrt{T}, \qquad (25.3)$$

where $\eta^*(\varepsilon)$, $\kappa^*(\varepsilon)$, $\mu^*(\varepsilon)$ and $\zeta^*(\varepsilon)$ are constants that depend only on ε (see also (23.4)).

Exercise 25.1 *Find expressions for the coefficients $\eta^*(\varepsilon)$, $\kappa^*(\varepsilon)$, $\mu^*(\varepsilon)$ and $\zeta^*(\varepsilon)$ using the relations for the kinetic coefficients and the cooling rate from Chapter 16!*

The set of equations (24.17) can be simplified using the relations

$$\tau_* \left(T \frac{\partial \zeta^{(0)}}{\partial T} \right)_h = \frac{1}{2} \tau_* \zeta_h^{(0)}, \qquad \tau_* \left(n \frac{\partial \zeta^{(0)}}{\partial n} \right)_h = \tau_* \zeta_h^{(0)}, \qquad (25.4)$$

which are obtained by means of (25.3). To simplify the analysis of the linearized equations (24.17) we chose the (still undefined) unit of time τ_* such that the coefficients in this equation become time independent. Let us analyse the coefficients in (24.17). The most complicated coefficient is

LINEARIZED HYDRODYNAMIC EQUATIONS FOR ε = CONST.   231

$$\frac{\eta}{2n_h T_h \tau_*} = \frac{\eta_0}{2n_h T_h \tau_*}\eta^*, \qquad (25.5)$$

where η^* is a pure number (see (25.1)). Hence, with the choice of τ_*,

$$\frac{\eta_0}{2n_h T_h \tau_*} = 1 \quad \text{or} \quad \tau_* \equiv \frac{\eta_0}{2n_h T_h}, \qquad (25.6)$$

this coefficient becomes time independent. With this choice of τ_* the coefficient $\tau_* \zeta_h^{(0)}$ reads [see (25.3)]

$$\tau_* \zeta_h^{(0)} = \frac{\eta_0}{2n_h T_h} \frac{n_h T_h}{\eta_0} \zeta^* = \frac{1}{2}\zeta^*. \qquad (25.7)$$

All other coefficients in (24.17) become also time independent and we arrive at the following form of the hydrodynamic equations:

$$\frac{\partial \rho}{\partial \tau} + \hat{\vec{\nabla}}\vec{w} = 0$$

$$\frac{\partial \vec{w}}{\partial \tau} - \frac{1}{4}\zeta^*\vec{w} = -\frac{1}{2}\left(\hat{\vec{\nabla}}\rho + \hat{\vec{\nabla}}\theta\right) + \eta^*\left[\hat{\nabla}^2\vec{w} + \frac{1}{3}\hat{\vec{\nabla}}\left(\hat{\vec{\nabla}}\cdot\vec{w}\right)\right] \qquad (25.8)$$

$$\frac{\partial \theta}{\partial \tau} + \frac{1}{4}\zeta^*\theta + \frac{1}{2}\zeta^*\rho = \frac{5}{2}\left(\kappa^*\hat{\nabla}^2\theta + \mu^*\hat{\nabla}^2\rho\right) - \frac{2}{3}\left(\hat{\vec{\nabla}}\cdot\vec{w}\right).$$

Note that the chosen unit of time τ_* depends on time itself:

$$\tau_*(t) \equiv \frac{\eta_0(t)}{2n_h T_h(t)} = \frac{5}{8}\left(4\sqrt{\pi}\sigma^2 g_2(\sigma)n_h \sqrt{\frac{T_h(t)}{m}}\right)^{-1} = \frac{5}{8}\tau_c(t), \qquad (25.9)$$

where $\tau_c(t)$ is the mean collision time in the homogeneous cooling gas at time t (see the footnote on page 138). From (24.16, 25.9) follows

$$d\tau = \frac{8}{5}\frac{dt}{\tau_c(t)} \quad \text{and} \quad \tau = \frac{8}{5}\int^t \frac{dt'}{\tau_c(t')}, \qquad (25.10)$$

that is, the dimensionless time τ is (up to the factor 8/5) measured by the accumulated number of collisions per particle. In contrast, the unit length is independent of time:

$$l_* = v_T \tau_* = \frac{5}{8}v_T(t)\tau_c(t) = \frac{5}{8}l_c, \qquad (25.11)$$

that is, the dimensionless length is measured (up to the factor 5/8) in units of the mean free path of a particle in a gas of homogeneous density n_h:

$$l_c = \left[2\sqrt{2\pi}\sigma^2 g_2(\sigma)n_h\right]^{-1}. \qquad (25.12)$$

25.2 Hydrodynamic modes

The Fourier transform of a field $\vec{a}(\vec{r}, \tau)$ is defined by

$$\vec{a}_{\vec{k}}(\tau) = \frac{1}{\sqrt{V}} \int d\vec{r} \; e^{-i\vec{k}\cdot\vec{r}} \vec{a}(\vec{r}, \tau)$$
$$\vec{a}(\vec{r}, \tau) = \frac{1}{\sqrt{V}} \sum_{\vec{k}} e^{i\vec{k}\cdot\vec{r}} \vec{a}_{\vec{k}}(\tau),$$
(25.13)

where $V = L^3$ is the volume of the granular gas. The waves $1/\sqrt{V} \exp\left(i\vec{k} \cdot \vec{r}\right)$, also called *modes*, form a complete set of orthonormal functions. Any field $\vec{a}(\vec{r}, \tau)$ may be represented as a superposition of these plane waves, where $\vec{a}_{\vec{k}}(\tau)$ is the amplitude of the mode $\vec{k}$. The summation over $\vec{k}$ in (25.13) runs over

$$\vec{k} = \frac{2\pi}{L} \begin{pmatrix} n_x \\ n_y \\ n_z \end{pmatrix} \quad \text{with} \quad n_x, n_y, n_z = \{0, \pm 1, \pm 2, \dots\} \quad (25.14)$$

Obviously, only modes with wave vectors $k_l \geq 2\pi/L$ (except for the trivial mode with $k = 0$) may exist in the system. Therefore, for the mode analysis not only the properties of the granular gas are relevant, but also the system size. The important property of the Fourier transform is that derivatives in the $\vec{r}$-space convert into simple multiplications in the $\vec{k}$-space:

$$\vec{\nabla} \cdot \vec{a}(\vec{r}, \tau) \to i\vec{k} \cdot \vec{a}_{\vec{k}}(\tau)$$
$$\vec{\nabla}^2 \vec{a}(\vec{r}, \tau) \to \left(i\vec{k}\right)^2 \vec{a}_{\vec{k}}(\tau).$$
(25.15)

By performing the Fourier transformation the set (25.8) of partial differential equations turns into a set of ordinary differential equations for the Fourier modes of the fields:

$$\frac{\partial \rho_{\vec{k}}}{\partial \tau} = -i\vec{k} \cdot \vec{w}_{\vec{k}}$$
$$\frac{\partial \vec{w}_{\vec{k}}}{\partial \tau} = \frac{1}{4}\zeta^* \vec{w}_{\vec{k}} - \frac{1}{2}i\vec{k}\left(\rho_{\vec{k}} + \theta_{\vec{k}}\right) - \eta^* \left[k^2 \vec{w}_{\vec{k}} + \frac{1}{3}\vec{k}\left(\vec{k} \cdot \vec{w}_{\vec{k}}\right)\right]$$
$$\frac{\partial \theta_{\vec{k}}}{\partial \tau} = -\frac{1}{4}\zeta^*\left(\theta_{\vec{k}} + 2\rho_{\vec{k}}\right) - \frac{5}{2}k^2\left(\kappa^* \theta_{\vec{k}} + \mu^* \rho_{\vec{k}}\right) - \frac{2}{3}\left(i\vec{k} \cdot \vec{w}_{\vec{k}}\right).$$
(25.16)

Hence, from the initial set of nonlinear partial differential equation with time-dependent coefficients we obtain a much simpler set of linear ordinary differential equations with constant coefficients. These equations for the Fourier components of the dimensionless fields depend on the dimensionless time and space coordinates. Their solution is straightforward. Writing (25.16) in matrix form

$$\frac{\partial \vec{\Psi}}{\partial \tau} = \hat{\mathbf{M}} \vec{\Psi} \quad (25.17)$$

the five-component vector $\vec{\Psi}$ contains $\rho_{\vec{k}}$, $\theta_{\vec{k}}$ and the three components of the vector $\vec{w}_{\vec{k}}$. The equations (25.16) for $\rho_{\vec{k}}$, $\theta_{\vec{k}}$ and $\vec{w}_{\vec{k}}$ are coupled. The eigenvectors of the 5×5 matrix $\hat{\mathbf{M}}$ are defined by

$$\hat{\mathbf{M}}\vec{\Psi}_l = \lambda_l(\vec{k})\vec{\Psi}_l \qquad l = 1, \ldots, 5, \qquad (25.18)$$

These eigenvectors are called *hydrodynamic modes*; in the framework of linearized hydrodynamics they are not coupled and evolve independently:

$$\vec{\Psi}_l(\tau) = \vec{\Psi}_l(0) \exp\left[\lambda_l(\vec{k})\tau\right]. \qquad (25.19)$$

If $\operatorname{Re}\left[\lambda_l(\vec{k})\right] \leq 0$, the concerned mode is stable, otherwise it is unstable, that is, for any initial conditions with $\Psi_l(0) \neq 0$ the amplitudes $\rho_{\vec{k}}$, $\theta_{\vec{k}}$, $\vec{w}_{\vec{k}}$ grow exponentially.

25.3 Vortex formation due to the instability of the transverse modes

We apply standard matrix algebra to find the eigenvalues and eigenmodes of the hydrodynamic equations. We choose the longitudinal component $\vec{w}_{\vec{k}\|} \equiv w_{\vec{k}\|}\vec{k}/k$ of the field $\vec{w}_{\vec{k}}$ along $\vec{k}$ and the perpendicular (transverse) component $\vec{w}_{\vec{k}\perp} \equiv \vec{w}_{\vec{k}} - \vec{w}_{\vec{k}\|}$. Since

$$\mathrm{i}\vec{k} \cdot \vec{w}_{\vec{k}\perp} = 0, \qquad \mathrm{i}\vec{k} \cdot \vec{w}_{\vec{k}\|} = \mathrm{i}k w_{\vec{k}\|}, \qquad (25.20)$$

the second equation in (25.16) yields an equation for the transverse component:

$$\frac{\partial \vec{w}_{\vec{k}\perp}}{\partial \tau} = \left(\frac{1}{4}\zeta^* - \eta^* k^2\right)\vec{w}_{k\perp}, \qquad (25.21)$$

which decouples from the other hydrodynamic equations. Its solution

$$\vec{w}_{\vec{k}\perp}(\tau) = \vec{w}_{\vec{k}\perp}(0) \exp\left[\left(\frac{1}{4}\zeta^* - \eta^* k^2\right)\tau\right], \qquad (25.22)$$

has the form (25.19), that is, the form of the hydrodynamic mode with the eigenvalue

$$\lambda_\perp\left(\vec{k}\right) = \frac{1}{4}\zeta^* - \eta^* k^2, \qquad (25.23)$$

which depends only on the absolute value of the wave vector $\vec{k}$. The direction of the vector $\vec{w}_{\vec{k}\perp}$ is not specified, the only requirement is that it must be perpendicular to $\vec{k}$. In a plane, perpendicular to $\vec{k}$ there exist two linearly independent vectors which we choose for the directions of the two transverse modes.[29]

---

[29] Usually these two vectors are chosen to be perpendicular to each other.

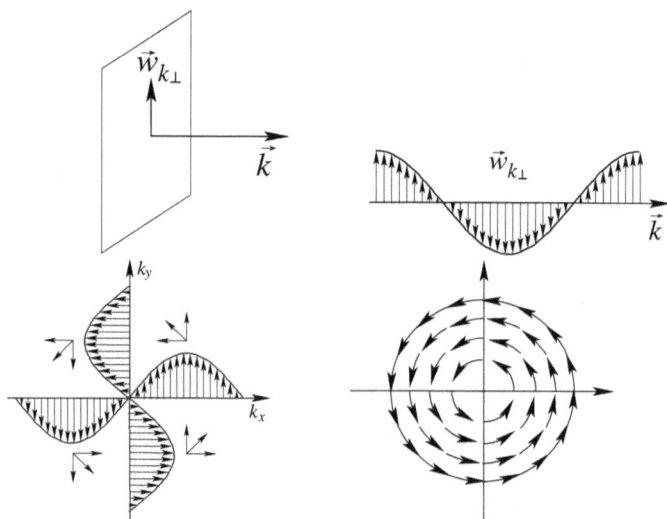

FIG. 25.1. Formation of a vortex by shear modes: in a shear mode the flow vector $\vec{w}_{\vec{k}}$ is perpendicular to the wave vector $\vec{k}$. Two shear modes with perpendicular wave vectors, for example, with wave vectors $\vec{k} = (k_x, 0, 0)$ and $\vec{k} = (0, k_y, 0)$ give rise to a vortex.

Thus the eigenvalue $\lambda_\perp(\vec{k})$ is twice degenerate. Within the linear hydrodynamics the transverse modes evolve independently from the other hydrodynamic modes and their stability is determined by the value of the wave vector $\vec{k}$. Since the eigenvalue (25.23) is real, these hydrodynamic modes, which are also called *shear modes*, do not oscillate. The marginal wave vector

$$k_\perp^*(\varepsilon) = \frac{1}{2}\sqrt{\frac{\zeta^*(\varepsilon)}{\eta^*(\varepsilon)}} \propto \sqrt{1-\varepsilon^2}, \qquad (25.24)$$

which depends only on ε, separates stable shear modes with $k \geq k_\perp^*(\varepsilon)$ from unstable modes.

Exercise 25.2 *Derive the function $k_\perp^*(\varepsilon)$ and show that it depends only on the coefficient of restitution! For $\varepsilon \lesssim 1$ show that $k_\perp^*(\varepsilon) \propto \sqrt{1-\varepsilon^2}$!*

Hence, short-wave shear modes (large k) are stable and decay rapidly, while long-wave modes grow exponentially. The physical meaning of the hydrodynamic shear modes is illustrated in Fig. 25.1, where it is shown that two transverse modes with perpendicular wave vectors give rise to a vortex. Therefore, long-wave modes with $k < k_\perp^*(\varepsilon)$ grow exponentially and lead to the formation of vortices in the initially flow-free granular gas. The well-developed vortex structure in Fig. 25.2 corresponds to a spatially almost uniform gas, that is, the formation of vortices precedes clustering. An exponential growth occurs for the reduced

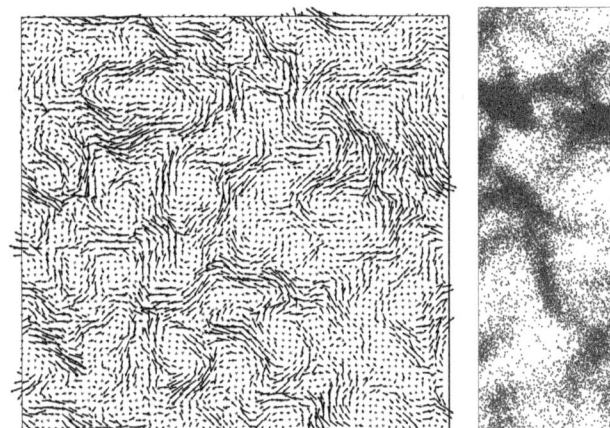
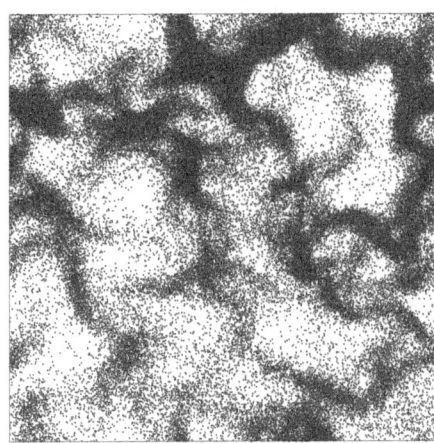

FIG. 25.2. Molecular dynamics simulations of a two-dimensional granular gas of 50,000 particles with $\varepsilon = 0.9$ and packing fraction 0.245 starting at a flow-free homogeneous state. The flow field $\vec{u}(\vec{r})$ after 80 collisions per particle (left), and the density field $n(\vec{r})$ after 160 collisions per particle (right). The figure has been taken from Brito and Ernst (1998a).

amplitude of the mode, which is measured in units of the thermal velocity. Its absolute amplitude decays with time.

Exercise 25.3 *Find the function $\vec{w}_{\vec{k}\perp}(t)$, using the relation between the laboratory time t and the time τ measured in the accumulated number of collisions! Find the function $\vec{u}_{\vec{k}\perp}(t)$ and show that the unscaled velocity field decays always with time!*

The vortex formation may be explained by a *thermal noise reduction process* (Brito and Ernst, 1998b): almost all initial conditions correspond to finite amplitudes of all transverse modes. At early times, these amplitudes are very small as compared with the random thermal velocities of the particles. As the system evolves, the amplitudes of the transverse modes with $k \geq k_\perp^*(\varepsilon)$ decay faster than the thermal velocity, while the amplitudes of the modes with $k < k_\perp^*(\varepsilon)$ decay slower than $v_T(t)$. Consequently, the long-wave transverse modes which initially have been hidden by the thermal noise (characterized by $T(t)$) become visible as a vortex structure.

For the formation of vortices it is necessary that modes with $k < k_\perp^*(\varepsilon)$ exist in the system, which implies that the characteristic size of the system L must exceed $2\pi/k_\perp^*(\varepsilon) \propto 1/\sqrt{1-\varepsilon^2}$. If the system is smaller, all shear modes are stable and no vortices can develop. As ε increases, the critical size of the system, which allows for the formation of vortices, grows.

The discussion of the minimal system size which allows for the formation of structures is controversial: on the one hand, the mathematical analysis yields a minimum system size below which structures cannot develop. On the other hand, throughout this book we describe granular gases in the absence of boundaries, that is, we describe a finite sub-volume of size L^3 containing N particles of a infinitely extended granular gas. The volume must be large enough to neglect fluctuations of the particle number. So if we say that the system is too small to develop structures we have three possible points of view:

1. We take the system size L literally, that is, the gas is confined in a container with certain boundary conditions. This assumption is problematic since the boundaries have crucial influence on the dynamics of granular gases. The description of the gas dynamics neglecting the boundaries is incomplete.
2. We consider a sub-space of size L of a much larger system with the conditions described above. This point of view is problematic too since there are no means to avoid clusters which range over the border of the system. In this sense the limitation to the sub-system of size L^3 becomes meaningless.
3. We consider a periodic system of period L. If there were structures in such systems whose extension is of the order L, the structures would interact with themselves. Such effects are not covered by our theory. Moreover, periodic systems which exist in nature, for example, planetary dust rings of the large planets of our solar system, require an external gravitational force, which confines these systems. The action of such forces is neither described by the present theory.

Therefore, the condition $k < k_\perp^*(\varepsilon)$ is controversially discussed and there is no generally acknowledged interpretation yet.

25.4 Cluster formation due the instability of the other hydrodynamic modes

We analyse now the remaining three equations for $\rho_{\vec{k}}$, $\theta_{\vec{k}}$ and $w_{\vec{k}\|}$, which can be written in the form

$$\frac{\partial}{\partial \tau}\begin{pmatrix}\rho_{\vec{k}}\\ \theta_{\vec{k}}\\ w_{\vec{k}\|}\end{pmatrix} = \hat{M}\left(\vec{k}\right)\begin{pmatrix}\rho_{\vec{k}}\\ \theta_{\vec{k}}\\ w_{\vec{k}\|}\end{pmatrix}, \qquad (25.25)$$

with the hydrodynamic matrix[30]

$$\hat{M}\left(\vec{k}\right) = \begin{pmatrix} 0 & 0 & -ik \\ -\tfrac{1}{2}\zeta^* - \tfrac{5}{2}\mu^* k^2 & -\tfrac{1}{4}\zeta^* - \tfrac{5}{2}\kappa^* k^2 & -\tfrac{2}{3}ik \\ -\tfrac{1}{2}ik & -\tfrac{1}{2}ik & \tfrac{1}{4}\zeta^* - \tfrac{4}{3}\eta^* k^2 \end{pmatrix}. \qquad (25.26)$$

To find the eigenvalues and eigenfunctions we solve the secular equation

---

[30] For simplicity we use the same notations for the reduced hydrodynamic matrix (25.25) as previously in (25.17).

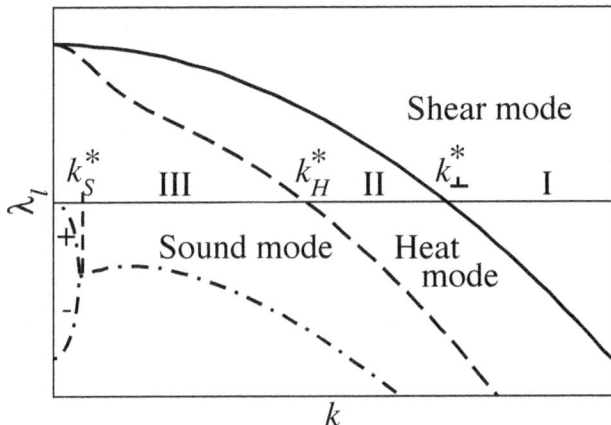

FIG. 25.3. Dependence of the real part of the eigenvalues of the hydrodynamic modes on the wave vector k. The marginal values $k_\perp^*$, k_H^* and k_S^* indicate, respectively, the regions of stability of the shear mode, of the heat mode and of the propagating sound modes.

$$\det \left| \hat{\mathbf{M}}\left(\vec{k}\right) - \lambda\left(\vec{k}\right) \hat{\mathbf{I}} \right| = 0, \qquad (25.27)$$

where $\hat{\mathbf{I}}$ is the unit matrix. For a 3×3 matrix (25.27) is a cubic equation, which may be solved analytically. The three roots of this equation correspond to the *heat mode* with a real eigenvalue $\lambda_H(\vec{k})$ and two *sound modes*. For small $k < k_S^*$ the sound modes have different real eigenvalues $\lambda_{S\,1/2}(\vec{k})$, while for $k \geq k_S^*$ the real parts of their eigenvalues coincide except for by their conjugate imaginary parts. This means that for $k \geq k_S^*$ the sound modes are *propagating modes* [which propagate as waves, see (25.19)], whereas the other modes (shear and heat mode) are *non-propagating*. The analysis of the hydrodynamic eigenvalues and eigenmodes may be found in detail in (Brito and Ernst, 1998b; van Noije and Ernst, 2000). Figure 25.3 illustrates the dependence of the hydrodynamic eigenvalues $\lambda_\perp(\vec{k})$, $\lambda_H(\vec{k})$ and $\lambda_{S\,1/2}(\vec{k})$ on the wave vector k as obtained in (Brito and Ernst, 1998b; van Noije and Ernst, 2000). The wave vector k_H^* separates the region of stability of the heat mode, $k \geq k_H^*$, from the region where the heat mode growths exponentially. Since the heat mode is a linear combination of $\rho_{\vec{k}}$, $\theta_{\vec{k}}$ and $w_{\vec{k}\|}$ (Brito and Ernst, 1998b; van Noije and Ernst, 2000), an exponential growth of the heat mode implies an exponential growth of the density mode $\rho_{\vec{k}}$, that is, clustering. Similarly as the shear mode, the heat mode can grow only if modes with $k < k_H^*$ exist in the system. Hence clustering may occur only if the system is large enough, so that its characteristic size is larger than $2\pi/k_H^*$ (see also the discussion on page 235).

The stability threshold for the heat mode reads $\lambda_H(\vec{k}) = 0$. It is obtained from the secular equation (25.27)

$$\det \left| \hat{\mathbf{M}} \left(\vec{k} \right) \right| = \frac{1}{2} k^2 \left(\frac{1}{2} \zeta^* + \frac{5}{2} \mu^* k^2 \right) - \frac{1}{2} k^2 \left(\frac{1}{4} \zeta^* + \frac{5}{2} \kappa^* k^2 \right) = 0 . \qquad (25.28)$$

The result

$$k_H^*(\varepsilon) = \sqrt{\frac{\zeta^*(\varepsilon)}{10 \left[\kappa^*(\varepsilon) - \mu^*(\varepsilon) \right]}} \propto \sqrt{1 - \varepsilon^2} \qquad (25.29)$$

shows that the smaller the dissipation $1 - \varepsilon^2$, the larger the system must be to reveal clusters.

Exercise 25.4 *Derive the dependence of the marginal wave vector on the coefficient of restitution: $k_H^*(\varepsilon) \propto \sqrt{1 - \varepsilon^2}$ for $1 - \varepsilon^2 \ll 1$!*

In the evolution of an initially flow-free homogeneous granular gas, clusters appear significantly later than vortices. In molecular dynamics simulations pronounced vortices have been observed in a spatially homogeneous system (see Fig. 25.2). Clusters can be expected to appear in regions of large flow velocity which corresponds to low pressure. In these regions, part of the energy of the initial random motion of the particles has been transformed into energy of the flux which corresponds to aligned motion of the particles. Hence, the temperature and pressure become smaller there than in the other regions. Therefore, clusters are formed mainly close to the borders of the vortices where the flow velocity is maximal (see Fig. 25.2). This effect may explain the typical filament structure of clusters. The similarity of the spatial structure of the velocity and density fields was first reported by Brey et al. (1999a).

The structure formation scenario is determined by the size of the system and by the value of the coefficient of restitution. If L is the characteristic length of the system, the following scenarios may occur:

- $L < 2\pi/k_\perp^*(\varepsilon)$: No vortices and no clusters are formed. The system remains in the regime of homogeneous cooling.
- $2\pi/k_\perp^*(\varepsilon) < L < 2\pi/k_H^*(\varepsilon)$: Vortices develop, but no clusters. The shear mode of maximal wavelength takes over finally and dominates the system.
- $L > 2\pi/k_H^*(\varepsilon)$: First vortices and later clusters are formed. The final state of the system is strongly inhomogeneous and system spanning clusters appear.

For systems with $L > 2\pi/k_S^*(\varepsilon)$ there exist non-propagating sound modes with $k < k_S^*(\varepsilon)$. A linear analysis as given above indicates that neither the unscaled velocity field $\vec{u}(\vec{r})$ nor the temperature fluctuations $\delta T(\vec{r})$ grow with time. These fields decay always due to dissipative particle collisions. Only $\delta n(\vec{r})$ increases with time if the initial perturbations excite the heat mode (Brey et al., 1999a).

The described scenarios of the evolution of a granular gas with the simplified collision model ($\varepsilon = \text{const.}$) have been confirmed by numerous numerical experiments. The behaviour of granular gases of viscoelastic particles deviates qualitatively from these results. Let us consider this problem in more detail in the next chapter.

26

STRUCTURE FORMATION IN GRANULAR GASES OF VISCOELASTIC PARTICLES

The stability analysis of the hydrodynamic equations for a gas of viscoelastic particles shows that the conditions of instability for the shear and heat modes depend explicitly on time. Hence, structure formation in granular gases occurs only as transient processes. After long time the granular gas eventually returns to the homogeneous cooling state.

26.1 Linearized equations for the hydrodynamic modes

In contrast to granular gases of particles whose interaction is described by the simplified collision model ($\varepsilon =$ const.) the coefficients of the hydrodynamic equations for granular gas of viscoelastic particles do not have the same temperature dependence. We write them in the form

$$\eta = \eta_0 \eta^* \left(\delta'\right) = \eta_0 \left(1 + \delta' \tilde{\eta}_1 + \delta'^2 \tilde{\eta}_2 + \cdots\right)$$
$$\kappa = \kappa_0 \kappa^* \left(\delta'\right) = \kappa_0 \left(1 + \delta' \tilde{\kappa}_1 + \delta'^2 \tilde{\kappa}_2 + \cdots\right),$$
(26.1)

where η_0 and κ_0 are given by (23.9), $\delta' \equiv \delta \left[2T(t)/T_0\right]^{1/10}$ is a small dissipative parameter, and the numerical coefficients $\tilde{\eta}_{1/2}$ and $\tilde{\kappa}_{1/2}$ are given by (21.19) and (21.30). We consider only the leading terms of the coefficients with respect to δ'. If we approximate $\eta^* \approx 1$, $\kappa^* \approx 1$ the coefficients of viscosity and thermal conductivity depend on temperature in the same way as for the case of $\varepsilon =$ const.:

$$\eta \propto \sqrt{T}, \qquad \kappa \propto \sqrt{T}.$$
(26.2)

The other two coefficients in the hydrodynamic equations are

$$\frac{\mu}{T} = \frac{\kappa_0}{n} \mu^* \left(\delta'\right) = \frac{\kappa_0}{n} \left(\delta' \tilde{\mu}_1 + \delta'^2 \tilde{\mu}_2 + \cdots\right)$$
$$\zeta^{(0)} = \frac{nT}{\eta_0} \zeta^* \left(\delta'\right) = \frac{nT}{\eta_0} \left(\delta' \omega_0 - \delta'^2 \omega_2 + \cdots\right) \frac{5}{24}\sqrt{\frac{2}{\pi}},$$
(26.3)

with the numerical coefficients $\tilde{\mu}_{1/2}$ and $\omega_{0/2}$ given by (23.12) and (21.6), respectively. In first-order approximation with respect to δ' we obtain

$$\mu^* \simeq \delta' \tilde{\mu}_1, \qquad \zeta^* = \frac{5}{24}\sqrt{\frac{2}{\pi}} \delta' \omega_0.$$
(26.4)

239

Thus, the dependence of these coefficients on temperature differs significantly from the dependence of η and κ:

$$\zeta^{(0)} \propto T^{3/5}, \qquad \frac{\mu}{T} \propto T^{3/5}. \tag{26.5}$$

The temperature dependence of the cooling rate $\zeta^{(0)}$ leads to a qualitative difference in the global behaviour of granular gases of viscoelastic particles as compared with gases of simplified particles ($\varepsilon = $ const.). As a consequence, for gases of viscoelastic particles it is not possible to derive a set of linearised hydrodynamic equations with time-independent coefficients unlike the set (25.8) in the previous chapter. Using (21.3) and keeping only the leading term in δ' we write

$$\tau_* \left(T \frac{\partial \zeta^{(0)}}{\partial T} \right)_h = \frac{3}{5} \tau_* \zeta^{(0)}, \qquad \tau_* \left(n \frac{\partial \zeta^{(0)}}{\partial n} \right)_h = \tau_* \zeta^{(0)}. \tag{26.6}$$

If we choose the unit of time

$$\tau_* = \frac{\eta_0}{2 n_h T_h} \qquad \text{with} \qquad \tau_* \zeta^{(0)} = \frac{1}{2} \zeta^* \tag{26.7}$$

we obtain, for the Fourier modes, the same set of equations (25.16) as before, but with $\eta^* = 1$, $\kappa^* = 1$ and with $\frac{3}{10}\zeta^*\theta$ instead of $\frac{1}{4}\zeta^*\theta$ in the second term of the last equation. With the choice (26.7) lengths are measured in units of the mean free path, just as in (25.11), and time in units of the accumulated number of collisions, as in (25.10):

$$\tau = \frac{8}{5} \int_0^t dt' \tau_c^{-1}(t') = \frac{8}{5} \int_0^t dt' \tau_c^{-1}(0) \sqrt{\frac{T_h}{T_0}} = \frac{8}{5\tau_c(0)} \int_0^t dt' \left(1 + \frac{t'}{\tau_0}\right)^{-5/6}$$

$$= \frac{48 \tau_0}{5\tau_c(0)} \left[\left(1 + \frac{t}{\tau_0}\right)^{1/6} - 1 \right] = \frac{3}{q_0 \delta} \left[\left(\frac{T_h}{T_0}\right)^{-1/10} - 1 \right], \tag{26.8}$$

where we use the law of temperature decay (9.40) for the homogeneous cooling state, $T_h/T_0 = (1 + t/\tau_0)^{-5/3}$ and (9.29) which relates τ_0 and $\tau_c(0)$. The constant $q_0 \equiv 2^{1/10} \omega_0/(16\sqrt{2\pi})$ is defined by (9.30). From (26.8) follows the law of temperature decay as a function of the rescaled time τ:

$$\frac{T_h}{T_0} = \left(1 + \frac{1}{3} q_0 \delta \tau\right)^{-10} \tag{26.9}$$

and correspondingly the dependence of δ' on τ:

$$\delta'(\tau) = 2^{1/10} \delta \left(1 + \frac{1}{3} q_0 \delta \tau\right)^{-1}. \tag{26.10}$$

26.2 Stability analysis of the hydrodynamic modes and structure formation

As for the case $\varepsilon = $ const., the equations for the shear modes decouple from the other equations. They have the same form

$$\frac{\partial \vec{w}_{k\perp}}{\partial \tau} = \left(\frac{1}{4}\zeta^* - \eta^* k^2\right) \vec{w}_{k\perp} = \left[b\delta'(\tau) - k^2\right] \vec{w}_{k\perp}, \quad (26.11)$$

where we use the above expression for ζ^* and introduce the numerical coefficient

$$b \equiv \frac{5}{48} \frac{\omega_0}{\sqrt{2\pi}}. \quad (26.12)$$

Note that the coefficient $\delta'(\tau)$ in (26.11) depends on the reduced time τ and, hence, leads to a qualitatively different evolution for the transverse modes. The solution of this equation reads

$$\vec{w}_{k\perp}(\tau) = \vec{w}_{k\perp}(0) \exp\left(-k^2\tau\right) \left(1 + \frac{1}{3}q_0\delta\tau\right)^{3 \cdot 2^{1/10} b/q_0}$$

$$= \vec{w}_{k\perp}(0) \left(1 + \frac{1}{3}q_0\delta\tau\right)^5 \exp\left(-k^2\tau\right), \quad (26.13)$$

where $\vec{w}_{k\perp}(0)$ is the initial amplitude of the shear mode and the above expressions for q_0 and b have been used. In Fig. 26.1 the dependence $\vec{w}_{k\perp}(\tau)$ is shown for different wave vectors k. From the analysis of the function $\vec{w}_{k\perp}(\tau)$ follows the marginal wave vector

$$k_\perp^* \equiv \sqrt{\frac{5q_0\delta}{3}}, \quad (26.14)$$

which distinguishes two different scenarios of the gas evolution: shear modes with $k \geq k_\perp^*$ decay, while those with $k < k_\perp^*$ grow initially and reach their maximal amplitudes

$$\vec{w}_{k\perp}^{\max} = \vec{w}_{k\perp}(0) \left(\frac{5q_0\delta}{3k^2 e}\right)^5 \exp\left(\frac{3k^2}{q_0}\delta\right) \quad (26.15)$$

at times

$$\tau_{k\perp}^{\max} = \frac{5}{k^2} - \frac{3}{q_0\delta}. \quad (26.16)$$

After reaching their maximum values they start to decrease and vanish eventually. The increase of the amplitudes of the shear modes corresponds to the formation of vortices. According to the linear stability analysis the vortices of size $\propto 1/k$ decay after a transient time $\propto \tau_{k\perp}^{\max}$. In a system of characteristic size L the smallest wave vector is of the order $2\pi/L$, thus in this system all shear modes will decay after the time $\propto (5L^2/4\pi^2 - 3/q_0\delta)$. This means that the vortex formation in granular gases of viscoelastic particles is a transient process, that is, the vortices tend to dissolve as the granular gas evolves further. This conclusion

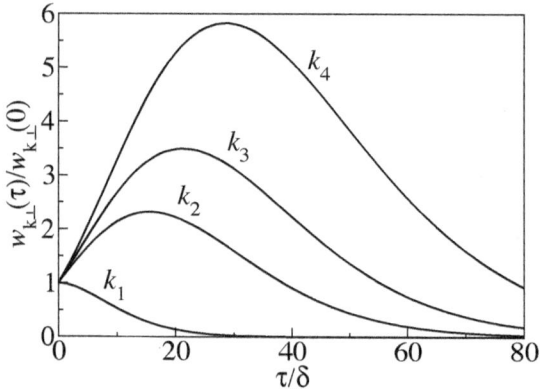

FIG. 26.1. The amplitude of the shear mode $\vec{w}_{k\perp}(\tau)$ as a function of the reduced time τ. Starting with an initial amplitude of the shear mode $\vec{w}_{k\perp}(0)$, for $k \geq k_\perp^*$ all shear modes decay, while for $k < k_\perp^*$ they initially grow up to a maximum amplitude $\vec{w}_{k\perp}^{\max}$ at time $\tau_{k\perp}^{\max} = 5/k^2 - 3/q_0\delta$ and decay for later times τ. The wave vectors of the modes are chosen due to $k_1 > k_2 > k_3 > k_4$ with $k_1 \geq k_\perp^*$ and $k_{2/3/4} < k_\perp^*$.

(based on the linear analysis) is in contrast to the results for granular gases with $\varepsilon = \text{const.}$: once created, the vortices remain persistent throughout the evolution of the granular gas.

The same analysis may be applied to the other three modes. The hydrodynamic matrix reads for the case of viscoelastic particles

$$\hat{M}(\vec{k},\tau) = \begin{pmatrix} 0 & 0 & -ik \\ -(2b + \tfrac{5}{2}\tilde{\mu}_1 k^2)\delta' & -\tfrac{6}{5}b\delta' - \tfrac{5}{2}k^2 & -\tfrac{2}{3}ik \\ -\tfrac{1}{2}ik & -\tfrac{1}{2}ik & b\delta' - \tfrac{4}{3}k^2 \end{pmatrix}, \quad (26.17)$$

where we take into account $\eta^* = 1$, $\kappa^* = 1$, $\zeta^* = 4b\delta'$ and $\mu^* = \tilde{\mu}_1 \delta'$. Contrary to the previous case ($\varepsilon = \text{const.}$), $\hat{M}$ depends now on time. Therefore, its eigenvalues and hydrodynamic modes also depend on time. However, for each time instant the structure of the hydrodynamic matrix is the same as for the case $\varepsilon = \text{const.}$ Therefore, for each time instant the structure of the hydrodynamic modes and the dependence of the eigenvalues on the wave vector k is the same as for $\varepsilon = \text{const.}$ As previously, we obtain the time-dependent marginal wave vectors $k_H^*(\tau)$ and $k_S^*(\tau)$, which separate regions of positive and negative eigenvalues for the heat modes and propagating and non-propagating regions for the sound modes (see Fig. 25.3). The marginal wave vector $k_H^*(\tau)$ may be found from the equation $\det \hat{M}(\vec{k},\tau) = 0$:

$$k_H^*(\tau) = \frac{2}{5}\sqrt{\frac{2b\delta'}{1 - \tilde{\mu}_1 \delta'}} \simeq \frac{2}{5}\sqrt{\frac{2^{11/10}\, b\delta}{1 + \tfrac{1}{3}q_0 \delta \tau}}. \quad (26.18)$$

In the second equation we neglect the small term $\tilde{\mu}_1 \delta'$. As for gases with $\varepsilon =$ const. the real part of the eigenvalue for the sound modes is negative, that is, all these modes decay. The heat modes of short wavelength with

$$k \geq k_H^*(0) = \frac{2}{5}\sqrt{2^{11/10}\, b\delta} \qquad (26.19)$$

decay too, whereas the amplitudes of the heat modes of long wavelengths with $k < k_H^*(0)$ grow initially.

The value of the marginal wave vector $k_H^*(\tau)$ decreases persistently with time. If initially it was larger than the minimal wave number of the system $2\pi/L$, it becomes smaller after some time τ_H, hence, over a period of time, all eigenvalues of all modes become negative. The time τ_H after which the amplitudes of all heat modes start to decay, can be estimated by equating the length of the marginal wavelength and the system size L, that is, from $k_H^*(\tau_H) = 2\pi/L$. We obtain

$$\tau_H = \frac{3}{q_0 \delta}\left(\frac{2^{11/10}\, b\delta L^2}{25\pi^2} - 1\right). \qquad (26.20)$$

Again we see that within the linear analysis the same scenario as for the shear modes takes place. Any initial fluctuation with $k < k_H^*(0)$ which refers to the heat mode (i.e. to some combination of $\theta_{\vec{k}}$, $\rho_{\vec{k}}$ and $w_{\vec{k}\parallel}$) first grows and then decays. The interval of time when the heat mode grows corresponds to an enhancement of the density inhomogeneities, that is, to clustering. Thus, we conclude that clusters of sizes larger than $\propto 1/k_H^*(0)$ start to emerge in the system, but at later time they eventually dissolve.

26.3 Structure formation as a transient process

By linear stability analysis we have shown that for a granular gas of viscoelastic particles all modes start to decay after some time lag, which depends on the coefficient of restitution of the particles and on the size of the system. Since vortex formation and clustering are attributed to the shear and heat mode instabilities, respectively, we conclude that for a gas of viscoelastic particles those structures may arise only as transient phenomena. Asymptotically the system tends to the flow-free state of homogeneous density.

Starting with a flux-free granular gas of viscoelastic particles of homogeneous density the system stays in this state and develops its velocity distribution (which deviates from the Maxwell distribution, see Chapter 9) while the temperature decreases steadily due to inelastic collisions. This state was termed homogeneous cooling state.

After a certain time the gas develops vortex structures while the density is still uniform. Later density inhomogeneities appear with denser regions close to the borders of the vortices and clusters start to appear. At the same time as clusters continue to grow, the vortices dissolve and finally the flux-free stage of the system is approached. Similarly, clusters become more and more nebulous

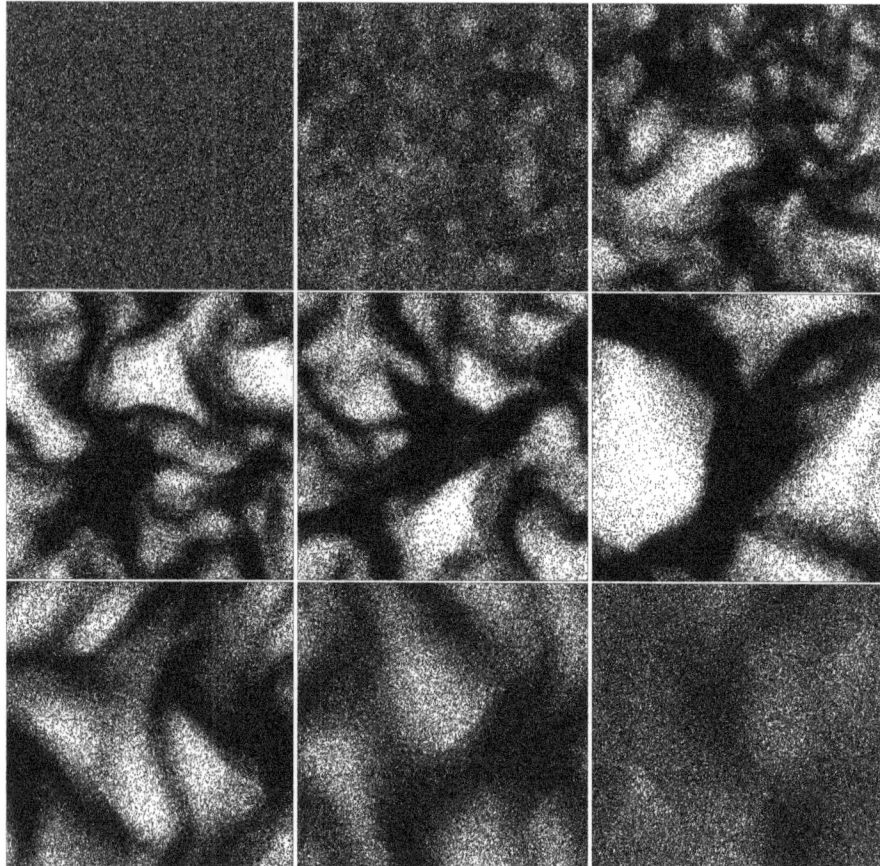

FIG. 26.2. The evolution of a granular gas of $N = 10^5$ viscoelastic particles. The snapshots are taken after 0, 200, 400, 600, 1000, 2000, 10,000, 20,000 and 40000 collisions per particle (top left to bottom right). Starting with homogenous density the particles develop density inhomogeneities and pronounced clusters which dissolve during the further evolution. Finally, the system approaches the homogeneous state.

with more and more diffusive boundaries and eventually they disappear too. Although we cannot judge whether the omitted nonlinear terms stabilise clusters and vortices, physical arguments support the above scenario: since the collisions become more and more elastic with increasing time, the system tends to behave as an assembly of elastic particles, that is, as a molecular gas. During the first stage of its evolution, however, the gas exhibits structure formation – the salient feature of dissipative systems. This behaviour of a granular gas is qualitatively different from the evolution of a gas of particles with a simplified collision model $\varepsilon = $ const., where the shear and heat modes grow infinitely for all wave numbers below a certain threshold. As a result, clusters and vortices once formed continue persistently to develop. In granular gases of viscoelastic particles these structures occur only as transient structures.

The transient nature of density inhomogeneities can also be observed in molecular dynamics simulations. Figure 26.2 shows snapshots of a granular gas of $N = 10^5$ particles after an increasing number of collisions. For the numerical method see Appendix B and Pöschel and Schwager (2004). We wish to remark that at intermediate times the system reveals system spanning clusters where the assumption of periodic boundary conditions becomes questionable. Nevertheless, this sequence of snapshots illustrates that cluster formation in granular gases of viscoelastic particles is a transient phenomenon (Brilliantov et al., 2004).

27

NONLINEAR MECHANISMS OF STRUCTURE FORMATION

The enslaving of temperature by the shear mode leads to an inverse dependence of pressure on density, that is, to a clustering instability. For a granular gas of viscoelastic particles, we show that the conditions of the temperature enslaving by the shear mode are not satisfied after some period of the gas evolution

Up to now we have analysed structure formation using linearized hydrodynamic equations. A nonlinear mechanism of cluster formation has been proposed by Goldhirsch and Zanetti (1993). Here we give a simplified description of this mechanism which is called *mode enslaving mechanism*.

Consider a shear mode with wave number $\vec{k}$. It may be shown that for $1-\varepsilon^2 \ll 1$ this mode evolves according to the linear theory as

$$\vec{u}_{\vec{k}\perp}(t) = \vec{u}_{\vec{k}\perp}(0) \left(1 + \frac{t}{\tau_0}\right)^{-48k^2/5(1-\varepsilon^2)} \tag{27.1}$$

with $\vec{u}_{\vec{k}\perp}(0)$ being the initial amplitude of the shear mode due to fluctuations.

Exercise 27.1 *Find the dependence of the non-reduced shear mode $\vec{u}_{\vec{k}\perp}$ on the non-reduced (laboratory) time t for $1 - \varepsilon^2 \ll 1$! Hint: use the dependence of temperature on time in the homogeneous cooling state and the relation between the laboratory time t and the reduced time τ.*

The temperature in the homogeneous cooling state decays as $T(t)/T_0 = (1 + t/\tau_0)^2$. Therefore, for

$$\frac{48}{5(1-\varepsilon^2)}k^2 \ll 2, \qquad \text{that is,} \qquad k^2 \ll \frac{5}{24}\left(1-\varepsilon^2\right), \tag{27.2}$$

where k is measured in units $l_*^{-1} = (8/5)l_c^{-1}$, the decay rate of the shear mode is much smaller than that of temperature. For these values of k the amplitude of the shear mode remains for a long time almost constant, while temperature may decrease significantly according to Haff's law.

Consider now the hydrodynamic equation (24.5). If we neglect all terms due to the field gradients, the remaining term, which scales as $\zeta T \propto T^{3/2}$ yields Haff's law for the homogeneous cooling state. On the other hand, due to fluctuations there always exist small-amplitude modes, even in the flow-free homogeneous cooling state. At early times these terms are negligible as compared with the dominant term ζT. As temperature decreases these terms become more and more important. According to the above analysis for the shear mode, the most important gradient terms in (24.5) are related to the shear modes, whose k satisfy the condition (27.2). A more detailed analysis confirms this expectation (Goldhirsch and Zanetti, 1993). The shear mode is perpendicular to the wave vector $\vec{k}$, that is, $\vec{k} \cdot \vec{u}_{\vec{k},\perp} = 0$. Since the application of the gradient $\vec{\nabla}$ for the Fourier components corresponds to a multiplication by $\vec{k}$ [see (25.15)], we conclude that $\vec{\nabla} \cdot \vec{u}_{\vec{k},\perp} = 0$. Then the non-vanishing terms on the RHS of (24.5) for the shear mode may be estimated by

$$\frac{2\eta}{3n}\left[(\nabla_i u_j)(\nabla_j u_i) + (\nabla_j u_i)(\nabla_j u_i)\right] \sim \frac{\eta}{n} k^2 \left[\vec{u}_{\vec{k},\perp}(0)\right]^2. \tag{27.3}$$

These terms correspond to the viscous heating. With the estimate (27.3) we obtain the equation for temperature,

$$\frac{\partial T}{\partial t} = -\zeta T + \frac{\eta}{n} k^2 \left[\vec{u}_{\vec{k}\perp}(0)\right]^2. \tag{27.4}$$

The first term on the RHS scales as $T^{3/2}$, while the second one as $\sqrt{T}$ as it follows from the temperature dependence of the viscosity. Initially the first (negative) term dominates, but when temperature decays, the second (positive) term takes over and the temperature saturates at some value. At this stage of the evolution the cooling due to the dissipative collision, given by the term ζT is compensated by the heating due to the nonlinear viscous term $\propto (\eta/n)k^2 \left[\vec{u}_{\vec{k},\perp}(0)\right]^2$. The evolution of temperature is now determined by the viscous heating. The described mechanism is called *mode enslaving*. Once the temperature field is enslaved, its time dependence follows the decay of the shear mode. The condition of the mode enslaving is obtained by equating the viscous heating and cooling term:

$$\zeta T = \frac{\eta}{n} k^2 \vec{u}_{\vec{k}\perp}(0)^2. \tag{27.5}$$

Since $\zeta \propto (1-\varepsilon^2) n\sqrt{T}$ and $\eta \propto \sqrt{T}$ we obtain for the enslaved temperature

$$T \propto \frac{1}{n^2(1-\varepsilon^2)}. \tag{27.6}$$

The balance between cooling and viscous heating is a local property, hence, the local temperature depends on the local density of the gas according to (27.6). From this relation one can obtain the local pressure

$$p = nT \propto \frac{1}{n}. \tag{27.7}$$

The higher the density, the lower the pressure which corresponds to negative compressibility, that is, the system is unstable with respect to density fluctuations. This is the nonlinear clustering mechanism based on the temperature enslaving phenomenon. Presently, it is not clear whether this mechanism is dominant in the process of cluster formation (Brey et al., 1999a).

The detailed analysis of all terms in the hydrodynamic equation for temperature (Goldhirsch and Zanetti, 1993) has shown that the discussed mechanism of the temperature field enslaving by the shear mode which leads to cluster formation requires the condition $kl < \sqrt{1-\varepsilon^2}$, where l is the mean free path.

Similar analysis of the mode enslaving mechanism may be performed for granular gases of viscoelastic particles. It has been shown, however, that the mode enslaving mechanism applies for the initial stages of the gas evolution only. As the gas cools the thermal velocity decays and, therefore, the typical coefficient of restitution (which corresponds to the thermal velocity) increases. Therefore, from a certain time on the condition for the mode enslaving mechanism $kl < \sqrt{1-\varepsilon^2}$ is not anymore fulfilled for any given k which may exist in a system of finite size. For a granular gas with a constant coefficient of restitution, the existence of the mode enslaving phenomenon was related to the possibility of a local cooling, which can overcome the thermal conductivity. For gases of viscoelastic particles the rate of cooling decreases due to increasing elasticity. As a result, after some time, when the elasticity becomes large enough, thermal conductivity takes over and destroys the mode enslaving. Finally, as it has been already demonstrated within the linear theory, the clusters dissolve and the system returns to the state of homogeneous density. Again, we come to the conclusion that structure formation in a gas of viscoelastic particles occurs only as a transient phenomenon.

28

TWO-DIMENSIONAL GRANULAR GASES

In general dimension d the linearized hydrodynamic equations read

$$\frac{\partial \rho}{\partial \tau} + \hat{\vec{\nabla}} \vec{w} = 0$$

$$\frac{\partial \vec{w}}{\partial \tau} - \frac{1}{2} \tau_* \zeta^{(0)} \vec{w} = -\frac{1}{2} \left(\hat{\vec{\nabla}} \rho + \hat{\vec{\nabla}} \theta \right) + \frac{\eta}{2 n_h T_h \tau_*} \left[\hat{\nabla}^2 \vec{w} + \frac{d-2}{d} \hat{\vec{\nabla}} \left(\hat{\vec{\nabla}} \cdot \vec{w} \right) \right]$$

$$\frac{\partial \theta}{\partial \tau} + \theta \tau_* \left(T \frac{\partial \zeta^{(0)}}{\partial T} \right)_h + \rho \tau_* \left(n \frac{\partial \zeta^{(0)}}{\partial n} \right)_h$$

$$= \frac{2}{d v_T^2 \tau_*} \left(\frac{\kappa}{n_h} \hat{\nabla}^2 \theta + \frac{\mu}{T_h} \hat{\nabla}^2 \rho \right) - \frac{2}{d} \left(\hat{\vec{\nabla}} \cdot \vec{w} \right). \tag{28.1}$$

If we choose the units of time $\tau_*(t)$ and length $l_* = \tau_* v_T(t)$ according to (25.6) and use the expressions (23.5) and (23.6) for the d-dimensional kinetic coefficients, we obtain equations for the Fourier modes for the case $\varepsilon = $ const.:

$$\frac{\partial \rho_{\vec{k}}}{\partial \tau} = -i \vec{k} \cdot \vec{w}_{\vec{k}}$$

$$\frac{\partial \vec{w}_{\vec{k}}}{\partial \tau} = \frac{1}{4} \zeta^* \vec{w}_{\vec{k}} - \frac{1}{2} i \vec{k} \left(\rho_{\vec{k}} + \theta_{\vec{k}} \right) - \eta^* \left[k^2 \vec{w}_{\vec{k}} + \frac{(d-2)}{d} \vec{k} \left(\vec{k} \cdot \vec{w}_{\vec{k}} \right) \right] \tag{28.2}$$

$$\frac{\partial \theta_{\vec{k}}}{\partial \tau} = -\frac{1}{4} \zeta^* \theta_{\vec{k}} - \frac{1}{2} \zeta^* \rho_{\vec{k}} - \frac{(d+2)}{(d-1)} k^2 \left(\kappa^* \theta_{\vec{k}} + \mu^* \rho_{\vec{k}} \right) - \frac{2}{d} \left(i \vec{k} \cdot \vec{w}_{\vec{k}} \right).$$

The variables in this equation depend on the reduced time and length, measured in units of

$$\tau_*(t) = \frac{(d+2)}{8} \tau_c(t) \quad \text{and} \quad l_* = \frac{(d+2)}{8} l_c, \tag{28.3}$$

with $\tau_c(t)$ and $l_c = \tau_c(t) v_T(t)$ being, respectively, the mean collision time and the mean free path (11.11). The equation for the $(d-1)$ shear modes has the same form, independently of the dimension. Therefore, the above analysis for the three-dimensional system may be directly applied to systems in d dimensions (using, of course, the corresponding kinetic coefficients). The analysis of the other modes is also similar as before. Instead of (25.28) we obtain the equation

$$\det\left|\hat{\mathbf{M}}\left(\vec{k}\right)\right| = \frac{1}{2}k^2\left(\frac{1}{2}\zeta^* + \frac{(d+2)}{(d-1)}\mu^*k^2\right) - \frac{1}{2}k^2\left(\frac{1}{4}\zeta^* + \frac{(d+2)}{(d-1)}\kappa^*k^2\right) = 0, \tag{28.4}$$

and correspondingly, the relation

$$k_H^*(\varepsilon) = \sqrt{\frac{(d-1)\zeta^*(\varepsilon)}{4(d+2)\left[\kappa^*(\varepsilon) - \mu^*(\varepsilon)\right]}} \tag{28.5}$$

for the marginal wave number which separates the unstable region of the heat mode from the stable region.

For the granular gas of viscoelastic particles the same equations as (28.2) may be obtained with the only difference that, instead of the numerical coefficient $1/4$, in the first term of the RHS of the equation for temperature (cooling rate term) reads $3/10$. The linear stability analysis of these equations for the d-dimensional case is analogous to the three-dimensional gas. For a two-dimensional gas of viscoelastic particles we obtain the reduced-time-dependent dissipation δ', similar to (26.10):

$$\delta'(\tau) = 2^{1/10}\delta\left(1 + \frac{2}{5}q_0\delta\tau\right)^{-1}, \tag{28.6}$$

and the corresponding temperature in the homogeneous cooling state as a function of the reduced time

$$\frac{T_h}{T_0} = \left(1 + \frac{2}{5}q_0\delta\tau\right)^{-10}. \tag{28.7}$$

The shear modes evolve as

$$\vec{w}_{k\perp}(\tau) = \vec{w}_{k\perp}(0)\left(1 + \frac{2}{5}q_0\delta\tau\right)^5 \exp\left(-k^2\tau\right) \tag{28.8}$$

and undergo the same evolution: the modes with $k > k_\perp^* \equiv \sqrt{2q_0\delta}$ decay always, while those with $k < k_\perp^*$ grow initially but then they decay as well. This indicates that vortices appear as transient structures. The stability analysis of the heat mode yields the marginal wave vector (with the previous definition (26.12) of b):

$$k_H^*(\tau) \simeq \frac{1}{5}\sqrt{\frac{3 \cdot 2^{11/10} b\delta}{1 + \frac{2}{5}q_0\delta\tau}}. \tag{28.9}$$

All heat modes with $k > k_H^*$ are stable, that is, they always decay. Since $k_H^*(\tau)$ decreases with time, all heat modes of any finite-size system become eventually stable. Hence, we again conclude that clustering occurs in a granular gas of viscoelastic particles only as a transient phenomenon.

SUMMARY

An initially homogeneous flux-free granular gas which is cooling in the absence of external forces, stays homogeneous and flux free during the first stage of its evolution. However, the homogeneous cooling state is inherently unstable and self-organized spatial structures appear at later stages.

The basic mechanism of structure formation can be easily understood: dissipative collisions reduce only the normal component of the relative velocities of the particles. The aligned traces of the particles lead to vortex formation. The temperature and, thus, pressure decay more rapidly in dense regions due to higher collision frequency. Therefore, these dense regions attract further particles and form eventually clusters.

For gases of viscoelastic particles this pressure instability is balanced by a superimposed effect: The effective elasticity of these particles is a function of the collision velocity and, therefore, increases as the gas cools down. Hence, particles in dense regions collide more elastically than those in dilute regions. A more quantitative analysis has shown that spontaneous structure formation in granular gases of viscoelastic particles is a transient phenomenon: vortices and clusters arise after the initial stage of homogeneous cooling, but they completely disappear at later stages. Eventually, the gas evolves back to a flux-free homogeneous state. The take-home message is:

1. The flux-free homogeneous state of granular gases is not stable under certain conditions. These conditions can be expressed in terms of the coefficient of restitution, the density and the system size.

2. For a granular gas of particles that collide with a simplified collision model $\varepsilon = \text{const.}$, there exist two characteristic wave numbers $k_\perp^*(\varepsilon) > k_H^*(\varepsilon)$. For $L < 2\pi/k_\perp^*(\varepsilon)$ no clusters and no vortices appear in the system, which always remains in the homogeneous cooling state. For $2\pi/k_\perp^*(\varepsilon) < L < 2\pi/k_H^*(\varepsilon)$ only vortices can arise while the density is kept homogeneous. Finally, for $2\pi/k_H^*(\varepsilon) < L$ first vortices and later clusters form; the final stage of the system is characterized by pronounced, system spanning clusters. Both quantities $k_\perp^*(\varepsilon)$ and $k_H^*(\varepsilon)$ are proportional to $\sqrt{1-\varepsilon^2}$ in the limit of small dissipation.

3. The process of structure formation in a granular gas of viscoelastic particles is qualitatively different. The characteristic wave numbers $k_\perp^*(\delta)$ and $k_H^*(\delta)$ are time-dependent functions of the dissipative parameter $\delta \propto 1 - \varepsilon$. For the system of viscoelastic particles, the structure formation is a transient process. At late stages of the gas evolution, all structures dissolve and the gas approaches the flow-free homogeneous state.

APPENDIX A

FUNCTIONS OF THE COLLISION INTEGRAL

A.1 Kinetic integrals and basic integrals

In many places throughout the book we face the problem to evaluate integrals of the type

$$\int d\vec{c}_1 \int d\vec{c}_2 \int d\vec{e}\, \Theta(-\vec{c}_{12} \cdot \vec{e}) |\vec{c}_{12} \cdot \vec{e}|\, \phi(c_1)\phi(c_2) \text{Expr}(\cdots)\,, \tag{A.1}$$

where $\vec{c}_1$, $\vec{c}_2$ and $\vec{e}$ are, respectively, the reduced velocities of two colliding spheres and a unit vector which specifies the geometry of the collision (see Fig. 6.1). The standard Maxwell distribution is denoted by

$$\phi(c) \equiv \pi^{-d/2} \exp\left(-c^2\right)\,. \tag{A.2}$$

We call integrals of the type (A.1) *Kinetic Integrals*. The kernel of the integral, **Expr**, may contain the pre- and after-collision velocities, $\vec{c}_{1/2}$, $\vec{c}\,'_{1/2}$, the Sonine polynomials S_i of these velocities, the Sonine coefficients a_i, the unit vector $\vec{e}$, the relative velocity $\vec{c}_{12}$, the centre of mass velocity $\vec{C}$, the coefficient of restitution, either $\varepsilon = \text{const.}$ or $\varepsilon = \varepsilon(g)$ for viscoelastic particles, scalar products of the mentioned vectors and other variables. Some examples are

$$\begin{aligned}
\text{Expr} &= \Delta\left(c_1^2 + c_2^2\right)\left[-\frac{1}{2}\left(1 + a_2\left(S_2(c_1^2) + S_2(c_2^2)\right)\right)\right] & \text{see (8.43)} \\
\text{Expr} &= (\vec{c}_{12} \cdot \vec{e})^2 (1 + \varepsilon)\left\{1 + a_2(t)\left[S_2\left(c_1^2\right) + S_2\left(c_2^2\right)\right]\right. \\
&\quad \left. + a_2^2(t) S_2\left(c_1^2\right) S_2\left(c_2^2\right)\right\} & \text{see (14.9)} \quad \text{(A.3)} \\
\text{Expr} &= \left[(\vec{c}\,'_1 \cdot \vec{c}_2)^2 + (\vec{c}\,'_2 \cdot \vec{c}_2)^2 - (\vec{c}_1 \cdot \vec{c}_2)^2 - (\vec{c}_2 \cdot \vec{c}_2)^2 \right. \\
&\quad \left. - \frac{1}{3} c_2^2 \Delta\left(c_1^2 + c_2^2\right)\right] & \text{see (19.27)}
\end{aligned}$$

Kinetic integrals may be transformed into sums of standard type integrals, this procedure is described in detail in Section 8.3.1. These *Basic Integrals* are of the form

$$J_{k,l,m,n,p,\alpha} = \int d\vec{c}_{12} \int d\vec{C} \int d\vec{e}\, \Theta\left(-\vec{c}_{12} \cdot \vec{e}\right) |\vec{c}_{12} \cdot \vec{e}|^{1+\alpha}$$
$$\phi(c_{12})\, \phi(C)\, C^k c_{12}^l \left(\vec{C} \cdot \vec{c}_{12}\right)^m \left(\vec{C} \cdot \vec{e}\right)^n (\vec{c}_{12} \cdot \vec{e})^p\,, \tag{A.4}$$

where $\phi(\vec{c}_{12})$ and $\phi(\vec{C})$ are respectively Maxwell distributions for the relative velocity and for the centre of mass velocity of a pair of particles. They read for the d-dimensional case

$$\phi(\vec{c}_{12}) = \frac{1}{(2\pi)^{d/2}} \exp\left(-\frac{1}{2}c_{12}^2\right), \quad \phi(\vec{C}) = \left(\frac{2}{\pi}\right)^{d/2} \exp\left(-2C^2\right). \quad (A.5)$$

There is a general solution of the Basic Integral which is described in the following section. Using this solution, therefore, any Kinetic Integral can be solved by a regular procedure. However, the transformation of a Kinetic Integral may lead to a sum of a large number of Basic Integrals, which may be as large as several thousands. Hence, we apply computational formula manipulation to evaluate Kinetic Integrals. The corresponding Maple program is explained in detail in Section A.3.

A.2   Evaluation of the basic integral

The solution of the Basic Integral (A.4) is straightforward for $n = 0$, however, for $n = 1, 2$ it requires some tricks (see, e.g. Resibois and de Leener, 1977). The presented approach may be generalised to $n > 2$, here we need only $n = 0, 1, 2$.

In the d-dimensional space the Cartesian coordinates of a vector $\vec{r} = (x_1, \ldots, x_d)$ are related to its spherical coordinates, expressed by the absolute value of the vector $r \geq 0$ and $d - 1$ angles $\theta_1, \theta_2, \ldots, \theta_{d-1}$, by the transformation

$$x_1 = r \sin\theta_1 \sin\theta_2 \cdots \sin\theta_{d-1}$$

$$x_m = r \cos\theta_{m-1} \prod_{k=m}^{d-1} \sin\theta_k, \quad m = 2, 3, \ldots, d-1 \quad (A.6)$$

$$x_d = r \cos\theta_{d-1},$$

where

$$\begin{array}{ll} 0 \leq \theta_1 \leq 2\pi \\ 0 \leq \theta_m \leq \pi, & m = 2, 3, \ldots, d-1. \end{array} \quad (A.7)$$

Hence, the infinitesimal solid angle reads

$$d\vec{e} = \prod_{k=1}^{d-1} \sin^{k-1}\theta_k \, d\theta_k. \quad (A.8)$$

Correspondingly we obtain for the surface area of a d-dimensional unit sphere

$$\Omega_d = \int d\vec{e} = \int_0^{2\pi} d\theta_1 \int_0^{\pi} \sin\theta_2 d\theta_2 \int_0^{\pi} \sin^2\theta_3 d\theta_3 \cdots \int_0^{\pi} \sin^{d-2}\theta_{d-1} d\theta_{d-1}$$

$$= 2\pi \frac{\Gamma(1)\Gamma\left(\frac{1}{2}\right)}{\Gamma\left(\frac{3}{2}\right)} \frac{\Gamma\left(\frac{3}{2}\right)\Gamma\left(\frac{1}{2}\right)}{\Gamma(2)} \frac{\Gamma(2)\Gamma\left(\frac{1}{2}\right)}{\Gamma\left(\frac{5}{2}\right)} \cdots \frac{\Gamma\left(\frac{d-1}{2}\right)\Gamma\left(\frac{1}{2}\right)}{\Gamma\left(\frac{d}{2}\right)}$$

$$= 2\pi \left[\Gamma\left(\frac{1}{2}\right)\right]^{d-2} \frac{1}{\Gamma\left(\frac{d}{2}\right)} = \frac{2\pi^{d/2}}{\Gamma\left(\frac{d}{2}\right)},$$

(A.9)

where we use the properties of the Gamma-function, $\Gamma(1) = 1$, $\Gamma\left(\frac{1}{2}\right) = \sqrt{\pi}$ and the trigonometric integral

$$\int_0^{\pi} \sin^l \theta d\theta = \frac{\Gamma\left(\frac{l+1}{2}\right)\Gamma\left(\frac{1}{2}\right)}{\Gamma\left(\frac{l}{2}+1\right)}. \tag{A.10}$$

Let us show how this kind of integrals may be evaluated. First, we notice that

$$\int_0^{\pi} \sin^l \theta d\theta = 2 \int_0^{\pi/2} \sin^l \theta d\theta = 2 \int_0^{\pi/2} \cos^l \theta d\theta \tag{A.11}$$

and evaluate with the substitute $y = \cos^2 \theta$ and the definition of the Beta-function $B(x,y)$ the integral

$$\int_0^{\pi/2} \sin^l \theta \cos^n \theta d\theta = \frac{1}{2} \int_0^1 (1-y)^{\frac{l-1}{2}} y^{\frac{n-1}{2}} dy$$

$$= \frac{1}{2} B\left(\frac{l+1}{2}, \frac{n+1}{2}\right) = \frac{1}{2} \frac{\Gamma\left(\frac{l+1}{2}\right)\Gamma\left(\frac{n+1}{2}\right)}{\Gamma\left(\frac{l+n}{2}+1\right)}. \tag{A.12}$$

(A.10) follows from (A.12, A.11).

Before calculating the Basic Integral we evaluate some more expressions. We will need the angular integral

$$\beta_m = \int d\vec{e}\, \Theta\left(\vec{e}\cdot\hat{g}\right)\left(\vec{e}\cdot\hat{g}\right)^m = (-1)^m \int d\vec{e}\, \Theta\left(-\vec{e}\cdot\hat{g}\right)\left(\vec{e}\cdot\hat{g}\right)^m, \tag{A.13}$$

where $\hat{g} \equiv \vec{g}/g$ is the unit vector directed along $\vec{g}$. In (A.13) we change the variables $\vec{e} \to -\vec{e}$ to get the expression for β_m which will be mainly used in the calculations. Let us choose the coordinate axis OX_d along $\vec{g}$, so that $\vec{e}\cdot\hat{g} = \cos\theta_{d-1}$. Using again (A.10, A.12), the above properties of the Gamma function and (A.8) for the infinitesimal solid angle $d\vec{e}$, we obtain

EVALUATION OF THE BASIC INTEGRAL

$$\beta_m = \int d\vec{e}\,\Theta\left(\vec{e}\cdot\hat{\vec{g}}\right)\left(\vec{e}\cdot\hat{\vec{g}}\right)^m$$

$$= \int_0^{2\pi} d\theta_1 \int_0^\pi \sin\theta_2 d\theta_2 \int_0^\pi \sin^2\theta_3 d\theta_3 \cdots \int_0^{\pi/2} \sin^{d-2}\theta_{d-1}\cos^m\theta_{d-1} d\theta_{d-1}$$

$$= 2\pi \frac{\Gamma(1)\Gamma\left(\frac{1}{2}\right)}{\Gamma\left(\frac{3}{2}\right)} \frac{\Gamma\left(\frac{3}{2}\right)\Gamma\left(\frac{1}{2}\right)}{\Gamma(2)} \cdots \frac{1}{2}\frac{\Gamma\left(\frac{d-1}{2}\right)\Gamma\left(\frac{m+1}{2}\right)}{\Gamma\left(\frac{d+m}{2}\right)}$$

$$= \pi^{(d-1)/2}\frac{\Gamma\left(\frac{m+1}{2}\right)}{\Gamma\left(\frac{m+d}{2}\right)}, \quad\text{(A.14)}$$

where we take into account that the integration over θ_{d-1} is to be performed in the interval $[0, \pi/2]$ due to the factor $\Theta\left(\vec{e}\cdot\hat{\vec{g}}\right)$. Note that β_m does not depend on the vector $\hat{\vec{g}}$. Similarly we obtain,

$$\int d\vec{e}\left(\vec{e}\cdot\hat{\vec{g}}\right)^m = \int d\vec{e}\left[\Theta\left(\vec{e}\cdot\hat{\vec{g}}\right)+\Theta\left(-\vec{e}\cdot\hat{\vec{g}}\right)\right]\left(\vec{e}\cdot\hat{\vec{g}}\right)^m$$

$$= [1+(-1)^m]\,\beta_m \quad\text{(A.15)}$$

$$= \int d\hat{\vec{g}}\left(\hat{\vec{g}}\cdot\hat{\vec{C}}\right)^m,$$

where in the last equation we take into account that $\vec{e}$ is the integration variable, and that β_m does not depend on $\hat{\vec{g}}$. This allows to change the notations for the integration variable and for $\hat{\vec{g}}$ as $\vec{e}\to\hat{\vec{g}}$ and $\hat{\vec{g}}\to\hat{\vec{C}}$, where $\hat{\vec{C}}=\vec{C}/C$ is a unit vector. We also need the integrals

$$\gamma_r \equiv \int d\vec{C}\,C^r \phi(C) = \int_0^\infty C^{d-1+r}\left(\frac{2}{\pi}\right)^{d/2}\exp\left(-2C^2\right)dC\int d\hat{\vec{C}}$$

$$= \left(\frac{2}{\pi}\right)^{d/2}\frac{1}{4}\frac{\sqrt{2}\,\Gamma\left(\frac{d+r}{2}\right)}{\sqrt{2^{d+r-1}}}\Omega_d = 2^{-r/2}\frac{\Gamma\left(\frac{d+r}{2}\right)}{\Gamma\left(\frac{d}{2}\right)} \quad\text{(A.16)}$$

and

$$\int d\vec{g}\,g^r\,\phi(g) = \int_0^\infty g^{d-1+r}\left(\frac{1}{2\pi}\right)^{d/2}\exp\left(-\frac{g^2}{2}\right)dg\int d\hat{\vec{g}}$$

$$= 2^{r/2}\frac{\Gamma\left(\frac{d+r}{2}\right)}{\Gamma\left(\frac{d}{2}\right)} = 2^r \gamma_r, \quad\text{(A.17)}$$

with

$$\int d\hat{\vec{g}} = \int d\hat{\vec{C}} = \int d\vec{e} = \Omega_d \quad\text{(A.18)}$$

according to (A.9).

For $n=0$, the definition of the Basic Integral (A.4) reads (with $\vec{g} \equiv \vec{c}_{12}$)

$$J_{k,l,m,0,p,\alpha} = \int d\vec{g} \int d\vec{C} \phi(g)\phi(C) C^k g^l (\vec{C}\cdot\vec{g})^m K_{\alpha,p}(g), \quad (A.19)$$

where we introduce the integral

$$K_{\alpha,p}(g) \equiv \int d\vec{\mu}\, |\vec{e}\cdot\vec{g}|^\alpha (\vec{e}\cdot\vec{g})^p, \quad (A.20)$$

with the short-hand notation

$$d\vec{\mu} \equiv d\vec{e}\, \Theta(-\vec{e}\cdot\vec{g})\, |\vec{e}\cdot\vec{g}|. \quad (A.21)$$

Similarly, for $n=1$:

$$J_{k,l,m,1,p,\alpha} = \int d\vec{g} \int d\vec{C}\, \phi(g)\phi(C) C^k g^l (\vec{C}\cdot\vec{g})^m \left[\vec{C}\cdot\vec{I}_{\alpha,p}(g)\right], \quad (A.22)$$

with the vectorial integral

$$\vec{I}_{\alpha,p}(g) \equiv \int d\vec{\mu}\, \vec{e}\, |\vec{e}\cdot\vec{g}|^\alpha (\vec{e}\cdot\vec{g})^p. \quad (A.23)$$

Finally, for $n=2$:

$$J_{k,l,m,2,p,\alpha} = \int d\vec{g} \int d\vec{C}\, \phi(g)\phi(C) C^k g^l (\vec{C}\cdot\vec{g})^m\, \vec{C}\cdot\hat{H}_{\alpha,p}(g)\cdot\vec{C} \quad (A.24)$$

where the dyad $\hat{H}_{\alpha,p}(g)$ is given by

$$\hat{H}_{\alpha,p}(g) \equiv \int d\vec{\mu}\, \vec{e}\,\vec{e}\, |\vec{e}\cdot\vec{g}|^\alpha (\vec{e}\cdot\vec{g})^p \quad (A.25)$$

and where we take into account that according to the vector-dyadic product

$$\vec{C}\cdot\vec{e}\vec{e}\cdot\vec{C} = \left(\vec{C}\cdot\vec{e}\right)\left(\vec{e}\cdot\vec{C}\right) = \left(\vec{C}\cdot\vec{e}\right)^2. \quad (A.26)$$

With the quantities introduced above we can express the Basic Integral step by step. The integral $K_{\alpha,p}(g)$ reads then

$$\begin{aligned} K_{\alpha,p}(g) &= \int d\vec{e}\, \Theta(-\vec{e}\cdot\vec{g})\, |\vec{e}\cdot\vec{g}|^{\alpha+1} (\vec{e}\cdot\vec{g})^p \\ &= (-1)^p g^{p+\alpha+1} \int d\vec{e}\, \Theta(\vec{e}\cdot\hat{\vec{g}})\, (\vec{e}\cdot\hat{\vec{g}})^{p+\alpha+1} \quad (A.27)\\ &= (-1)^p g^{p+\alpha+1} \beta_{p+\alpha+1} \end{aligned}$$

with the definition (A.13) of β_m.

EVALUATION OF THE BASIC INTEGRAL

Consider now the vectorial integral $\vec{I}_{\alpha,p}(g)$. The result of the integration over $\vec{\mu}$, that is, over $\vec{e}$ (see (A.21)), must be a vector. Due to the symmetry of the problem, this vector should be directed along the vector $\vec{g}$ since there is no other preferred direction. Thus,

$$\vec{I}_{\alpha,p}(g) = \vec{g}\, G(g) \tag{A.28}$$

with the function $G(g)$ to be found. We multiply (A.23) by $\vec{g}$ and obtain

$$\vec{g}\cdot\vec{I}_{\alpha,p}(g) = g^2 G(g) = \int d\vec{e}\,\Theta(-\vec{e}\cdot\vec{g})\,|\vec{e}\cdot\vec{g}|^{\alpha+1}(\vec{e}\cdot\vec{g})^{p+1} = K_{\alpha,p+1}(g)\,. \tag{A.29}$$

Therefore, $G(g) = g^{-2} K_{\alpha,p+1}(g)$ and thus with (A.27) follows

$$\vec{I}_{\alpha,p}(g) = (-1)^{p+1} \beta_{p+\alpha+2}\, g^{p+\alpha}\,\vec{g}\,. \tag{A.30}$$

The integrand in (A.25) is a dyad proportional to the dyad $\vec{e}\vec{e}$. Since integration over $\vec{e}$ does not change the symmetry, we expect that the components of the resultant dyad $\hat{H}_{\alpha,p}(g)$ are composed of the components of the vector $\vec{g}$, which is the only vectorial quantity in (A.25). Since the result may also have a non-vectorial part, $\hat{H}_{\alpha,p}(g)$ is of the general form

$$\hat{H}(g) = A(g)\,\vec{g}\vec{g} + B(g)\,g^2\,\hat{U}\,, \tag{A.31}$$

where $\hat{U}$ is a unit dyad (i.e. the unit diagonal matrix) and the functions $A(g)$ and $B(g)$ are to be determined. Multiplying $\hat{H}$ (scalar product) from both sides by $\vec{g}$ we obtain the first equation for $A(g)$ and $B(g)$:

$$\begin{aligned}\vec{g}\cdot\hat{H}\cdot\vec{g} &= A(g)g^4 + B(g)g^4 \\ &= \int d\vec{e}\,\Theta(-\vec{e}\cdot\vec{g})\,|\vec{e}\cdot\vec{g}|^{\alpha+1}(\vec{e}\cdot\vec{g})^{p+2} = K_{\alpha,p+2}(g)\,,\end{aligned} \tag{A.32}$$

where we use

$$\vec{g}\cdot\vec{g}\vec{g}\cdot\vec{g} = (g^2)^2 = g^4\,, \qquad \vec{g}\cdot\hat{U}\cdot\vec{g} = \vec{g}\cdot\vec{g} = g^2\,. \tag{A.33}$$

Then taking the trace of $\hat{H}$ (i.e. the sum of diagonal elements) we arrive at the second equation

$$\mathrm{Tr}\,\hat{H} = A(g)g^2 + d\,B(g)g^2 = \int d\vec{e}\,\Theta(-\vec{e}\cdot\vec{g})\,|\vec{e}\cdot\vec{g}|^{\alpha+1}(\vec{e}\cdot\vec{g})^{p} = K_{\alpha,p}(g)\,. \tag{A.34}$$

where we take into account (with the summation convention)

$$\mathrm{Tr}\,\vec{g}\vec{g} = g_i g_i = g^2\,, \qquad \mathrm{Tr}\,\hat{U} = \delta_{ii} = d\,, \qquad \mathrm{Tr}\,\vec{e}\vec{e} = e_i e_i = e^2 = 1\,. \tag{A.35}$$

Solving the set of equations (A.32, A.34) for the functions $A(g)$ and $B(g)$ we obtain with (A.27) for $K_{\alpha,p}(g)$

$$\hat{H} = \frac{(-1)^p\, g^{p+\alpha-1}}{(d-1)} \left[(d\beta_{p+\alpha+3} - \beta_{p+\alpha+1})\, \vec{g}\vec{g} + (\beta_{p+\alpha+1} - \beta_{p+\alpha+3})\, g^2 \hat{U} \right]. \tag{A.36}$$

With the obtained results for the scalar integral $K_{\alpha,p}(g)$, for the vectorial integral $\vec{I}_{\alpha,p}(g)$ and for the tensorial integral $\hat{H}_{\alpha,p}(g)$ we can evaluate the Basic Integrals. For the first Basic Integral with $n=0$ we obtain with (A.27)

$$\begin{aligned}
J_{k,l,m,0,p,\alpha} &= (-1)^p \beta_{p+\alpha+1} \int d\vec{g} \int d\vec{C}\, \phi(g)\phi(C) C^{k+m} g^{l+m+p+\alpha+1} (\hat{\vec{C}} \cdot \hat{\vec{g}})^m \\
&= (-1)^p \beta_{p+\alpha+1} \int d\vec{C}\, C^{k+m}\phi(C) \int d\vec{g}\, g^{l+m+p+\alpha+1}\phi(g) \int d\hat{\vec{g}}\, \left(\hat{\vec{g}}\cdot\hat{\vec{C}}\right)^m \Omega_d^{-1},
\end{aligned} \tag{A.37}$$

where we write the vectors $\vec{g}$ and $\vec{C}$ in the form $\vec{g} = g\hat{\vec{g}}$ and $\vec{C} = C\hat{\vec{C}}$ and for the integration over $d\vec{g}$ we use the identity

$$\begin{aligned}
\int d\vec{g}\, F_1(g) F_2(\hat{\vec{g}}) &= \int_0^\infty g^{d-1} F_1(g)\, dg \int d\hat{\vec{g}}\, F_2(\hat{\vec{g}}) \\
&= \int_0^\infty g^{d-1} F_1(g)\, dg \int d\hat{\vec{g}}\, \frac{\int d\hat{\vec{g}}\, F_2(\hat{\vec{g}})}{\int d\hat{\vec{g}}} \\
&= \int d\vec{g}\, F_1(g) \int d\hat{\vec{g}}\, F_2(\hat{\vec{g}})\, \Omega_d^{-1}
\end{aligned} \tag{A.38}$$

(recall that $\int d\hat{\vec{g}} = \Omega_d$). The integral over $d\hat{\vec{g}}$ in (A.37) is $[1+(-1)^m]\beta_m$, according to (A.15), while the integrals over $d\vec{C}$ and over $d\vec{g}$ are given by (A.16) and (A.17), respectively. Finally, we obtain the Basic Integral for $n=0$:

$$\boxed{\begin{aligned}
J_{k,l,m,0,p,\alpha} &= (-1)^p \left[1+(-1)^m\right] 2^{l+m+p+\alpha+1} \Omega_d^{-1} \\
&\quad \times \beta_{p+\alpha+1}\beta_m \gamma_{k+m}\gamma_{l+m+p+\alpha+1}
\end{aligned}} \tag{A.39}$$

The evaluation of the Basic Integral for $n=1$ and $n=2$ may be performed by substituting (A.30, A.36), respectively, into (A.22, A.24). The integrals that are obtained by this substitute are precisely of the same structure as in (A.37). Hence, the calculations for $n=1$ and $n=2$ are analogous to the case $n=0$. The final result reads for $n=1$

$$\boxed{\begin{aligned}
J_{k,l,m,1,p,\alpha} &= (-1)^{p+1} \left[1+(-1)^{m+1}\right] 2^{l+m+p+\alpha+1} \Omega_d^{-1} \\
&\quad \times \beta_{p+\alpha+2}\beta_{m+1}\gamma_{k+m+1}\gamma_{l+m+p+\alpha+1}
\end{aligned}} \tag{A.40}$$

and for $n = 2$

$$\begin{aligned}J_{k,l,m,2,p,\alpha} = &(-1)^p \left[1 + (-1)^m\right] 2^{l+m+p+\alpha+1} \left[(d-1)\Omega_d\right]^{-1} \\ &\times \gamma_{k+m+2}\, \gamma_{l+m+p+\alpha+1}\, \left[(d\beta_{p+\alpha+3} - \beta_{p+\alpha+1})\beta_{m+2}\right. \\ &\left. + (\beta_{p+\alpha+1} - \beta_{p+\alpha+3})\beta_m\right]\end{aligned} \quad \text{(A.41)}$$

For convenience we list here the constants:

$$\beta_m = \pi^{(d-1)/2}\frac{\Gamma\left(\frac{m+1}{2}\right)}{\Gamma\left(\frac{m+d}{2}\right)}, \quad \gamma_m = 2^{-m/2}\frac{\Gamma\left(\frac{m+d}{2}\right)}{\Gamma\left(\frac{d}{2}\right)}, \quad \Omega_d = \frac{2\pi^{d/2}}{\Gamma\left(\frac{d}{2}\right)}$$

(A.42)

For the three-dimensional case explicit expressions for the Basic Integral are given in Chapter 4.2.3 by (8.31–8.33).

A.3 Computational formula manipulation of kinetic integrals

Kinetic Integrals as specified by (A.1)

$$\int d\vec{c}_1 \int d\vec{c}_2 \int d\vec{e}\, \Theta\left(-\vec{c}_{12} \cdot \vec{e}\right) |\vec{c}_{12} \cdot \vec{e}|\, \phi(c_1)\, \phi(c_2)\, \text{Expr}(\cdots)\,, \quad \text{(A.43)}$$

can be evaluated in an algorithmic way using the method of Basic Integrals. For practical applications, however, even simple looking expressions, Expr, may result in sums of thousands of terms of the type $J_{k,l,m,n,p,\alpha}$. Therefore, we apply a formula manipulation program, such as Maple.

Before giving a systematic description we wish to discuss a simple example: we want to compute the fourth moment of the collision integral in linear approximation with respect to the second Sonine coefficient. The formula is given by (8.23):

$$\begin{aligned}\mu_4 = -\frac{1}{2}\int d\vec{c}_1 \int d\vec{c}_2 \int d\vec{e}\, \Theta\left(-\vec{c}_{12}\cdot \vec{e}\right) |\vec{c}_{12}\cdot\vec{e}|\, \phi(c_1)\,\phi(c_2) \\ \times \left\{1 + a_2\left[S_2(c_1^2) + S_2(c_2^2)\right]\right\}\Delta\left(c_1^4 + c_2^4\right)\,.\end{aligned} \quad \text{(A.44)}$$

Comparing this expression with our standard form (A.43) we obtain

$$\text{Expr} = -\frac{1}{2}\left\{1 + a_2\left[S_2(c_1^2) + S_2(c_2^2)\right]\right\}\Delta\left(c_1^4 + c_2^4\right)\,. \quad \text{(A.45)}$$

The listing demonstrates the application of the package `KineticIntegral.m`, which will be described in detail below.

```
>  restart;
>  libname:=libname,'/home/kies/SMGG/':
>  with(KineticIntegral);
```
        $[DefDimension, DefJ, DefS, KIinit, getJexpr, unDefJ]$
```
>  KIinit();
>  DefDimension(3);
```

Dimension = 3

Sonine polynomials defined
```
>  DefJ();
```

Basic Integrals will be evaluated
```
>  Expr:=Delta4*(-1/2*(1+a2*(S(2,c1p)+S(2,c2p)))):
>  mu4:=getJexpr(0,Expr,0);
```

expressing 90 terms in form of Basic Integrals

$$\mu 4 := -\frac{1}{32}(1+\varepsilon)\sqrt{\pi}\sqrt{2}$$
$$(30\,\varepsilon^3\,a2 + 32\,\varepsilon^3 - 30\,a2\,\varepsilon^2 - 32\,\varepsilon^2 + 144\,\varepsilon + 207\,a2\,\varepsilon - 144 - 271\,a2)$$

We start the computation with `restart` to clean the memory of the computer from earlier results, then with `libname...` we extend the library path of Maple. In our case the library `KineticIntegral.m` is located in the directory `/home/kies/SMGG/`. It is loaded by `with....` The next line is an output produced by Maple, indicating that the commands `DefDimension`, `DefJ`, `DefS`, `KIinit`, `getJexpr` and `unDefJ` have been provided by the package `KineticIntegral.m`. Here starts the evaluation of our integral: first we initialize the computation by `KIinit()` which defines some internal variables. Then we specify the dimension of our problem. Here we wish to evaluate the integral in three dimensions. With the line `DefJ()` we require that the Basic Integrals will be evaluated, that is, in the final result we generate real numbers according to (A.39)-(A.41), e.g., $J_{0,0,2,0,2,0} = 6\sqrt{2\pi}$. The line `Expr:=...` defines the kernel of the integral due to (A.44). Finally `mu4:=getJexpr(0,Expr,0)` starts the computation. The arguments of `getJexpr()` as well as the syntax of `Expr` are described in detail below. The Maple output (which is the solution of Exercise 8.3) coincides with (8.38).

Since the first part of the program is identical in all applications, only the main part is given in the text. The program above then reads

```
1  DefDimension(3);
2  DefJ();
3  Expr:=Delta4*(-1/2**(1+a2*(S(2,c1p)+S(2,c2p)))):
4  mu4:=getJexpr(0,Expr,0);
```

We wish to explain the Maple package `KineticIntegral.m` in detail. To this end we comment the program line by line. If we refer to a certain line number L of the code we use the notation $\boxed{\text{L}}$. There is no need to retype the program, it can be found at the address http://www.oup.co.uk/isbn/0-19-853038-2. The program uses internal variables whose names follow a regular scheme: c1p, c2p, c2a and c2p stand for the pre-collision and after-collision velocities of the particles. The unit vector e$\equiv E \equiv \vec{r}_{12}/r_{12}$ and the dimensionality DIM are further variables. Scalar products are named by xDOTy, where x and y are vectors. An example is c1pDOTc2p$\equiv \vec{c}_1 \cdot \vec{c}_2$. A full list of the internal variables is given in the table below. All these symbols can be used to specify the term Expr which is the kernel of the integral.

c1p	$\vec{c}_1$:	pre-collision velocity of particle 1
c2p	$\vec{c}_2$:	pre-collision velocity of particle 2
c1a	$\vec{c}_1'$:	after-collision velocity of particle 1
c2a	$\vec{c}_2'$:	after-collision velocity of particle 2
c12	$\vec{c}_{12}$:	relative velocity
C	$\vec{C}$:	centre of mass velocity
c12DOTe=eDOTc12	$\vec{c}_{12} \cdot \vec{e}$	
CDOTc12=c12DOTC	$\vec{C} \cdot \vec{c}_{12}$	
c1pDOTc1p	$(\vec{c}_1)^2 = \vec{C}^2 + \vec{c}_{12}^2/4 + \vec{C} \cdot \vec{c}_{12}$	
c2pDOTc2p	$(\vec{c}_2)^2 = \vec{C}^2 + \vec{c}_{12}^2/4 - \vec{C} \cdot \vec{c}_{12}$	
c1pDOTc2p =c2pDOTc1p	$\vec{c}_1 \cdot \vec{c}_2 = \vec{C}^2 - \vec{c}_{12}^2/4$	
c1aDOTc1a	$(\vec{c}_1')^2 = \vec{C}^2 + \vec{c}_{12}^2/4 - (1-\varepsilon^2)/4 (\vec{c}_{12} \cdot \vec{e})^2$ $+ \vec{C} \cdot \vec{c}_{12} - (1+\varepsilon)(\vec{C} \cdot \vec{e})(\vec{c}_{12} \cdot \vec{e})$	
c2aDOTc2a	$(\vec{c}_2')^2 = \vec{C}^2 + \vec{c}_{12}^2/4 - (1-\varepsilon^2)/4 (\vec{c}_{12} \cdot \vec{e})^2$ $- \vec{C} \cdot \vec{c}_{12} + (1+\varepsilon)(\vec{C} \cdot \vec{e})(\vec{c}_{12} \cdot \vec{e})$	
c1aDOTc2a=c2aDOTc1a	$\vec{c}_1' \cdot \vec{c}_2' = \vec{C}^2 - \vec{c}_{12}^2/4 + (1-\varepsilon^2)/4 (\vec{c}_{12} \cdot \vec{e})^2$	
c1pDOTc1a=c1aDOTc1p	$\vec{c}_1 \cdot \vec{c}_1' = \vec{C}^2 + \vec{c}_{12}^2/4 + \vec{C} \cdot \vec{c}_{12} - (1+\varepsilon)/4 (\vec{c}_{12} \cdot \vec{e})^2$ $-(1+\varepsilon)/2(\vec{C} \cdot \vec{e})(\vec{c}_{12} \cdot \vec{e})$	
c2pDOTc2a=c2aDOTc2p	$\vec{c}_2 \cdot \vec{c}_2' = \vec{C}^2 + \vec{c}_{12}^2/4 - \vec{C} \cdot \vec{c}_{12} - (1+\varepsilon)/4 (\vec{c}_{12} \cdot \vec{e})^2$ $+(1+\varepsilon)/2(\vec{C} \cdot \vec{e})(\vec{c}_{12} \cdot \vec{e})$	
c1aDOTc2p=c2pDOTc1a	$\vec{c}_1' \cdot \vec{c}_2 = \vec{C}^2 - \vec{c}_{12}^2/4 - (1+\varepsilon)/2(\vec{C} \cdot \vec{e})(\vec{c}_{12} \cdot \vec{e})$ $+(1+\varepsilon)/4 (\vec{c}_{12} \cdot \vec{e})^2$	
c1pDOTc2a=c2aDOTc1p	$\vec{c}_1 \cdot \vec{c}_2' = \vec{C}^2 - \vec{c}_{12}^2/4 + (1+\varepsilon)/2(\vec{C} \cdot \vec{e})(\vec{c}_{12} \cdot \vec{e})$ $+(1+\varepsilon)/4 (\vec{c}_{12} \cdot \vec{e})^2$	
Delta2, Delta4	see (A.46, A.47)	
DIM	dimension	
C1, dprime	describes the coefficient of restitution of viscoelastic particles, see (A.49)	
absc12DOTe5	$\|\vec{c}_{12} \cdot \vec{e}\|^{1/5}$, see (A.49)	

Two more variables, Delta2 and Delta4, are defined in the initialization program KIinit():

FUNCTIONS OF THE COLLISION INTEGRAL

────────── KIinit ──────────
```
 1  KIinit:=proc()
 2    global Delta2,Delta4:
 3    unprotect('Delta2','Delta4'):
 4    Delta2:=(1/2*epsilon^2-1/2)*c12DOTe^2:
 5    Delta4:=(-4*epsilon-4)*CDOTe*c12DOTe*CDOTc12+(1/8-1/4*epsilon^2
 6      +1/8*epsilon^4)*c12DOTe^4+(-1/4+1/4*epsilon^2)*c12^2*c12DOTe^2
 7      +(2*epsilon^2+2+4*epsilon)*c12DOTe^2*CDOTe^2
 8      +(epsilon^2-1)*C^2*c12DOTe^2:
 9    protect('Delta2','Delta4'):
10  end:
```

These variables express the change of $\vec{c}_1^p + \vec{c}_2^p$, $p = \{2, 4\}$, due to a collision. They are evaluated in (8.28, 8.29):

$$\texttt{Delta2} \equiv \Delta\left(c_1^2 + c_2^2\right) = \left(\frac{1}{2}\varepsilon^2 - \frac{1}{2}\right)(\vec{c}_{12} \cdot \vec{e})^2 \tag{A.46}$$

$$\texttt{Delta4} \equiv \Delta\left(c_1^4 + c_2^4\right) = 2\left(1+\varepsilon\right)^2 (\vec{c}_{12} \cdot \vec{e})^2 \left(\vec{C} \cdot \vec{e}\right)^2 + \frac{1}{8}\left(1-\varepsilon^2\right)^2 (\vec{c}_{12} \cdot \vec{e})^4$$
$$-\frac{1}{4}\left(1-\varepsilon^2\right)(\vec{c}_{12} \cdot \vec{e})^2 \vec{c}_{12}^{\,2} - \left(1-\varepsilon^2\right)\vec{C}^2 (\vec{c}_{12} \cdot \vec{e})^2$$
$$-4\left(1+\varepsilon\right)(\vec{C} \cdot \vec{c}_{12})(\vec{C} \cdot \vec{e})(\vec{c}_{12} \cdot \vec{e}) \ . \tag{A.47}$$

The kernel Expr may contain Sonine polynomials. They are defined by DefS().

────────── DefS ──────────
```
11  DefS:=proc()
12    global S:
13    if(assigned(DIM)) then
14      unprotect('S'):
15      S:=proc(a::integer,expr::algebraic)
16        local r,xx:
17        xx:=subs(c1p=C+1/2*c12,expr):
18        xx:=subs(c2p=C-1/2*c12,xx):
19        xx:=subs(c1a=C+1/2*c12-1/2*(1+epsilon)*c12DOTe*e,xx):
20        xx:=subs(c2a=C-1/2*c12+1/2*(1+epsilon)*c12DOTe*e,xx):
21        xx:=expand(xx^2):
22        xx:=algsubs(C*c12=CDOTc12,xx):
23        if(a=0)then
24          r:=1:
25        elif (a=1) then
26          r:=-xx+DIM/2:
27        elif (a=2) then
28          r:=xx^2/2 - (DIM+2)*xx/2 +(DIM+2)*DIM/8:
29        else
30          ERROR('invalid argument',a):
31        fi:
32        r:=factor(r):
33      end:
34      protect('S'):
35      printf("Sonine polynomials defined\n"):
36    else
37      ERROR('Dimension undefined'):
38    fi:
39  end:
```

First at $\boxed{13}$ it is checked whether the dimensionality DIM is specified, otherwise DefS reports an error at $\boxed{37}$. The syntax of the Sonine polynomials is S(a,expr), a specifies the order of the polynomial and expr is its algebraic argument which may contain the pre- and after-collision particle velocities c1p, c1a, c2p and c2a. In $\boxed{17\text{-}20}$ these velocities are translated into expressions of the centre of mass velocity C, the relative velocity c12, the coefficient of restitution epsilon, the unit vector e and the scalar product c12DOTe, due to (8.27). Note the different notation of the Sonine polynomials as defined by (7.7) (for three dimensions) (11.3) (for general dimension) and in the Maple program:

$$\text{S(a,expr)} \equiv S_a\left(\text{expr}^2\right). \qquad (A.48)$$

This difference of the notation is taken into account by $\boxed{21}$ where the entire term expr with all the described substitutions is squared. In the squared expressions there appears the term $\vec{C}\cdot\vec{c}_{12}$, which is again substituted by CDOTc12. The program evaluates Sonine polynomials of the order $a = \{0,1,2\}$ which is done in $\boxed{23\text{--}28}$. Any incorrect input a is acknowledged by an error message $\boxed{30}$.

All of the defined internal variables, including the Sonine polynomials, are accessible within the Maple program, for example,

```
>   restart;
>   libname:=libname,'/home/kies/SMGG/':
>   with(KineticIntegral):
>   KIinit();
>   DefDimension(3);
Dimension = 3
Sonine polynomials defined
>   S(2,c1a);
```
$-\frac{5}{2}CDOTc12 - \frac{5}{8}c12^2 - \frac{5}{2}C^2 - \frac{5}{8}c12DOTe^2 - \frac{5}{4}c12DOTe^2\varepsilon - \frac{5}{8}c12DOTe^2\varepsilon^2$
$+\frac{5}{2}C\,c12DOTe\,e + \frac{5}{4}c12\,c12DOTe\,e + \frac{5}{2}C\,c12DOTe\,e\varepsilon + \frac{5}{4}c12\,c12DOTe\,e\varepsilon$
$-C^3\,c12DOTe\,e\varepsilon - \frac{1}{2}C^2\,c12\,c12DOTe\,e + \frac{3}{4}C^2\,c12DOTe^2\varepsilon + \frac{3}{4}C^2\,c12DOTe^2\varepsilon^2$
$-\frac{1}{4}C\,c12DOTe\,e\,c12^2 + \frac{1}{2}C\,c12DOTe^2\,c12 - \frac{3}{4}C\,c12DOTe^3\,e^3\varepsilon - \frac{3}{4}C\,c12DOTe^3\,e^3\varepsilon^2$
$-\frac{1}{4}C\,c12DOTe^3\,e^3\varepsilon^3 - \frac{1}{8}c12^3\,c12DOTe\,e\varepsilon + \frac{3}{8}c12^2\,c12DOTe^2\varepsilon - C^3\,c12DOTe\,e$
$+\frac{3}{16}c12^2\,c12DOTe^2\varepsilon^2 + \frac{3}{4}C^2\,c12DOTe^2 - \frac{1}{4}C\,c12DOTe^3\,e^3 - \frac{1}{8}c12^3\,c12DOTe\,e$
$+\frac{3}{16}c12^2\,c12DOTe^2 - \frac{1}{8}c12\,c12DOTe^3\,e^3 + \frac{1}{8}c12DOTe^4\,e^4\varepsilon + \frac{3}{16}c12DOTe^4\,e^4\varepsilon^2$
$+\frac{1}{8}c12DOTe^4\,e^4\varepsilon^3 + \frac{1}{32}c12DOTe^4\,e^4\varepsilon^4 + \frac{1}{4}CDOTc12\,c12DOTe^2 + C^2\,CDOTc12$
$+\frac{15}{8} + \frac{1}{4}C^2\,c12^2 + \frac{1}{2}CDOTc12^2 + \frac{1}{4}CDOTc12\,c12^2 + \frac{1}{32}c12DOTe^4\,e^4 + \frac{1}{32}c12^4$
$+\frac{1}{2}C^4 - \frac{1}{2}C^2\,c12\,c12DOTe\,e\varepsilon + C\,c12DOTe^2\,c12\varepsilon - \frac{1}{4}C\,c12DOTe\,e\varepsilon\,c12^2$
$+\frac{1}{2}C\,c12DOTe^2\varepsilon^2\,c12 - \frac{3}{8}c12\,c12DOTe^3\,e^3\varepsilon - \frac{3}{8}c12\,c12DOTe^3\,e^3\varepsilon^2$
$-\frac{1}{8}c12\,c12DOTe^3\,e^3\varepsilon^3 - CDOTc12\,C\,c12DOTe\,e - \frac{1}{2}CDOTc12\,c12\,c12DOTe\,e$
$+\frac{1}{2}CDOTc12\,c12DOTe^2\varepsilon + \frac{1}{4}CDOTc12\,c12DOTe^2\varepsilon^2 - CDOTc12\,C\,c12DOTe\,e\varepsilon$
$-\frac{1}{2}CDOTc12\,c12\,c12DOTe\,e\varepsilon$

In general, it is of not much value to list them since according to the substitution of variables even small terms lead to impressive formulae, for example, For debugging purposes, however, this feature may be useful.

There are two different output modes: first we may be interested in the result of an integration in terms of Basic Integrals $J_{k,l,m,n,p,\alpha}$ and second (usually) we are interested in obtaining real numbers. The translation of the Basic Integrals into numbers is given by (A.39–A.41). These equations are implemented in DefJ():

─────────────────────────── DefJ ───────────────────────────
```
40  DefJ:=proc()
41    global J:
42    unprotect(J):
43    if(assigned(DIM)) then
44      J:=proc(k::integer,l::integer,m::integer,n::integer,p::integer,
45              a::integer)
46        local J0, J1, J2, beta, gamma, omega:
47        beta:=m->Pi^((DIM-1)/2)*GAMMA((m+1)/2)/GAMMA((m+DIM)/2):
48        gamma:=m->2^(-m/2)*GAMMA((m+DIM)/2)/GAMMA(DIM/2):
49        omega:=2*Pi^(DIM/2)/GAMMA(DIM/2):
50        J0:=(k,l,m,p,a)->(-1)^p*(1+(-1)^m)*2^(1+m+p+a+1)/omega*beta(p+a+1)
51            *beta(m)*gamma(k+m)*gamma(1+m+p+a+1):
52        J1:=(k,l,m,p,a)->(-1)^(p+1)*(1+(-1)^(m+1))
53            *2^(1+m+p+a+1)/omega*beta(p+a+2)*beta(m+1)*gamma(k+m+1)
54            *gamma(1+m+p+a+1):
55        J2:=(k,l,m,p,a)->(-1)^p*(1+(-1)^m)*2^(1+m+p+a+1)/omega*gamma(k+m+2)
56            *gamma(1+m+p+a+1)*((DIM*beta(p+a+3)-beta(p+a+1))*beta(m+2)
57            + (beta(p+a+1)-beta(p+a+3))*beta(m))/(DIM-1):
58        simplify(1/2*J0(k,l,m,p,a/5) * (n-1)*(n-2) -J1(k,l,m,p,a/5)*n*(n-2)
59            + 1/2*J2(k,l,m,p,a/5)*n*(n-1)):
60      end:
61      protect('J'):
62      printf("Basic Integrals will be evaluated\n"):
63    else
64      ERROR('Dimension undefined'):
65    fi:
66  end:
```

After calling defJ all results are given as real numbers by evaluating the Basic Integrals, for example,

```
>    restart;
>    libname:=libname,'/home/kies/SMGG/':
>    with(KineticIntegral);
         [DefDimension, DefJ, DefS, KIinit, getJexpr, unDefJ]
>    KIinit();
>    DefDimension(3);

Dimension = 3

Sonine polynomials defined
>    DefJ();

Basic Integrals will be evaluated
>    BIdemo:=J(0,0,1,1,1,1);
```

$$BIdemo := \frac{11}{100} \frac{2^{(3/5)} \pi^{(3/2)}}{\sin(\frac{1}{10}\pi) \Gamma(\frac{9}{10})}$$

```
>    evalf(BIdemo);
                                2.811423645
```

The counterpart of DefJ() is unDefJ(). After calling unDefJ, any result is given in terms of Basic Integrals.

```
                              ─── unDefJ ───
67  unDefJ:=proc()
68    global J:
69    if (assigned(J)) then
70      unprotect(J):
71      unassign('J');
72    fi;
73    printf("Basic Integrals will not be evaluated\n"):
74  end:
```

In the described programs DefS and DefJ we have used the variable DIM for the dimensionality of the Basic Integral. This variable is set by DefDimension():

```
                            ─── DefDimension ───
75  DefDimension:=proc(d::integer)
76    global DIM:
77    unprotect('DIM'):
78    DIM:=d:
79    printf("Dimension = %d\n",d):
80    protect('DIM'):
81    DefS():
82    if (assigned(J)) then
83      DefJ():
84    fi:
85  end:
```

After changing the variable DIM in line $\boxed{78}$ the Sonine polynomials which are specific for any dimension have to be redefined $\boxed{81}$. For the case that DefJ() has been called before, the Basic Integrals have to be redefined as well, $\boxed{83}$.

The heart of the Maple package is the program getJexpr(). This program has three arguments: epstype specifies the type of the particle interaction, epstype=0 stands for constant coefficient of restitution and epstype=1 stands for viscoelastic particles. The second argument is the kernel of the integral expr which is a function of the particle velocities before and after the collision, the dimension, the Sonine polynomials and their arguments, the Sonine coefficients and scalar products of the vectorial values. In fact expr may contain all variables which are listed in the table on page 261. The third argument describing the output mode is explained later. The text of getJexpr() is listed below:

```
                                       getJexpr
 86  getJexpr:=proc(epstype::integer,expr::algebraic,ou::integer)
 87    global di,ddi:
 88    local xx,termJ,it,term,ko,epow,aa,k,l,m,n,p,a2pow,vareps,C1expl,
 89       C1omega0,C12omega1:
 90    xx:=subs(c2pDOTc1p=c1pDOTc2p,expr);
 91    xx:=subs(c2aDOTc1a=c1aDOTc2a,xx);
 92    xx:=subs(c1pDOTc1a=c1aDOTc1p,xx);
 93    xx:=subs(c2pDOTc2a=c2aDOTc2p,xx);
 94    xx:=subs(c2pDOTc1a=c1aDOTc2p,xx);
 95    xx:=subs(c2aDOTc1p=c1pDOTc2a,xx);
 96    xx:=subs(c1pDOTc1p=(C*C+1/4*c12*c12+CDOTc12),xx);
 97    xx:=subs(c2pDOTc2p=(C*C+1/4*c12*c12-CDOTc12),xx);
 98    xx:=subs(c1pDOTc2p=(C*C-1/4*c12*c12),xx);
 99    xx:=subs(c1aDOTc1a=(C*C+1/4*c12*c12
100       -(1-epsilon*epsilon)/4*c12DOTe*c12DOTe+CDOTc12
101       -(1+epsilon)*CDOTe*c12DOTe),xx);
102    xx:=subs(c2aDOTc2a=(C*C+1/4*c12*c12
103       -(1-epsilon*epsilon)/4*c12DOTe*c12DOTe-CDOTc12
104       +(1+epsilon)*CDOTe*c12DOTe),xx);
105    xx:=subs(c1aDOTc2a=(C*C-1/4*c12*c12
106       +(1-epsilon*epsilon)/4*c12DOTe*c12DOTe),xx);
107    xx:=subs(c1aDOTc1p=(C*C+1/4*c12*c12+CDOTc12-(1+epsilon)/2*CDOTe*c12DOTe
108       -(1+epsilon)/4*c12DOTe*c12DOTe),xx);
109    xx:=subs(c2aDOTc2p=(C*C+1/4*c12*c12-CDOTc12+(1+epsilon)/2*CDOTe*c12DOTe
110       -(1+epsilon)/4*c12DOTe*c12DOTe),xx);
111    xx:=subs(c1aDOTc2p=(C*C-1/4*c12*c12-(1+epsilon)/2*CDOTe*c12DOTe
112       +(1+epsilon)/4*c12DOTe*c12DOTe),xx);
113    xx:=subs(c1pDOTc2a=(C*C-1/4*c12*c12+(1+epsilon)/2*CDOTe*c12DOTe
114       +(1+epsilon)/4*c12DOTe*c12DOTe),xx);
115    if(epstype=0) then
116       xx:=expand(xx):
117    elif(epstype=1) then
118       vareps:=1-C1*dprime*absc12DOTe5+3/5*C1^2*dprime^2*absc12DOTe5^2:
119       xx:=expand(algsubs(epsilon=vareps,xx)):
120    else
121       ERROR('invalid argument',epstype):
122    fi:
123    xx:=expand(simplify(xx)):
124    termJ:=0:
125    printf("expressing %d terms in form of Basic Integrals\n",nops(xx)):
126    for it from 1 to nops(xx) do
127       term:=op(it,xx):
128       ko:=coeffs(term,{epsilon,absc12DOTe5,C,c12,CDOTc12,CDOTe,c12DOTe,a2}):
```

COMPUTATIONAL FORMULA MANIPULATION   267

```
129       epow:=degree(term,epsilon):
130       aa:=degree(term,absc12DOTe5):
131       k:=degree(term,C):
132       l:=degree(term,c12):
133       m:=degree(term,CDOTc12):
134       n:=degree(term,CDOTe):
135       p:=degree(term,c12DOTe):
136       a2pow:=degree(term,a2):
137       termJ:=termJ+epsilon^epow*ko*a2^a2pow*J(k,l,m,n,p,aa):
138       di[it]:=ko*epsilon^epow*c12DOTe^aa*C^k*c12^l*CDOTc12^m*CDOTe^n*
139          c12DOTe^p*a2^a2pow:
140       ddi[it]:=term:
141       if(simplify(di[it]-ddi[it])!=0) then
142          ERROR('error evaluating term',it,' . see vectors di[], ddi[]'):
143       fi:
144    od:
145    termJ:=simplify(termJ):
146    if (ou=1) then
147       C1expl:=GAMMA(3/5)*sqrt(Pi)/(2^(1/5)*5^(2/5)*GAMMA(21/10));
148       termJ:=subs(C1=C1expl,termJ):
149       expand(evalf(termJ)):
150    elif (ou=0) then
151       factor(simplify(termJ)):
152    elif (ou=2) then
153       C1omega0:=omega0/(2*sqrt(2*Pi)*2^(1/10)*GAMMA(21/10)):
154       C12omega1:=omega1/(sqrt(2*Pi)*2^(1/5)*GAMMA(16/5)):
155       termJ:=taylor(termJ,dprime=0,3):
156       termJ:=normal(simplify(termJ,GAMMA,trig));
157       termJ:=subs(C1^2=C12omega1,termJ);
158       termJ:=subs(C1=C1omega0,termJ);
159       termJ:=normal(simplify(termJ));
160    else
161       ERROR('invalid argument',ou):
162    fi:
163 end:
```

First, in lines 90–114, the kernel of the integral expr is processed. All variables are expressed in terms of epsilon, C, c12, CDOTc12, CDOTe, c12DOTe, a2. If we deal with viscoelastic particles (epstype=1), the coefficient of restitution is a function of the relative velocity. It is expressed in our variables by (9.19):

$$\varepsilon = 1 - C_1 \delta'(t) \left| \vec{c}_{12} \cdot \vec{e} \right|^{1/5} + \frac{3}{5} C_1^2 \delta'^2(t) \left| \vec{c}_{12} \cdot \vec{e} \right|^{2/5} . \quad (A.49)$$

The coefficients δ' describes the material properties of the viscoelastic particles and C_1 is a numerical constant, see (3.22). Therefore, for this case ε has to be replaced by an expression in our variables which is done in lines 117–119. We introduce a new internal variable absc12DOTe5 $\equiv \left| \vec{c}_{12} \cdot \vec{e} \right|^{1/5}$.

The entire kernel of the integral is then in 123 represented as a sum of polynoms in the variables C, c12, CDOTc12, CDOTe, c12DOTe, a2, epsilon (for ε = const.) and absc12DOTe5 (for viscoelastic particles). These are precisely the variables which define our Basic Integrals for which we know the analytic solution. For any of these terms we determine in the loop $\boxed{126\text{--}144}$ the indices of the according Basic Integral. The terms are then successively picked out of the sum by $\boxed{127}$ and stored in `term`. Then the numerical coefficient of this term is determined, `ko`. In lines $\boxed{129\text{--}136}$ the powers of the variables in the present term are determined, i.e., `epow` is the power of `epsilon` in `term`, `aa` is the power of `absc12D0Te5`, etc. These powers determine the indices of the Basic Integral.

In this way, all terms of the sum are transformed in line $\boxed{137}$ into expressions of the Basic Integrals and the according prefactors. By processing the terms, `termJ` grows and finally it contains the full sum in terms of Basic Integrals.

The vector `di[]` in line $\boxed{138}$ is for testing purposes. It reconstructs the present term by using the powers of the variables as computed above and its prefactors. If the reconstructed term differs from the original term $\boxed{142}$ an error has occurred. This may happen by an typing error for the kernel of the integral, e.g. if erroneously an undefined variable would be used in the `Expr` term of a Kinetic Integral. In this case an error is reported.

In line $\boxed{145}$ the integral is completely expressed in terms of Basic Integrals. There are different output modes, specified by the third argument `ou` of the program `getJexpr()`. If `ou=0` the output is given in symbolic form, using the Gamma–function and its properties. The result is analytically exact, however, the output may become very large. For `ou=1`, the result is given by a floating point number. Before, the numerical constant C_1 is replaced in $\boxed{147}$ by its analytical expression (3.45):

$$C_1 = \frac{\sqrt{\pi}}{2^{1/5}5^{2/5}} \frac{\Gamma(3/5)}{\Gamma(21/10)} \,. \tag{A.50}$$

For the choice `ou=2` the result is given as a Taylor expansion using the numerical constants $\omega_{1/2}$ defined by (9.22).

Finally the defined commands have to be bundled to a package and stored with the name `KineticIntegral.m`. This is done by the last lines of the program:

```
                             makePackage
164  KineticIntegral:=table():
165  KineticIntegral[KIinit]:=eval(KIinit):
166  KineticIntegral[DefJ]:=eval(DefJ):
167  KineticIntegral[unDefJ]:=eval(unDefJ):
168  KineticIntegral[getJexpr]:=eval(getJexpr):
169  KineticIntegral[DefDimension]:=eval(DefDimension):
170  KineticIntegral[DefS]:=eval(DefS):
171  save KineticIntegral, 'KineticIntegral.m';
```

The presented program generates the maple library `KineticIntegral.m`. To evaluate integrals of the type (A.43) the library can be invoked and used as demonstrated by the example in the beginning of this section.

A summary of the available commands can be found on page 77. In many cases the results of this package have to be further processed by Maple. There are many applications of this program in this book. For practical applications it may be useful to study these examples on pages 76, 142, 206, 209, 215 and 216.

APPENDIX B

MOLECULAR DYNAMICS OF GRANULAR GASES

B.1 Event-driven molecular dynamics

The idea of molecular dynamics simulations is to solve Newton's equation of motion

$$\ddot{\vec{r}}_i = \frac{1}{m_i} \vec{F}_i \left(\vec{r}_1, \dot{\vec{r}}_1, \ldots, \vec{r}_N, \dot{\vec{r}}_N \right), \qquad i = 1, \ldots, N \qquad (B.1)$$

simultaneously for all particles which belong to the system. Starting the simulation with certain initial conditions, this system of equations can be integrated by a numerical algorithm such as a Runge–Kutta scheme or a predictor-corrector algorithm for any time t, provided the interaction force $\vec{F}_i$ is known as a function of the particle positions and velocities. Moreover, we have to specify the interaction forces at the system boundaries. In the case of viscoelastic particles the interaction force is given by (3.2, 3.4). This scheme, which we call *force-based molecular dynamics*, is very general and has been applied to many particle system, among them also granular systems (e.g. Herrmann et al., 1998).

As any many-particle system granular gases may be simulated by force-based molecular dynamics, however, there is a much more efficient algorithm which we call *event-driven molecular dynamics*. Granular gases are dilute systems, that is, the typical duration of a collision is much shorter than the mean time in between successive collisions of a particle. Therefore, collisions in which more than two particles are involved, are very rare events. Such many-particle interactions can be neglected as it is done throughout this book. The harder the particles, that is, the steeper the interaction potential, the better is this approximation. Hence, most of the time each of the particles propagates along a straight line, interrupted by collisions with other particles of very short duration. Therefore, the pairwise collisions of particles may be considered as instantaneous events and each of these events may be treated separately.

Assume two particles collide with the velocities $\vec{v}_1$ and $\vec{v}_2$. We can now integrate Newton's equation of motion for this collision and obtain the velocities after the collision, $\vec{v}_1'$ and $\vec{v}_2'$. As shown in Chapter 3, the integration of the equations of motion leads to the coefficient of restitution ε which relates the velocities before and after the collision. For the case of viscoelastic particles the coefficient of restitution is a function of the normal component of the relative velocity. Frequently, it is assumed that $\varepsilon = $ const., for a critical discussion of this approximation see Section 3.2.1.

The main idea of event-driven molecular dynamics is that in the entire system at any time instant there occurs at most one collision of infinitesimal duration. This collision alters the velocities of the involved particles according to a collision law which is characterized by the coefficient of restitution. During the time intervals in between collisions the particles move along ballistic trajectories. Hence, in the absence of external fields the particles move with constant velocities along straight lines. Obviously, this algorithm is much more efficient than force-based algorithms since the time consuming numerical integration of the particle interaction is avoided. Instead the dynamics of the system is determined by a series of discrete events.

The enormous increase of the simulation speed is the main motivation for applying event-driven algorithms. Therefore it has been applied not only to granular gases but also to other granular systems. The application of event-driven simulations is, however, not always justified. A heap of steel spheres, simulated with event-driven molecular dynamics could not conduct electrical current since at any time at most two of the spheres can be in touch. Obviously, this is not an adequate description.

Using modern algorithms we are presently able to simulate on a personal computer (Pentium 4, 2.5 GHz, 2 GBytes memory) several million particles over billions of collisions. Such algorithms are described in detail by Pöschel and Schwager (2004). Here we discuss a very simple algorithm which may be used to simulate small systems of up to about 10,000 particles in two dimensions.

B.2   A simple event-driven algorithm

The motion of particles in between successive collisions at time $t_k < t_{k+1}$ follows straight lines, that is,

$$\vec{r}_i(t_{k+1}) = \vec{r}_i(t_k) + \vec{v}_i(t_k)(t_{k+1} - t_k), \qquad i = 1, \ldots, N, \qquad (B.2)$$

where $v(t_k)$ denotes the velocity *after* the collision which occurred at time t_k.

The change of the velocities due to a collision of particles i and j at time t_{k+1} is determined by the collision rule (2.7):

$$\vec{v}_l(t_{k+1}) = \vec{v}_l(t_k) \qquad \text{for} \qquad l = 1, \ldots, N, \quad l \neq i, \quad l \neq j,$$

$$\vec{v}_i(t_{k+1}) = \vec{v}_i(t_k) - \frac{1}{2}(1+\varepsilon)\{[\vec{v}_i(t_k) - \vec{v}_j(t_k)] \cdot [\vec{r}_i(t_{k+1}) - \vec{r}_j(t_{k+1})]\}$$
$$\times \frac{\vec{r}_i(t_{k+1}) - \vec{r}_j(t_{k+1})}{[\vec{r}_i(t_{k+1}) - \vec{r}_j(t_{k+1})]^2} \qquad (B.3)$$

$$\vec{v}_j(t_{k+1}) = \vec{v}_j(t_k) + \frac{1}{2}(1+\varepsilon)\{[\vec{v}_i(t_k) - \vec{v}_j(t_k)] \cdot [\vec{r}_i(t_{k+1}) - \vec{r}_j(t_{k+1})]\}$$
$$\times \frac{\vec{r}_i(t_{k+1}) - \vec{r}_j(t_{k+1})}{[\vec{r}_i(t_{k+1}) - \vec{r}_j(t_{k+1})]^2},$$

where we take into account $m_i = m$ for $i = 1, \ldots, N$. The coefficient of restitution may be either a constant or a function of the normal component of the relative velocity $\varepsilon = \varepsilon(|\vec{v}_{ij}(t_k) \cdot \vec{e}_{ij}|)$.

Based on the rules (B.2, B.3) we can construct a simple algorithm for the simulation of a force-free granular gas:

1. Initialize the positions $\vec{r}_i$ and velocities $\vec{v}_i$ of all particles $i = 1, \ldots, N$ at time $t_k = t_0 = 0$
2. Determine the time of the following collision $t_{k+1} > t_k$ of two particles of the gas, that is, the time when the distance of the centres of the particles approaches the sum of their radii:

$$t_{k+1} = \min \left(t_{ij}^* > t_k : \left| \vec{r}_i \left(t_{ij}^* \right) - \vec{r}_j \left(t_{ij}^* \right) \right| = R_i + R_j \, ; \, i,j = 1, \ldots, N \right) \tag{B.4}$$

3. Compute the positions of all particles at time t_{k+1} by (B.2).
4. Compute the new velocities of the particles i and j, after their collision at time t_{k+1} using (B.3).
5. Increment $k := k+1$ and proceed with step 2 of the algorithm for the next collision.

Since the action of a uniform external force such as gravity does not affect the collision times, the generalization to this case is straightforward.

For the case that the granular gas is confined by a container we have to specify the interaction of the particles with the walls. We will discuss this issue in the next section. If we chose periodic boundary condition, the periodicity has to be taken into account for both, the propagation between the collisions and the computation of the next collision time.

The main precondition for this algorithms is the assumption that there are exclusively pairwise particle collisions. This assumption is equivalent to the assumption of instantaneous collisions, that is, the duration of a collision is zero. This condition contradicts basic mechanics since even the collision of elastic spheres ($\varepsilon = 1$) lasts for a finite time (Hertz, 1882)

$$t_{\text{coll}} \propto v_{12}^{-1/5} \tag{B.5}$$

and a dissipative collision for the same elastic constant lasts even longer (Schwager and Pöschel, 1998). An algorithm which considers the finite duration of collisions has been described in (Luding and McNamara, 1998). Here we assume that the system is dilute enough to disregard the duration of collisions.

B.3 A simple program for event driven simulations

The design of an efficient molecular dynamics program is a difficult problem which is beyond the scope of this book. Efficient implementations can be found, for example, in Pöschel and Schwager (2004). For the simple algorithm that is described here we follow exactly the scheme of the previous section. The C++ program code is available in the Internet at http://www.oup.co.uk/isbn/0-19-853038-2.

We restrict ourselves to identical spheres, i.e., $R_i = R$ and $m_i = m$, the generalization is obvious. The gas is confined by a quadratic box of size 2×box. The corners of the box are located at (-box,-box), (-box,box), (box,box) and (box,-box). First, we describe some useful subroutines which are assembled later in the main program.

Motion of the particles in between collisions
In between the collisions the particles follow straight lines. During the time tprop they progress by $\dot{\vec{r}}_i\, t_{\text{prop}}$.

```
                            propagation
1  void propagation(double tprop){
2    for(int i=0; i<N; i++){
3      x[i]+=tprop*vx[i];
4      y[i]+=tprop*vy[i];}}
```

Collision of particles
The collision rule (B.3) is implemented in the function collision():

```
                            collision
5  void collision(pair<int,int> ij){
6    int i=ij.first, j=ij.second;
7    double dx = x[i]-x[j], dy = y[i]-y[j];
8    double dist=sqrt(dx*dx+dy*dy);
9    double ndx=dx/dist;
10   double ndy=dy/dist;
11   double h=(1+eps)*((vx[i]-vx[j])*ndx+(vy[i]-vy[j])*ndy)/2;
12   vx[i]-=h*ndx; vy[i]-=h*ndy;
13   vx[j]+=h*ndx; vy[j]+=h*ndy;}
```

The arguments of this function are the indices i and j of the colliding particles in the form pair<int,int>. It modifies the vector elements vx[i], vy[i], vx[j] and vy[j] according to (B.3).

Collision time of two particles
According to step 2 of the algorithm described in the previous section, we have to determine the time of collision for any pair of particles i and j which progress along straight lines due to their present velocities. To this end we solve the equation

$$\|[\vec{r}_i + (t^* - t)\,\vec{v}_i] - [\vec{r}_j + (t^* - t)\,\vec{v}_j]\| = R_i + R_j, \quad (\text{B.6})$$

for t^*, where t is the present time and t^* the time of the collision, hence there is a side condition $t^* > t$. We obtain

$$(t^* - t)^2 + 2\,(t^* - t)\,\frac{(\vec{r}_i - \vec{r}_j)\cdot(\vec{v}_i - \vec{v}_j)}{(\vec{v}_i - \vec{v}_j)^2} + \frac{(\vec{r}_i - \vec{r}_j)^2 - (R_i + R_j)^2}{(\vec{v}_i - \vec{v}_j)^2} = 0. \quad (\text{B.7})$$

A necessary condition for a collision is

$$(\vec{r}_i - \vec{r}_j)\cdot(\vec{v}_i - \vec{v}_j) < 0, \quad (\text{B.8})$$

that is, the particles have to approach. This condition can be checked easily by the algorithm. The particles collide if the quadratic equation (B.7) has real roots, that is, if

$$\left[\frac{(\vec{r}_i - \vec{r}_j) \cdot (\vec{v}_i - \vec{v}_j)}{(\vec{v}_i - \vec{v}_j)^2}\right]^2 + \frac{(R_i + R_j)^2 - (\vec{r}_i - \vec{r}_j)^2}{(\vec{v}_i - \vec{v}_j)^2} > 0. \qquad (B.9)$$

Given both conditions (B.8) and (B.9) are fulfilled, the particles will collide at time t^*.

The listed function ppcoll() computes the collision time t^* in case the particles collide, otherwise it delivers infty, that is, a predefined very large number which is regarded as infinity.

ppcoll
```
14  double ppcoll(int i, int j, double nada){
15    double dx,dy,dvx,dvy,xci,xcj,yci,ycj,scalar,h,p,q,w,ct,RR,dist;
16    dx=x[i]-x[j]; dy=y[i]-y[j]; dvx=vx[i]-vx[j]; dvy=vy[i]-vy[j];
17    scalar=dx*dvx+dy*dvy;
18    if(scalar>0){ return infty;}
19    else{
20      dist=dx*dx+dy*dy;
21      RR=(dist>=R*R ? R : R-nada);
22      h=1/(dvx*dvx+dvy*dvy);
23      p=scalar*h;
24      q=(dist-4*RR*RR)*h;
25      w=p*p-q;
26      if(w<0) return infty;
27      else{
28        ct=tim+q/(-p+sqrt(w));
29        xci=x[i]+(ct-tim)*vx[i]; xcj=x[j]+(ct-tim)*vx[j];
30        if((xci>box-R)||(xci<-box+R)||(xcj>box-R)||(xcj<-box+R))
31          return infty;
32        else{
33          yci=y[i]+(ct-tim)*vy[i]; ycj=y[j]+(ct-tim)*vy[j];
34          if((yci>box-R)||(yci<-box+R)||(ycj>box-R)||(ycj<-box+R))
35            return infty;
36          else return ct;}}}}
```

The third argument of this function nada is a predefined very small number of the order 10^{-10}, which is needed for the correction of numerical errors. For the choice nada=0, the function yields the correct collision time t^* according to (B.7).

With the condition (B.9) equation (B.7) has two real roots of which we need the smaller one. Later, we will collect all possible collision times of particle pairs in a list. To accelerate the computation and to save computer memory we wish to keep the list short by excluding all collisions which cannot occur. Therefore, ppcoll() returns also infty for the case that the place of the putative collision lies outside the simulation area.

The smaller root of the quadratic equation (B.7) reads according to the standard rule

$$t^* - t = -\frac{\vec{r}_{ij} \cdot \vec{v}_{ij}}{\vec{v}_{ij}^2} - \sqrt{\left(\frac{\vec{r}_{ij} \cdot \vec{v}_{ij}}{\vec{v}_{ij}^2}\right)^2 + \frac{R_{ij}^2 - \vec{r}_{ij}^2}{\vec{v}_{ij}^2}}, \quad \text{(B.10)}$$

where

$$\vec{r}_{ij} \equiv \vec{r}_i - \vec{r}_j, \qquad \vec{v}_{ij} \equiv \vec{v}_i - \vec{v}_j, \qquad R_{ij} \equiv R_i + R_j \quad \text{(B.11)}$$

and we take into account $\vec{r}_{ij} \cdot \vec{v}_{ij} < 0$ according to (B.8). The numerical evaluation of this expression is, however, rather problematic since for small distance of the particle surfaces ($R_{ij} \approx |\vec{r}_{ij}|$) the evaluation of (B.10) may lead to large numerical errors. The reason of this problem is that the difference of two almost equal large numbers has to be determined. Instead, we apply the mathematically equivalent expression

$$t^* - t = -\frac{\vec{r}_{ij}^2 - R_{ij}^2}{-\vec{r}_{ij} \cdot \vec{v}_{ij} + \sqrt{(\vec{r}_{ij} \vec{v}_{ij})^2 + \vec{v}_{ij}^2 \left(R_{ij}^2 - \vec{r}_{ij}^2\right)}} \quad \text{(B.12)}$$

which is numerically stable.

Wall collisions

We assume elastic walls, that is, particles which are colliding with the walls are elastically reflected. The change of the particle velocity due to a wall collision is performed by wcollision(), its argument is the index of the particle. For the case of an elastic collision the velocity component which is normal to the wall changes its sign.

```
                              wcollision
37  void wcollision(int i){
38    if( y[i]<0){
39      if((y[i] > x[i]) || (y[i] > -x[i])){ vx[i]=-vx[i];}
40      else{ vy[i]=-vy[i];}}
41    else{
42      if((y[i] < -x[i]) || (y[i] < x[i])){ vx[i]=-vx[i]; }
43      else{ vy[i]=-vy[i];}}}
```

Similar as a particle–particle collision, a collision of a particle with a wall is also an *event* in the sense of our event-driven algorithm. Therefore, for all particles we need the times when they will collide with any of the walls, provided it moves on with its present velocity. This time is given by

$$t^* = \min \begin{cases} t_{\text{left}} & \text{if } v_x(i) < 0, \quad \infty \text{ otherwise} \\ t_{\text{right}} & \text{if } v_x(i) > 0, \quad \infty \text{ otherwise} \\ t_{\text{top}} & \text{if } v_y(i) > 0, \quad \infty \text{ otherwise} \\ t_{\text{bottom}} & \text{if } v_y(i) < 0, \quad \infty \text{ otherwise} \end{cases} \quad \text{(B.13)}$$

This time is computed by pwcoll():

A SIMPLE PROGRAM FOR EVENT DRIVEN SIMULATIONS 275

```
           ─────────────── pwcoll ───────────────
44  double pwcoll(int i,double nada){
45    double tx, ty;
46    double RR=R;
47    if((box+x[i]>R)||(box-x[i]<R)||(box+y[i]>R)||(box-y[i]<R)) {RR=R-nada;}
48    if(vx[i]==0) tx=infty;
49    else{if(vx[i] >0) tx=(box-x[i]-RR)/vx[i];
50    else tx=(-box-x[i]+RR)/vx[i];}
51    if(vy[i]==0) ty=infty;
52    else{if(vy[i] >0) ty=(box-y[i]-RR)/vy[i];
53    else ty=(-box-y[i]+RR)/vy[i];}
54    return tim+min(tx,ty);}
```

As for particle–particle collisions, the parameter nada is used to correct numerical errors.

Initialization

The procedure init() initializes the particle positions and velocities in a way that there is no mutual penetration of particles. If we place particles of size R randomly, due to numerical errors there may occur tiny overlaps of particles, therefore, for the initialization we assume the radius R+null, where null is a very small constant. This constant is only used for the initialization, it does not affect the simulation. At the end of the initialization procedure the function ctime(i) is called for all particle indices i. This procedure generates the collision lists based on the particle's positions and velocities.

```
           ─────────────── init ───────────────
55  void init(double nada){
56    bool overlap;
57    int j;
58    x[0]=ranf(box-R-nada); y[0]=ranf(box-R-nada);
59    vx[0]=ranf(1); vy[0]=ranf(1);
60    for(int i=1; i<N; i++){
61      if(!(i % 100 )) cout << "Init " << i << endl;
62      do{
63        overlap=false;
64        x[i]=ranf(box-R-nada); y[i]=ranf(box-R-nada);
65        j=0;
66        do{overlap=((x[i]-x[j])*(x[i]-x[j])+(y[i]-y[j])*(y[i]-y[j])<
67                    4*(R+nada)*(R+nada));}
68        while((++j<i) && !overlap);}
69      while(overlap);
70      vx[i]=ranf(1); vy[i]=ranf(1);}
71    for(int i=0; i<N; i++) ctime(i);}
```

List of collision times

As described in the previous section, at any time instant (time of an event) t_k we need to know which event follows next at time t_{k+1}, that is, which pair of particles will collide next or which particle collides with one of the walls next, respectively. In principle it is possible to compute the times of all $N(N-1)/2$ putative particle–particle collisions using ppcoll() and N putative wall collisions by `pwcoll()` after each collision at which the velocities of one particle (wall collision) or two particles (particle-particle collision) have been modified. From all these event times we would have to select the smallest to decide which event follows next.

This procedure would work correctly, however, very inefficiently. Assume at time t_k we had a complete list of all possible events, that is, a list of all collisions (collision times and partners) which occur provided all particle move on with their present velocities. We assume further that at time t_k particles i and j collide, that is, the velocities of these particles change. This collision affects the collision list since some of the collisions which are contained in the list will not occur anymore, and there are new events which have to be registered in the list due to the new velocities $\vec{v}_i(t_k)$ and $\vec{v}_j(t_k)$. However, not the entire list of events are invalid. Instead only entries in which either i or j are involved as collision partners with other particles k have to be updated. All other entries are still valid.

To remove the invalid list entries, we have to search the list of collisions `cseq` for the particles i and j. A more efficient way is to assign each particle a vector `clist` which contains all times of putative events in which this particle is involved. These vectors allow for an efficient removal of entries from the collision list.

We introduce the collision list `cseq` as `map<double,pair<int,int>>`. Its first template parameter (the key) acts as the (real valued) index, the second argument is the value of the list. Hence, the index (the argument) of the list is the collision time and the value is the pair of the indices of the colliding particles. The function `ctime()` keeps the list `cseq` and the vectors `vector<double> clist[N]` updated after each collision between two particles or one particle with the wall.

─────────── ctime ───────────
```
72  void ctime(int i){
73    double ct;
74    for(unsigned int ii=0; ii!=clist[i].size(); ii++)
75      cseq.erase(clist[i][ii]);
76    clist[i].clear();
77
78    for(int j=0; j<N; j++){
79      if(i!=j){
80        ct=ppcoll(i,j,null);
81        if(ct < infty){
82          cseq[ct]=pair <int, int> (i,j);
83          clist[i].push_back(ct);
84          clist[j].push_back(ct);
85    }}}
86    ct=pwcoll(i,null);
87    if(ct < infty){
88      cseq[ct]=pair <int, int> (i,-1);
89      clist[i].push_back(ct);}}
```

The argument of the function is the index i of a particle whose velocity has changed. Hence, all events in `clist[i]` became invalid. Using the command cseq.erase(clist[i][ii]) we first delete all entries in cseq whose arguments (collision times) are contained in the vector clist[i]. Since we know that at each time there occurs at most one collision, there cannot be two entries with identical arguments (collision times) in the list. Therefore, this procedure is an elegant way to remove invalid entries from the list.

Since the velocity of the particle i has changed we need to recompute the new putative collision times with all other particles, that is, the old vector clist[i] became worthless. We delete this vector and reconstruct it. To this end the collision times of particle i with all other particles $j = 0, \ldots, N-1$ are calculated. If particles i and j collide (ct!=infty), we assemble the pair pa from i and j and insert this pair into the collision list cseq. The corresponding collision time ct is added to the vectors clist[i] and clist[j]. As described above these vectors simplify the removal of entries from the collision list which may be necessary later if either i or j collides with any other partner k.

In the same way the time of the next wall collision is inserted. To mark it as a wall-collision event the second collision partner pa.second has the value -1.

Whenever the velocity of any particle i is modified, the procedure call ctime(i) is sufficient to update the collision list. Hence, after the collision of particles i and j we have to call this function twice, once for i and once for j.

Main program

By now we have described all necessary subroutines. In the header of the program we define the constants and variables. Most of them do not need any further explanation, either they have been explained in the description of the subroutines above or their meaning is obvious from their names. We only wish to mention: tps – the interval (number of collisions) after which Postscript pictures of the system are printed, infty – a very large number which is interpreted by the program as ∞, null – a very small number to correct numerical errors.

```
                            prefix
 90  #include <iostream>
 91  #include <fstream>
 92  #include <vector>
 93  #include <map>
 94
 95  const int N=5000, itend=10000000, tps=10000;
 96  const double R=1, eps=0.98, box=200, pssize=500 ;
 97  const double infty=1e20, null=1e-10;
 98
 99  vector<double> x(N),y(N),vx(N),vy(N);
100  vector<double> clist[N];
101  double tim=0;
102  typedef map<double,pair<int,int> > ctype;
103  ctype cseq;
104  ofstream psout("MD.ps"), eout("energy");
105
106  void init(double);
107  void propagation(double);
108  void collision(pair<int,int>);
```

```
109  void ctime(int);
110  void wcollision(int);
111  void psplot(int, ofstream &);
112  bool checkoverlap();
113  double ppcoll(int,int,double);
114  double pwcoll(int,double);
115  double ranf(double x){return (2*double(rand())/(1+double(RAND_MAX))-1)*x;}
116  double kinenergy();
```

In each iteration `it` the main program determines the time of the next event `tnext` and the corresponding collision partners, that is, the pair `ijnext` which is located at the top of the collision list `cseq`. If the index of the second partner is -1, a wall collision (of the first partner) is described, otherwise a particle–particle collision is meant.

In case of a wall collision, the time of the collision `tn` is computed again by `pwcoll()`. In difference to the call of this function in `ctime()`, the third argument is now `null`. Then all particles are moved according to their present velocities by `propagation()`. Finally, the velocity of the colliding particle is changed in the wall collision function `wcollision()` and `ctime()` is called to update the collision list.

Particle–particle collisions are treated analogously where `ctime()` has to be called for both collision partners.

──── main ────
```
117  int main(){
118    double tnext, tn=0;
119    pair<int,int> ijnext;
120
121    init(null);
122    for(int it=0; it<itend; it++){
123      if(!(it % 10000)) checkoverlap();
124      if(!(it % 1000)) eout << it << " " << kinenergy() << endl;
125      tnext=cseq.begin()->first;
126      if(!(it%1000)) cout << "IT: " << it << "  time= " << tim << endl;
127      ijnext=cseq.begin()->second;
128      if(ijnext.second==-1){
129        tn=pwcoll(ijnext.first,null);
130        propagation(tn-tim);
131        tim=tnext;
132        wcollision(ijnext.first);
133        ctime(ijnext.first);}
134      else{
135        tn=ppcoll(ijnext.first,ijnext.second,null);
136        propagation(tn-tim);
137        tim=tnext;
138        collision(ijnext);
139        ctime(ijnext.first);
140        ctime(ijnext.second);}
141      if(!(it%tps)) psplot(it/tps,psout);
142    }
143  }
```

Why do we determine the time of the collision again, although it is already known from the entry in the collision table? Between the recording of the entry in the collision table for a certain collision between i and j and the time of the collision when the velocities of i and j are altered according to the collision rule there may occur thousands of collisions between other particles, where each time `propagation()` is called. Each time this function is called, the positions of all particles are changed which implies accumulating tiny numerical errors. When the collision i–j occurs, these errors may cause that particles overlap by a tiny distance, i.e., their distance is slightly smaller than $2R$. This tiny overlap does, in fact, not affect the dynamics of a granular gas, however, the algorithm may break down, in particlular, when such errors occur in dense regions since overlapping particles may then stay overlapping forever. Since numerical calculations are always affected by rounding errors we need to apply certain measures to avoid the described situation.

For reasons of principle there is no perfect protection to avoid overlapping particles due to numerical errors. An approved method is to shorten the propagation by a tiny amount so that the colliding particles preserve a tiny distance. For this reason, we use in `propagation()` the shortened time as obtained by `pwcoll()` or `ppcoll()`, respectively, called with the third argument `nada`.

Since there is no perfect protection against overlapping we check from time to time whether overlapping occurs. There are various methods to detect and to purge such illegal situations. A simple method is to store the coordinates and velocities of the particles in certain (large) intervals. If an overlap has occurred we can recover the previously stored situation, build up the collision list from scratch and proceed with the computation. The function `checkoverlap()` checks for illegal situations.

```
                           ───── checkoverlap ─────
144   bool checkoverlap(){
145     double dist;
146     for(int i=1; i<N; i++){
147       for(int j=0; j<i; j++){
148         dist=(x[i]-x[j])*(x[i]-x[j])+(y[i]-y[j])*(y[i]-y[j])-4*R*R;
149         if(dist<0){
150           cout << " OVERLAP " << i << " " << j << " " << dist << endl;
151           return true;}}}
152     for(int i=0; i<N; i++){
153       if((x[i]-R<-box)||(y[i]-R<-box)||(x[i]+R>box)||(y[i]+R>box)){
154         cout << "OVERLAP WALL" << i << " " << y[i]+R-box << endl;
155         return true;}}
156     return false;}
```

This function has to be called in certain intervals. It yields `true` if a illegal situation is encountered, otherwise `false`. Since illegal situations occur due to accumulated errors, for small particle number ($N \lesssim 10^4$) they are extremely rare events. Therefore, in the presented program we do not give the code for the correction as described above. Its implementation is simple.

Snapshots and measurements

Finally, we provide two small functions which demonstrate how to measure the desired data in a simulation. The first procedure `psplot()` produces Postscript snapshots of the particles in certain time intervals. They can be used to produce animations as shown at http://www.oup.co.uk/isbn/0-19-853038-2. Since this function does not directly belong to the molecular dynamics algorithm we do not wish to discuss it in detail here.

```
                            ─── psplot ───
157  void psplot(int page, ofstream & psout){
158    if(!page){
159      psout << "%!PS-Adobe-2.0" << endl;
160      psout << "%%BoundingBox: 0 0 "<<pssize+20<<" "<< pssize+20<< endl;
161      psout << "%%EndComments" << endl;
162      psout << "/frame {10 10 translate " <<box/pssize<<" setlinewidth ";
163      psout << pssize/(2*box) <<" "<< pssize/(2*box) <<" scale ";
164      psout << box<<" "<<box<<" translate newpath "<< -box<<" "<<-box
165             <<" moveto " << 2*box;
166      psout <<" 0 rlineto 0 "<<2*box<<" rlineto "<<-2*box
167             <<" 0 rlineto closepath stroke}def" << endl;
168      psout << "/c { 1 0 360 arc stroke} def" << endl;
169    }
170    psout << "%%Page: " << page << " " << page << endl;
171    psout << "frame " <<endl;
172    for(int i=0; i<N; i++) psout << x[i]<<" "<<y[i]<<" c" << endl;
173    psout << "stroke showpage " << endl;}
```

The function `kinenergy()` measures the total kinetic energy as an example of measuring data in a molecular dynamics simulation.

```
                            ─── kinenergy ───
174  double kinenergy(){
175    double ekin=0;
176    for(int i=0; i<N; i++) ekin+=vx[i]*vx[i]+vy[i]*vy[i];
177    return ekin;}
```

Thus, our simple program for event-driven molecular dynamics is now complete. It can be downloaded from http://www.oup.co.uk/isbn/0-19-853038-2 and directly used for simulations of small systems up to about 10,000 particles due to the limitations of computer time. For larger systems or long time simulations more sophisticated algorithms should be applied.

B.4 Efficient algorithms

The program that has been presented in the previous section performs precisely the algorithms which has been described on page 270. We wish to emphasize that this algorithm does not reflect the state of the art of molecular dynamics. Instead it is one of the most simple reasonable molecular dynamics programs, since although the program is algorithmically correct, it contains a series of deficiencies which may be improved. For example:

1. In the initialization the particles are distributed randomly. For any particle i we check for overlap with all $i-1$ previously placed particles.

2. During the initialization we compute N^2 collision times, among them also for pairs of particles which are very distant and whose collision probability is, therefore, very small.

3. After each wall collision we compute N new collision times and for a particle–particle collision $2N$ times. If the putative collision partners are very distant the probability for the collision to occur is very small.

4. In simulations of a large number of particles the collision list as well as the vectors of the collision times become very large. Most of these entries do not correspond to actual collisions but will be deleteded before they occur, due to collisions of one of the partners with other particles.

These problems may be solved by using more sophisticated algorithms which are described in the literature (e.g. Rapaport, 1980; Marín et al., 1993; Baeza-Yates et al., 1994; Marín, 1998; Pöschel and Schwager, 2004).

We wish to add a final comment regarding the inelastic collapse as described in Section 4.1. For event–driven simulations, it is assumed that the collisions are instantaneous, that is, $t_c = 0$. If one further assumes $\varepsilon = $ const., there may occur a pathological effect which is called *inelastic collapse* (Shida and Kawai, 1989; McNamara and Young, 1992) we have discussed this effect in detail in Section 4.1. During an inelastic collapse there occurs an infinite number of collisions in finite time, that is, our event-driven simulation stops to advance in time as soon as a collapse occurs. An inelastic collapse of three particles may occur if the coefficient of restitution obeys the condition $\varepsilon \leq 7 - 4\sqrt{3}$. For such choice of ε, a collapse occurs only if the initial velocities and positions are chosen appropriately which establishes a second condition.

If more than three particles move along a line, a collapse can occur in a wider range of ε (McNamara and Young, 1992), again provided the initial conditions are chosen appropriately. The more particles are involved, the wider is the range of ε. At the same time particles have to move coherently to perform an inelastic collapse. Therefore, for almost elastic particles $\varepsilon \to 1$ it is extremely unlikely to encounter an inelastic collapse.

Moreover, according to our understanding it is not clear yet whether a perfect collapse may occur in higher dimensions, since the subspace of appropriate initial velocities and positions may have a smaller dimension than the phase space. In this case the probability to find such a situation by chance is zero.

Nevertheless, even an incomplete collapse may lead to numerical problems since according to an exponential decrease of the intervals between collisions the real time of the simulation progresses extremely slow. From a certain moment on the relative velocities of the particles and their distances become so small in a certain region of the gas that numerical errors become arbitrarily large. Such situations have been observed in numerical simulations of granular gases in two and three dimensions, (McNamara and Young, 1993; Du et al., 1995; Zhou and Kadanoff, 1996).

There are several numerical tricks to avoid a collapse: Luding and McNamara (1998) assumed that there is a small but finite duration of a collision t_c. If during that time a particle suffers another collision, this collision is computed with $\varepsilon = 1$. This method prevents the collapse since the limit t_c is reached after a finite number of collisions and the infinite chain of collisions which would occur during a collapse is interrupted.

Another method has been proposed by Deltour and Barrat (1997) and Grossman (1997). As soon as an approaching collapse is detected the velocity vectors of the involved particles are rotated by a small amount, for example, by 5°. By this procedure, a part of the normal velocity is turned into tangential velocity which interrupts the sequence of an inelastic collapse.

We wish to mention that in real systems of force free gases the inelastic collapse cannot exist since the coefficient of restitution is not constant but depends on the relative normal velocity. For decaying velocity it approaches the elastic limit, $\varepsilon = 1$. Therefore, the condition for an inelastic collapse is violated from a certain time on. As soon as the relative velocity falls below a certain threshold the collision sequence is interrupted (Goldman et al., 1998).

APPENDIX C

SOLUTIONS TO PROBLEMS

Solution of problem 2.1
Derive the collision law (2.7)!

First, we change from the velocities $\vec{v}_1$ and $\vec{v}_2$ to the centre of mass and relative velocity:

$$\vec{v}_{\rm cm} = \frac{m_1}{M}\vec{v}_1 + \frac{m_2}{M}\vec{v}_2 \qquad (C.1)$$
$$\vec{v}_{12} = \vec{v}_1 - \vec{v}_2,$$

where $M = m_1 + m_2$. Then the inverse transformation reads

$$\vec{v}_1 = \vec{v}_{\rm cm} + \frac{m_2}{M}\vec{v}_{12}, \qquad \vec{v}_2 = \vec{v}_{\rm cm} - \frac{m_1}{M}\vec{v}_{12}. \qquad (C.2)$$

Consider the normal and transversal components of the relative velocity $\vec{v}_{12}$ at the point of contact of the colliding spheres:

$$\vec{v}_{12\|} = (\vec{v}_{12} \cdot \vec{e})\,\vec{e}, \qquad \vec{v}_{12\perp} = \vec{v}_{12} - (\vec{v}_{12} \cdot \vec{e})\,\vec{e}. \qquad (C.3)$$

We assume that the interaction force has only a normal component, that is, the tangential component is negligible. This assumption is approximatively valid for smooth spheres (but see the discussion on page 35). The collision affects, hence, only the normal component of the relative velocity while the tangential component of $\vec{v}_{12}$ does not change. The components of the relative velocity after the collision read

$$\vec{v}\,'_{12\|} = -\varepsilon \vec{v}_{12\|}, \qquad \vec{v}\,'_{12\perp} = \vec{v}_{12\perp}, \qquad (C.4)$$

so that

$$\vec{v}\,'_{12} = \vec{v}\,'_{12\perp} + \vec{v}\,'_{12\|} = \vec{v}_{12} - (\vec{v}_{12}\cdot\vec{e})\,\vec{e} - \varepsilon(\vec{v}_{12}\cdot\vec{e})\,\vec{e} = \vec{v}_{12} - (1+\varepsilon)(\vec{v}_{12}\cdot\vec{e})\,\vec{e}. \qquad (C.5)$$

Hence

$$\vec{v}\,'_1 = \vec{v}\,'_{\rm cm} + \frac{m_2}{M}\vec{v}\,'_{12} = \frac{m_1}{M}\vec{v}_1 + \frac{m_2}{M}\vec{v}_2 + \frac{m_2}{M}\vec{v}_1 - \frac{m_2}{M}\vec{v}_2 - \frac{m_2}{M}(1+\varepsilon)(\vec{v}_{12}\cdot\vec{e})\,\vec{e}$$
$$= \vec{v}_1 - \frac{m^{\rm eff}}{m_1}(1+\varepsilon)(\vec{v}_{12}\cdot\vec{e})\,\vec{e}. \qquad (C.6)$$

Similarly, we obtain

$$\vec{v}\,'_2 = \vec{v}_2 + \frac{m^{\rm eff}}{m_2}(1+\varepsilon)(\vec{v}_{12}\cdot\vec{e})\,\vec{e}. \qquad (C.7)$$

Solution of problem 2.2
Two spheres collide with an angle of their paths α. Express the angle of their traces after the collision as function of the coefficient of restitution!

The motion of two colliding spheres occurs always in a plane as it follows immediately from the conservation of momentum (show that!). Since the dissipation during a collision concerns only the velocity component in normal direction $\vec{e}_n = |\vec{r}_1 - \vec{r}_2| / (\vec{r}_1 - \vec{r}_2)$, the angle after the collision is always smaller than before for all $\varepsilon < 1$. From simple geometry the result $\tan \alpha' = \varepsilon \tan \alpha$ follows.

Solution of problem 2.3
Assume $\varepsilon = $ const. A particle falls from height H and rebounces recurrently from the floor. At which time t_∞ does it come to rest? Can t_∞ be finite?

At the first contact with the floor, the particle's velocity v is given by the equivalence of kinetic and potential energy $mv^2/2 = mgH$. So it reaches the floor with $v_0 = \sqrt{2gH}$ after time $t_0 = \sqrt{2H/g}$. The time between the first and the second bounce is $t_1 = 2\varepsilon\sqrt{2H/g}$. In general, the delay between the ith and $(i+1)$th bounce $(i \geq 1)$ is $t_i = 2\varepsilon^i \sqrt{2H/g}$. Hence, the total time is

$$t_\infty = \sqrt{\frac{2H}{g}} + \sum_{i=1}^{\infty} 2\sqrt{\frac{2H}{g}} \varepsilon^i = \sqrt{\frac{2H}{g}} \left(\frac{1+\varepsilon}{1-\varepsilon} \right). \tag{C.8}$$

This means that the bouncing sphere comes to rest after finite time.

Solution of problem 3.1
Assume ε depends on the impact velocity as $\varepsilon(v) = 1 - Cv^{1/5}$ (This formula is an approximation of (3.22)). A particle is again dropped from height H to a rigid floor and rebounces recurrently. After which time t_∞^v this sphere will come to rest? Compare the bouncing times for the cases $\varepsilon = $ const. and $\varepsilon = \varepsilon(v)$! Assume that at the first contact with the floor the velocity dependent coefficient of restitution and the constant one have the same value!

The velocity before the kth bounce is v_k, that is, $v_{k+1} = \varepsilon v_k$. The corresponding energies are $E_{k+1} = \varepsilon^2 E_k$. The time lag is $t_{k+1} - t_k = 2\sqrt{2E_k/mg^2}$. Therefore

$$\frac{E_{k+1} - E_k}{t_{k+1} - t_k} = \frac{(\varepsilon^2 - 1) E_k g \sqrt{m}}{2\sqrt{2}\sqrt{E_k}}. \tag{C.9}$$

Assuming small energy differences and time lags, we formulate the according differential equation

$$\frac{dE}{dt} = \frac{g\sqrt{m}}{2\sqrt{2}} (\varepsilon^2 - 1) \sqrt{E}. \tag{C.10}$$

For $\varepsilon = $ const. we abbreviate $\dot{E} = a\sqrt{E}$ with the solution $\sqrt{E} = at/2 + \sqrt{E_0}$, where $E_0 = mgH$ is the initial energy. The energy is completely dissipated after the time

$$t^c_\infty = -\frac{2}{a}\sqrt{E_0} = \frac{2\sqrt{2}}{g\sqrt{m}}\frac{1}{1-\varepsilon^2}2\sqrt{E_0} = \frac{4\sqrt{2}}{\sqrt{g}}\frac{\sqrt{H}}{1-\varepsilon^2}. \qquad (C.11)$$

Note that t^c_∞ corresponds to the starting time $t = 0$ at the ground. Since the ball is dropped from the height H the total time is by $t_0 = \sqrt{2H/g}$ larger. The result (C.11) and the solution of problem 2.3, equation (C.8) (with t_0 substracted for the comparison), differ from each other, although they describe the same physical situation. This difference comes from the approximation of continuous time which has been made in (C.10). This approximation is justified if the kinetic energy between successive bounces decays only by a small amount, that is, for $\varepsilon \to 1$. In this limit, both solutions, (C.8) and (C.11) coincide. However, (C.8) is rigorous since no assumption about ε has been made. For the velocity dependent coefficient of restitution, the summation in (C.8) cannot be performed.

For the velocity dependent coefficient of restitution we obtain $1-\varepsilon^2 = 2Cv^{1/5}$ with $v = \sqrt{2E/m}$. Equation (C.10) for this case reads

$$\dot{E} = -gCm^{2/5}2^{-2/5}E^{3/5} = bE^{3/5}, \qquad (C.12)$$

with the solution $E^{2/5} = 2bt/5 + E_0^{2/5}$. The energy is dissipated after

$$t^v_\infty = -\frac{5}{2b}E_0^{2/5} = \frac{1}{gC}\frac{5}{2^{3/5}}\frac{1}{m^{2/5}}E_0^{2/5}, \qquad (C.13)$$

again with $E_0 = mgH$. We wish to compare the times for the cases $\varepsilon = $ const. and $\varepsilon = 1-Cv^{1/5}$. To this end we assume that for the first bounce the coefficients of restitution are identical, that is,

$$1 - \varepsilon^2 = 2C\frac{2^{1/10}E_0^{1/10}}{m^{1/10}} \qquad (C.14)$$

and insert this quantity into (C.11):

$$t^c_\infty = -\frac{2^{7/5}}{gCm^{2/5}}E_0^{2/5}. \qquad (C.15)$$

With (C.13) we obtain

$$\frac{t^c_\infty}{t^v_\infty} = \frac{4}{5}, \qquad (C.16)$$

that is, the ball which bounces with $\varepsilon = $ const. dissipates its energy in 80% of the time the viscoelastic ball needs.

Solution of problem 3.2
Derive (3.36)!

For the inverse collision

$$\frac{dE(\hat{\xi})}{d\hat{\xi}} = +\varkappa\dot{\hat{\xi}}\sqrt{\hat{\xi}}, \qquad \hat{\xi}(0) = 0, \qquad \dot{\hat{\xi}}(0) = \varepsilon. \tag{C.17}$$

According to the definition of the inverse collision, which is just the direct collision in reverse time, the energy at the point of maximal compression (turning point) and at the beginning of the collision reads

$$E(\hat{\xi}_0) = \frac{1}{2}\hat{\xi}_0^{5/2}, \qquad E(0) = \frac{1}{2}\dot{\hat{\xi}}^2(0) = \frac{\varepsilon^2}{2}. \tag{C.18}$$

Then integrating the LHS of (C.17), we obtain

$$\int_0^{\hat{\xi}_0} d\hat{\xi} \frac{dE(\hat{\xi})}{d\hat{\xi}} = \frac{1}{2}\hat{\xi}_0^{5/2} - \frac{\varepsilon^2}{2}. \tag{C.19}$$

On the other hand, since

$$\dot{\hat{\xi}}(\hat{\xi}) \approx \varepsilon\sqrt{1 - (\hat{\xi}/\hat{\xi}_0)^{5/2}}, \tag{C.20}$$

integration of the RHS of (C.17) yields

$$\int_0^{\hat{\xi}_0} d\hat{\xi}\,\varkappa\dot{\hat{\xi}}\sqrt{\hat{\xi}} = \varkappa\varepsilon \int_0^{\hat{\xi}_0} d\hat{\xi}\sqrt{1 - (\hat{\xi}/\hat{\xi}_0)^{5/2}}\sqrt{\hat{\xi}}$$
$$= \varkappa\varepsilon\hat{\xi}_0^{3/2} \int_0^1 \sqrt{1 - x^{5/2}}\sqrt{x}\,dx = \varkappa\varepsilon\hat{\xi}_0^{3/2} d, \tag{C.21}$$

where we use the definition (3.33). Comparing (C.19) and (C.21) we arrive at (3.36).

Solution of problem 3.3
Derive the propagation rule (3.52) by means of (3.51, 3.49)!

Consider the collision of two spherical particles of radius R with velocities $\vec{v}_1$, $\vec{v}_2$ and angular velocities $\vec{\omega}_1$, $\vec{\omega}_2$. Let $\vec{r}_{12} = \vec{e}R$. The velocity of the surface of the first sphere at the point of contact is $\vec{v}_1 + R(\vec{e} \times \vec{\omega}_1)$ and that of the second sphere is $\vec{v}_2 - R(\vec{e} \times \vec{\omega}_2)$. Then the relative velociy at the point of contact reads

$$\vec{v}_1 + R(\vec{e} \times \vec{\omega}_1) - \vec{v}_2 + R(\vec{e} \times \vec{\omega}_2) = \vec{v}_{12} + R(\vec{e} \times \vec{\omega}_{12}), \tag{C.22}$$

where we introduce $\vec{\omega}_{12} \equiv \vec{\omega}_1 + \vec{\omega}_2$. The normal and transverse components of this velocity are given by (3.49). From the first two equations of (3.51) we obtain

$$\vec{\omega}'_{12} - \vec{\omega}_{12} = -\frac{Rm}{I}(\vec{e}\times\vec{v}'_{12}) + \frac{Rm}{I}(\vec{e}\times\vec{v}_{12}). \qquad (C.23)$$

Combining (3.48, 3.49), the tangential velocity after the collision reads

$$\vec{v}'_{12} - (\vec{g}^n)' + R(\vec{e}\times\vec{\omega}'_{12}) = \varepsilon^t \vec{g}^t. \qquad (C.24)$$

With $\vec{\omega}'_{12}$ given by (C.23) we write

$$\begin{aligned}\vec{v}'_{12} &= -\varepsilon^n \vec{g}^n + \varepsilon^t \vec{g}^t - R(\vec{e}\times\vec{\omega}_{12}) + \frac{R^2 m}{I}(\vec{e}\times\vec{e}\times\vec{v}'_{12}) - \frac{R^2 m}{I}(\vec{e}\times\vec{e}\times\vec{v}_{12}) \\ &= -\varepsilon^n \vec{g}^n + \varepsilon^t \vec{g}^t - R(\vec{e}\times\vec{\omega}_{12}) + \frac{(\vec{v}'_{12}\cdot\vec{e})\vec{e} - \vec{v}'_{12}}{q} - \frac{(\vec{v}_{12}\cdot\vec{e})\vec{e} - \vec{v}_{12}}{q} \\ &= -\varepsilon^n \vec{g}^n + \varepsilon^t \vec{g}^t - (\vec{g}^t - \vec{v}_{12} + \vec{g}^n) + \frac{-\varepsilon^n \vec{g}^n - \vec{v}'_{12}}{q} - \frac{\vec{g}^n - \vec{v}_{12}}{q},\end{aligned} \qquad (C.25)$$

where we use the relation $\vec{A}\times\vec{B}\times\vec{C} = \vec{B}(\vec{A}\cdot\vec{C}) - \vec{C}(\vec{A}\cdot\vec{B})$, introduce $q \equiv I/(mR^2)$ and take into account the definition of $\vec{g}^n$, (3.49) and (3.48), which relates $\vec{g}^n$ and $(\vec{g}^n)'$. We rewrite (C.25) in the form

$$\vec{v}'_{12}\left(1+\frac{1}{q}\right) = -\varepsilon^n \vec{g}^n \left(1+\frac{1}{q}\right) - (1-\varepsilon^t)\vec{g}^t + \left(1+\frac{1}{q}\right)(\vec{v}_{12}-\vec{g}^n), \qquad (C.26)$$

that is,

$$\vec{v}'_1 - \vec{v}'_2 = -\varepsilon^n \vec{g}^n - \frac{(1-\varepsilon^t)}{(1+q^{-1})}\vec{g}^t + \vec{v}_1 - \vec{v}_2 - \vec{g}^n. \qquad (C.27)$$

With the conservation of momentum

$$\vec{v}'_1 + \vec{v}'_2 = \vec{v}_1 + \vec{v}_2, \qquad (C.28)$$

we obtain the result

$$\vec{v}'_1 = \vec{v}_1 - \frac{1+\varepsilon^n}{2}\vec{g}^n - \frac{1-\varepsilon^t}{2(1+q^{-1})}\vec{g}^t, \qquad (C.29)$$

which coincides with the first equation in the system (3.52). If we subtract (C.27) from (C.28) we obtain the second equation in (3.52). From (3.51), the third equation in (3.52) follows:

$$\begin{aligned}\vec{\omega}'_1 &= \vec{\omega}_1 - \frac{Rm}{I}[\vec{e}\times(\vec{v}'_1 - \vec{v}_1)] \\ &= \vec{\omega}_1 - \frac{Rm}{I}(\vec{e}\times\vec{g}^n)\frac{1+\varepsilon^n}{2} + \frac{Rm}{I}(\vec{e}\times\vec{g}^t)\frac{1-\varepsilon^t}{2(1-q^{-1})} \qquad (C.30) \\ &= \vec{\omega}_1 + \frac{1-\varepsilon^t}{1-q^{-1}}\frac{1}{2R}(\vec{e}\times\vec{g}^t).\end{aligned}$$

The last equation in (3.52) may be obtained completely analogously.

Solution of problem 3.4
The reduced moment of inertia $q \equiv I/(mR^2)$ characterizes the distribution of particle material inside the grain. How does this quantity affect the coupling between the rotational and translational motion? Look at two opposite cases: (i) all mass is distributed in a very thin shell of radius R, and (ii) the mass is concentrated in a very small volume around the centre of the particle. What is the value of q for the grains of a uniform density? Does q depend on mass, density, or radius in the latter case?

For the case when all mass is distributed in a very thin shell $I = mR^2$ and $q = q_{max} = 1$; the coupling between the rotational and translational degrees of freedom is maximal, as it follows from (3.52). For the opposite case, when all mass is located at the centre of the sphere, $I = 0$ and $q = q_{min} = 0$, so that $1/(1 + 1/q) = 0$; (3.52) hence shows that there is no coupling between rotational motion and translational motion. If the grains are of uniform density $I = 2/5\, mR^2$ and $q = 2/5$, this quantity depends neither on the masses, the density nor on the radii of the particles.

Solution of problem 4.1
Derive (4.34) and (4.36)!

The extremal value of $n = n^*$ follows from the equation $dv'_n/dn = 0$ or from $d\log(v'_n)/dn = 0$, since the logarithm is a monotonic function. Hence

$$\frac{d\log(v'_n)}{dn} = \log(1+\varepsilon) - \log\left[1 + \left(\frac{m_n}{m_0}\right)^{1/n}\right]$$

$$+ \frac{n}{1+\left(\frac{m_n}{m_0}\right)^{1/n}} \left(\frac{m_n}{m_0}\right)^{1/n} \frac{1}{n^2} \log\left(\frac{m_n}{m_0}\right) = 0. \quad (C.31)$$

If we introduce $x_0 = (m_n/m_0)^{1/n}$ this equation turns into

$$\log\frac{(1+x_0)}{(1+\varepsilon)} = \frac{x_0}{1+x_0}\log x_0, \quad (C.32)$$

which is equivalent with (4.35):

$$(1+x_0) = (1+\varepsilon)x_0^{x_0/(1+x_0)}. \quad (C.33)$$

From the definition of x_0 follows (4.34) for n^*. Using the definition of x_0 we can also write

$$\left(\frac{v'_n}{v_0}\right)_{\text{extr}} = \left(\frac{1+\varepsilon}{1+\left(\frac{m_n}{m_0}\right)^{1/n^*}}\right)^{n^*} \quad v_0 = \left(\frac{1+\varepsilon}{1+x_0}\right)^{n^*}$$

$$= \left(\frac{1+x_0}{x_0^{x_0/(1+x_0)}(1+x_0)}\right)^{n^*} = \left[\left(\frac{m_0}{m_n}\right)^{1/n^*}\right]^{n^* x_0/(1+x_0)} = \left(\frac{m_0}{m_n}\right)^{x_0/(1-x_0)}, \tag{C.34}$$

which proves (4.36).

Solution of problem 5.1
Estimate the ratio of triple and binary collisions for a gas of soft particles of radius R and mass m which interact with a repulsive potential $\Phi(r) = A\xi^\alpha$, where ξ is the compression of particles (see Chapter 3)! The gas has temperature T and number density n.

To estimate the ratio of triple and binary collisions we consider a typical collision of particles with velocities close to a thermal velocity $v_T = \sqrt{2T/m}$. When particles collide, the typical kinetic energy of their relative motion, $m^{\text{eff}} v_T^2/2$ ($m^{\text{eff}} = m/2$) transforms partly into the potential energy of the elastic deformation so that the conservation of energy gives

$$\frac{m^{\text{eff}}}{2}\left(\frac{d\xi}{dt}\right)^2 + A\xi^\alpha = \frac{m^{\text{eff}} v_T^2}{2}. \tag{C.35}$$

The maximal deformation at the collision $\xi_{\max}$ then reads

$$\xi_{\max} = \left(\frac{m^{\text{eff}} v_T^2}{2A}\right)^{1/\alpha} = 2^{-1/\alpha} A^{-1/\alpha} T^{1/\alpha}, \tag{C.36}$$

and the collision duration is

$$\tau_{\text{coll}} = 2\int_0^{\xi_0} \frac{d\xi}{\sqrt{v_T^2 - (2A/m^{\text{eff}})\xi^\alpha}} = b\left(\frac{m^{\text{eff}}}{2A}\right)^{1/\alpha} v_T^{2/\alpha - 1}, \tag{C.37}$$

with the constant

$$b = 2\int_0^1 \frac{dx}{\sqrt{1-x^\alpha}} = \frac{2}{\alpha} B\left(\frac{1}{2}, \frac{1}{\alpha}\right) = \frac{2}{\alpha} \frac{\Gamma\left(\frac{1}{2}\right)\Gamma\left(\frac{1}{\alpha}\right)}{\Gamma\left(\frac{1}{2}+\frac{1}{\alpha}\right)}, \tag{C.38}$$

where $B(x,y)$ and $\Gamma(x)$ are, respectively, the Beta and the Gamma functions. Using the definitions of v_T and m^{eff} we can also write

$$\tau_{\text{coll}} = \frac{b}{\sqrt{2}} A^{-1/\alpha} m^{1/2} T^{1/\alpha - 1/2}. \tag{C.39}$$

The average time of the free flight (mean collision time) τ_c reads (see (5.3))

$$\tau_c^{-1} = \pi\sigma^2 \bar{v}_{12} n = 4\pi R^2 \sqrt{\frac{2T}{m^{\text{eff}}}} n = 8\pi R^2 m^{-1/2} T^{1/2} n. \qquad (C.40)$$

The ratio of the collision duration and the mean collision time may then be written using (C.36),

$$\frac{\tau_{\text{coll}}}{\tau_c} = 4\pi 2^{1/\alpha + 1/2} \, bn R^2 \xi_{\max} = 3 \cdot 2^{1/\alpha + 1/2} \, b\eta \left(\frac{\xi_{\max}}{R}\right), \qquad (C.41)$$

where $\eta = (4/3)\pi R^3 n$ is the packing fraction, which illustrates that the approximation of instantaneous collisions becomes better for harder particles and for more dilute gases.

The fraction of time which a particle spends in binary collisions may be estimated as a ratio of the collision duration and the time of the mean free flight, that is, by (C.41). Therefore, if the total number of particles is N, then at each time instant about $N\left(\tau_{\text{coll}}/\tau_c\right)$ of them are involved in a binary collisions. Thus, in a unit volume there exists $1/2\, n\, (\tau_{\text{coll}}/\tau_c)$ pairs of particles which are in contact, that is, the concentration of pairs n_{pair} is $1/2\, n\, (\tau_{\text{coll}}/\tau_c)$.

Now it is easy to calculate the collision frequency of a single particles with pairs, that is, the collision frequency of triple collisions, $\tau_{c,\text{trip}}^{-1}$. This may be done exactly in the same way as before for the case of binary collisions (see Chapter 5):

$$\tau_{c,\text{trip}}^{-1} = 4\pi D R^2 \sqrt{\frac{T}{m}} n_{\text{pair}}, \qquad (C.42)$$

where D is a numerical factor of the order of unity which accounts for the effective cross-section of the collision between a single particle and a pair and for the effective mass (equal to $2m/3$). Hence, we obtain for the ratio of the collision frequencies of triple and binary collisions,

$$\frac{\tau_{c\,\text{trip}}^{-1}}{\tau_c^{-1}} = \frac{4\pi D R^2 \sqrt{T/m} \, n_{\text{pair}}}{8\pi R^2 \sqrt{T/m} \, n} = \frac{D}{4} n \, \frac{\tau_{\text{coll}}}{\tau_c} \sim \eta \frac{\xi_{\max}}{R}, \qquad (C.43)$$

which again shows that the more dilute a gas and the harder the interaction potential, the more accurate is the binary collision approximation.

Solution of problem 5.2
Using the results of the Exercise 5.1 estimate the temperature dependence of the ratio of triple and binary collisions for a gas of viscoelastic particles!

From the Hertz law (3.2) follows the potential energy of compressed viscoelastic particles, $\Phi(\xi) = \frac{2}{5}\rho \xi^{5/2}$ and thus the maximal compression,

$$\xi_{\max} = \left(\frac{m^{\text{eff}} v_T^2}{2A}\right)^{1/\alpha} = 2^{-2/5} \left(\frac{2}{5}\rho\right)^{-2/5} T^{2/5}, \qquad (C.44)$$

where we assume that the dissipation is small so that the value of $\xi_{\max}$ may be approximated by the Hertz law. Similarly, we approximate the collision duration. Then the ratio of the collision frequencies of triple and binary collisions is given by (C.43), which we write using (C.44)

$$\frac{\tau_{c\,\text{trip}}^{-1}}{\tau_c^{-1}} \sim \eta \left(\frac{\xi_{\max}}{R}\right) \sim \eta T^{2/5} R^{-1}. \tag{C.45}$$

The last equation shows that this ratio scales with temperature as $T^{2/5}$, that is, the binary collision approximation becomes better as temperature decreases, that is, the particles become appearently harder for smaller collision velocities at low temperature.

Solution of problem 8.1
Recast the basic property of the collision integral (6.25) into its dimensionless form (8.4)!

Let ψ in (6.25) be a function of the dimensionless velocity c_1. Using the definition of the dimensionless collision integral (8.3) we write

$$\int d\vec{v}_1 \psi(\vec{c}_1) I(f,f) = \int v_T^3 d\vec{c}_1 \psi(\vec{c}_1) \sigma^2 n^2 v_T^{-2} \tilde{I}\left(\tilde{f},\tilde{f}\right)$$
$$= \sigma^2 n^2 v_T \int d\vec{c}_1 \psi(\vec{c}_1) \tilde{I}\left(\tilde{f},\tilde{f}\right), \tag{C.46}$$

which equals the RHS of (6.25) (without the omitted factor $g_2(\sigma)$):

$$\frac{\sigma^2}{2} \int v_T^3 d\vec{c}_1 v_T^3 d\vec{c}_2 \int d\vec{e}\,\Theta\left(-v_T \vec{c}_{12} \cdot \vec{e}\right) |v_T \vec{c}_{12} \cdot \vec{e}|$$
$$\times \frac{n}{v_T^3} \tilde{f}(\vec{c}_1) \frac{n}{v_T^3} \tilde{f}(\vec{c}_2) \Delta\left[\psi(\vec{c}_1) + \psi(\vec{c}_2)\right] \tag{C.47}$$
$$= \frac{1}{2} \sigma^2 n^2 v_T \int d\vec{c}_1 d\vec{c}_2 \int d\vec{e}\,\Theta\left(-\vec{c}_{12} \cdot \vec{e}\right) |\vec{c}_{12} \cdot \vec{e}|$$
$$\times \tilde{f}(\vec{c}_1) \tilde{f}(\vec{c}_2) \Delta\left[\psi(\vec{c}_1) + \psi(\vec{c}_2)\right].$$

Comparing (C.46) with (C.47), we obtain the basic property of the dimensionless collision integral (8.4).

Solution of problem 8.2
Derive the expression for μ_2, given by (8.23) in terms of Basic Integrals as defined by (8.30)!

Substituting (8.26, 8.28) into (8.23) for $p=2$ we obtain

$$\mu_2 = \frac{(1-\varepsilon^2)}{4} \int d\vec{c}_1 \int d\vec{c}_2 \int d\vec{e}\,\Theta\left(-\vec{c}_{12}\cdot\vec{e}\right)|\vec{c}_{12}\cdot\vec{e}|\,\phi(c_1)\,\phi(c_2)$$
$$\times \left[(\vec{c}_{12}\cdot\vec{e})^2 + a_2\left((\vec{c}_{12}\cdot\vec{e})^2 C^4 + (\vec{c}_{12}\cdot\vec{e})^2\left(\vec{C}\cdot\vec{c}_{12}\right)^2 + \frac{1}{16}(\vec{c}_{12}\cdot\vec{e})^2 c_{12}^4 \right.\right.$$
$$\left.\left. + \frac{1}{2}(\vec{c}_{12}\cdot\vec{e})^2 C^2 c_{12}^2 - 5(\vec{c}_{12}\cdot\vec{e})^2 C^2 - \frac{5}{4}(\vec{c}_{12}\cdot\vec{e})^2 c_{12}^2 + \frac{15}{4}(\vec{c}_{12}\cdot\vec{e})^2\right)\right]. \tag{C.48}$$

From the definition (8.30) of the Basic Integrals it follows that μ_2 in the last equation may be written in the form (8.36).

Solution of problem 8.3
Derive μ_4 given by (8.38) using Maple!

The corresponding Maple program reads.

```
1  DefDimension(3);
2  DefJ();
3  Expr:=Delta4*(-1/2*(1+a2*(S(2,c1p)+S(2,c2p)))):
4  mu4:=getJexpr(0,Expr,0);
5  T1:=factor(coeff(mu4,a2,0)/4/sqrt(2*Pi));
6  T2:=simplify(coeff(mu4,a2,1)/4/sqrt(2*Pi));
```

The solution of this problem is discussed in more detail on page 260. It is easy to check that the result given by Maple may be written in the form (8.38, 8.39).

Solution of problem 10.1
Find the solution of (10.22), that is, find the exponent ν and the prefactor for the Ansatz $\varphi \propto (1+t/\tau_0)^{\nu}$!

Substituting the Ansatz $\varphi = A\left(1+t/\tau_0\right)^{\nu}$ into equation (10.22) we obtain

$$A\nu\left(1+t/\tau_0\right)^{\nu-1}\tau_0^{-1} + \frac{5A}{6\tau_0}\left(1+\frac{t}{\tau_0}\right)^{\nu-1} = \sqrt{\frac{\pi}{8}}\tau_c(0)^{-1}\left(1+\frac{t}{\tau_0}\right)^{-5/6}. \tag{C.49}$$

Comparing the exponents in the LHS and RHS we conclude that $\nu - 1 = -5/6$, that is, $\nu = 1/6$. Substituting this exponent into (C.49) we obtain

$$A = \sqrt{\frac{\pi}{8}\frac{\tau_0}{\tau_c(0)}} = b\delta^{-1}, \tag{C.50}$$

where we use (9.29) for $\tau_0/\tau_c(0)$ and (10.24) for b.

Solution of problem 13.1
Derive (13.37) using the definition of the collision operator (13.35) and the collision rules (6.5)!

Since the indices i, j are arbitrary we choose $i, j = 1, 2$. From the definition of the binary collision operator (13.35) and of the operator $\hat{b}^{\vec{e}}_{ij}$ (13.36) follows

$$\hat{T}_{12}\vec{v}_2 = \sigma^2 \int d\vec{e}\,\Theta\left(-\vec{v}_{12} \cdot \vec{e}\right) |\vec{v}_{12} \cdot \vec{e}|\, \delta\left(\vec{r}_{12} - \sigma\vec{e}\right)\left(\vec{v}'_2 - \vec{v}_2\right), \tag{C.51}$$

where according to the collision rules (6.5)

$$\vec{v}'_2 - \vec{v}_2 = \frac{1+\varepsilon}{2}\left(\vec{v}_{12} \cdot \vec{e}\right)\vec{e} = -\left(\vec{v}'_1 - \vec{v}_1\right). \tag{C.52}$$

Thus,

$$\hat{T}_{12}\vec{v}_2 = -\sigma^2 \int d\vec{e}\,\Theta\left(-\vec{v}_{12} \cdot \vec{e}\right) |\vec{v}_{12} \cdot \vec{e}|\, \delta\left(\vec{r}_{12} - \sigma\vec{e}\right)\left(\vec{v}'_1 - \vec{v}_1\right) \tag{C.53}$$
$$= -\hat{T}_{12}\vec{v}_1.$$

Similarly, we can write

$$\hat{T}_{21}\vec{v}_1 = \sigma^2 \int d\vec{e}\,\Theta\left(-\vec{v}_{21} \cdot \vec{e}\right) |\vec{v}_{21} \cdot \vec{e}|\, \delta\left(\vec{r}_{21} - \sigma\vec{e}\right)\left(\hat{b}^{\vec{e}}_{21}\vec{v}_1 - \vec{v}_1\right)$$
$$= \sigma^2 \int d\vec{e}\,\Theta\left(+\vec{v}_{12} \cdot \vec{e}\right) |\vec{v}_{12} \cdot \vec{e}|\, \delta\left(\vec{r}_{12} + \sigma\vec{e}\right)\left(\hat{b}^{\vec{e}}_{12}\vec{v}_1 - \vec{v}_1\right), \tag{C.54}$$

where we use $\vec{v}_{21} = -\vec{v}_{12}$, $\vec{r}_{21} = -\vec{r}_{12}$ and take into account $\hat{b}^{\vec{e}}_{21} = \hat{b}^{\vec{e}}_{12}$, according to the definition (13.36) of this operator. If we change $\vec{e} \to -\vec{e}$ in the integrand and notice that $\hat{b}^{-\vec{e}}_{12} = \hat{b}^{\vec{e}}_{12}$, as it follows from the collision rule (6.5), we obtain

$$\hat{T}_{21}\vec{v}_1 = \sigma^2 \int d\vec{e}\,\Theta\left(-\vec{v}_{12} \cdot \vec{e}\right) |\vec{v}_{12} \cdot \vec{e}|\, \delta\left(\vec{r}_{12} - \sigma\vec{e}\right)\left(\hat{b}^{\vec{e}}_{12}\vec{v}_1 - \vec{v}_1\right) \tag{C.55}$$
$$= \hat{T}_{12}\vec{v}_1.$$

Solution of problem 13.2
Prove (13.39)!

Since

$$\mathcal{L}^0_i v_1^2 = \vec{v}_i \cdot \frac{\partial}{\partial \vec{r}_i} v_1^2 = 0, \tag{C.56}$$

the term with the free-streaming operators does not contribute. According to the definition (13.35) the expression $\hat{T}_{ij}v_1^2$ with $i < j$ contains the factor $(\hat{b}^{\vec{e}}_{ij} - 1)v_1^2$.

From the definition of $\hat{b}^e_{ij}$ which describes the collision between a pair (i,j) follows

$$\left(\hat{b}^e_{ij} - 1\right)\vec{v}_1^2 = \vec{v}_1'^2 - \vec{v}_1^2 = 0 \quad \text{if} \quad i \neq 1, \tag{C.57}$$

since if $i \neq 1$ the particle 1 is not involved in the collision (recall that $j > i$). If $i = 1$, the summation over the index $j > 1$ in (13.39) gives $(N-1)$ identical terms, the same for all j. With $j = 2$ we arrive at the final result (13.39).

Solution of problem 14.1
Prove the property (14.3) of the binary collision operator!

From the identity of the particles and (13.37) follows

$$\begin{aligned}\left\langle \vec{v}_1 \hat{T}_{12} \vec{v}_1 \right\rangle &= \frac{1}{2}\left(\left\langle \vec{v}_1 \hat{T}_{12} \vec{v}_1 \right\rangle + \left\langle \vec{v}_2 \hat{T}_{21} \vec{v}_2 \right\rangle\right) \\ &= \frac{1}{2}\left(\left\langle \vec{v}_1 \hat{T}_{12} \vec{v}_1 \right\rangle - \left\langle \vec{v}_2 \hat{T}_{12} \vec{v}_1 \right\rangle\right) = \frac{1}{2}\left\langle \vec{v}_{12} \hat{T}_{12} \vec{v}_1 \right\rangle.\end{aligned} \tag{C.58}$$

Solution of problem 14.2
Derive (14.6)!

We integrate over $d\vec{r}_3 \cdots d\vec{r}_N d\vec{v}_3 \cdots d\vec{v}_N$ by means of (13.42),

$$N(N-1)\int d\vec{r}_3 \cdots d\vec{r}_N \int d\vec{v}_3 \cdots d\vec{v}_N \rho(t) = f_2(\vec{r}_1, \vec{r}_2, \vec{v}_1, \vec{v}_2, t), \tag{C.59}$$

and use the approximation (13.43),

$$f_2(\vec{r}_1, \vec{r}_2, \vec{v}_1, \vec{v}_2, t) = g_2(r_{12}) f(\vec{v}_1, t) f(\vec{v}_2, t), \tag{C.60}$$

which corresponds to the assumption of molecular chaos. The integration over $d\vec{r}_1 d\vec{r}_2$ may be performed using the transformation of variables $d\vec{r}_1, d\vec{r}_2 \to d\vec{r}_1, d\vec{r}_{12}$ with the result

$$\begin{aligned}\frac{1}{N}\int d\vec{r}_1 d\vec{r}_2 g_2(r_{12}) f(\vec{v}_1, t) f(\vec{v}_2, t) \delta(\vec{r}_{12} - \sigma\vec{e}) \\ = f(\vec{v}_1, t) f(\vec{v}_2, t) \int d\vec{r}_{12} g_2(r_{12}) \delta(\vec{r}_{12} - \sigma\vec{e}) \frac{1}{N}\int d\vec{r}_1 \\ = f(\vec{v}_1, t) f(\vec{v}_2, t) g_2(\sigma) \frac{V}{N} = f(\vec{v}_1, t) f(\vec{v}_2, t) n^{-1} g_2(\sigma).\end{aligned} \tag{C.61}$$

Substituting this result into (14.5) we arrive at (14.6).

Solution of problem 14.3

Derive the velocity correlation time for a granular gas of particles that interact with $\varepsilon = $ const. in linear approximation with respect to a_2: first perform the manual calculations, then use Maple (see Appendix A) to get the expression for τ_v in terms of Basic Integrals. Finally, derive (14.22)!

In linear approximation with respect to a_2 one can write (14.9) using (14.11) and expressing the factor $\left(S_2(c_1^2) + S_2(c_2^2)\right)$ in terms of the new variables $\vec{c}_{12}$ and $\vec{C}$:

$$\tau_v^{-1}(t) = \frac{1+\varepsilon}{12\sqrt{2\pi}} \tau_c^{-1}(t) \int d\vec{c}_{12} \int d\vec{C} \phi(c_{12}) \phi(C)$$

$$\times \int d\vec{e}\, \Theta\left(-\vec{c}_{12} \cdot \vec{e}\right) |\vec{c}_{12} \cdot \vec{e}| \left(\vec{c}_{12} \cdot \vec{e}\right)^2 (1 + a_2 F) \quad \text{(C.62)}$$

where

$$F = C^4 + \left(\vec{C} \cdot \vec{c}_{12}\right)^2 + \frac{1}{16} c_{12}^4 + \frac{1}{2} C^2 c_{12}^2 - 5C^2 - \frac{5}{4} c_{12}^2 + \frac{15}{4}. \quad \text{(C.63)}$$

From the definition of the Basic Integrals (8.30), the expression for τ_v follows:

$$\tau_v^{-1} = \frac{1+\varepsilon}{12\sqrt{2\pi}} \tau_c^{-1} \Big[J_{0,0,0,0,2,0} + a_2 \Big(J_{4,0,0,0,2,0} + J_{0,0,2,0,2,0}$$

$$\frac{1}{6} J_{0,4,0,0,2,0} + \frac{1}{2} J_{2,2,0,0,2,0} - 5 J_{2,0,0,0,2,0} - \frac{5}{4} J_{0,2,0,0,2,0} + \frac{15}{4} J_{0,0,0,0,2,0} \Big) \Big]. \quad \text{(C.64)}$$

Substituting the values of the Basic Integrals, which may be found from (8.31), we obtain (14.22). The same result may be obtained by Maple:

```
1  DefDimension(3);
2  unDefJ();
3  Expr:=tau_c_inv/(12*sqrt(2*Pi))*(1+epsilon)*c12D0Te*c12D0Te
4      *(1+a2*(S(2,c1p)+S(2,c2p))):
5  tau_v_inv:=getJexpr(0,Expr,0);
6  DefJ();
7  tau_v_inv:=getJexpr(0,Expr,0);
```

which gives as expected (14.22).

Solution of problem 14.4

Find the self-diffusion coefficient up to the second order in a_2!

For the complete description of the self-diffusion on the level of the first non-vanishing Sonine coefficient a_2 (with all high-order coefficients neglected) one has to take into account the term $a_2^2 S_2\left(c_1^2\right) S_2\left(c_2^2\right)$ in (14.9) (which was omitted in the linear analysis) and use (8.65) for τ_0, which refers to the complete description of the evolution of temperature (on the level of a_2). Using Maple we obtain

$$\tau_v^{-1}(t) = \frac{(1+\varepsilon)}{3}\left(1 + \frac{3a_2}{32}\right)^2 \tau_c^{-1}(t) \tag{C.65}$$

and then with (14.17, 14.18, 14.20) we arrive at

$$D(t) = \frac{2D_0}{1+\varepsilon}\frac{1}{\left(1+\frac{3}{32}a_2\right)^2}\left(1+\frac{t}{\tau_0}\right)^{-1} \tag{C.66}$$

where in the last expression τ_0 from (8.65) is to be used. The Maple program for the evaluation of τ_v reads in this case

```
1  DefDimension(3);
2  unDefJ();
3  Expr:=tau_c_inv/(12*sqrt(2*Pi))*(1+epsilon)*c12D0Te*c12D0Te
4      *(1+a2*(S(2,c1p)+S(2,c2p))):
5  tau_v_inv:=getJexpr(0,Expr,0);
6  DefJ();
7  tau_v_inv:=getJexpr(0,Expr,0);
```

Solution of problem 14.5
Find the velocity correlation time τ_v up to the second order in δ and in a_2 by means of Maple!

The result of the Maple program

```
1  DefDimension(3);
2  DefJ();
3  Expr:=tau_c_inv/(12*sqrt(2*Pi))*(1+epsilon)*c12D0Te*c12D0Te
4      *(1+a2*(S(2,c1p)+S(2,c2p))+a2^2*S(2,c1p)*S(2,c2p)):
5  tau_v_inv:=getJexpr(1,Expr,2);
6  coeff(tau_v_inv, dprime,0)/tau_c_inv;
7  coeff(tau_v_inv, dprime,1)/tau_c_inv;
8  coeff(tau_v_inv, dprime,2)/tau_c_inv;
```

reads

$$\begin{aligned}\tau_v^{-1} &= \frac{2}{3}\tau_c^{-1}\left(1+\frac{3}{32}a_2\right)^2 - \frac{\sqrt{2\pi}\omega_0}{12\pi}\tau_c^{-1}\left(1+\frac{6}{25}a_2+\frac{21}{2500}a_2^2\right)\delta' \\ &+ \frac{\sqrt{2\pi}\omega_1}{22\pi}\tau_c^{-1}\left(1+\frac{119}{400}a_2+\frac{4641}{640000}a_2^2\right)\delta'^2.\end{aligned} \tag{C.67}$$

Solution of problem 15.1
Prove (15.22) *and* (15.23)!

Consider the vectorial integral

$$\vec{a} = \int d\vec{e}\,(\vec{g}\cdot\vec{e})|\vec{g}\cdot\vec{e}|\Theta(-\vec{g}\cdot\vec{e})\vec{e} \tag{C.68}$$

(here we will use $\vec{g}$ in place of $\vec{w}$ in (15.20, 15.21)). The result of the integration over $\vec{e}$ must be a vector. Due to the symmetry of the problem, this vector should be directed along $\vec{g}$ since there is no other preferred directions. Thus,

$$\vec{a} = G(g)\vec{g} \tag{C.69}$$

with the function $G(g)$ to be determined. We chose the coordinate system with the Z axis directed along $\vec{g}$ and multiply (C.68) by $\vec{g}$. With the definition (C.69) of $G(g)$ we find

$$G(g)g^2 = \int_{\pi/2}^{\pi} \sin\theta d\theta (g\cos\theta)^2 g|\cos\theta| \int_0^{2\pi} d\varphi = \frac{2\pi}{4}g^3, \tag{C.70}$$

that is, $G(g) = (\pi/2)g$ and thus

$$\vec{a} = \frac{\pi}{2}g\vec{g} \tag{C.71}$$

with (C.71) the and definition of $\vec{A}$, (15.20), we obtain (15.22). To prove the other relation (15.23) we consider the tensor with the components $\{i,j\} = \{x,y,z\}$:

$$n_{ij} = \int d\vec{e}\,e_i e_j (\vec{g}\cdot\vec{e})^2|\vec{g}\cdot\vec{e}|\Theta(-\vec{g}\cdot\vec{e}). \tag{C.72}$$

The integrand in (C.72) is a tensor with components given by products of the components of the vector $\vec{e}$. Since integration over $\vec{e}$ does not change the symmetry, we expect that the components of the resultant tensor are functions of the components of the vector $\vec{g}$, which is the only involved vector, plus the unit tensor:

$$n_{ij} = H(g)g^2\delta_{ij} + K(g)g_i g_j. \tag{C.73}$$

The trace (that is the sum of diagonal elements) of the tensor n_{ij} reads

$$3H(g)g^2 + K(g)g^2 = \int d\vec{e}\,(\vec{g}\cdot\vec{e})^2|\vec{g}\cdot\vec{e}|\Theta(-\vec{g}\cdot\vec{e})$$
$$= \int_{\pi/2}^{\pi} \sin\theta d\theta (g\cos\theta)^2 g|\cos\theta| \int_0^{2\pi} d\varphi = \frac{2\pi}{4}g^3, \tag{C.74}$$

where we take into account that $\delta_{ii} = 1+1+1 = 3$, $g_i g_i = g_x^2 + g_y^2 + g_z^2 = g^2$ and $e_i e_i = 1$. Multiplying n_{ij} by $g_i g_j$ and summing over i and j we obtain

$$H(g)g^4 + K(g)g^4 = \int d\vec{e}\,(\vec{g}\cdot\vec{e})^2(\vec{g}\cdot\vec{e})^2|\vec{g}\cdot\vec{e}|\Theta(-\vec{g}\cdot\vec{e})$$
$$= \int_{\pi/2}^{\pi} \sin\theta d\theta (g\cos\theta)^4 g|\cos\theta| \int_0^{2\pi} d\varphi = \frac{2\pi}{6}g^5 \qquad \text{(C.75)}$$

where we use $\delta_{ij}g_ig_j = g^2$ and $e_ig_i = e_jg_j = \vec{g}\cdot\vec{e}$. The system

$$3H(g) + K(g) = \frac{\pi}{2}g$$
$$H(g) + K(g) = \frac{\pi}{3}g \qquad \text{(C.76)}$$

which follows from (C.74, C.75), has the solution

$$H(g) = \frac{\pi}{12}g$$
$$K(g) = \frac{\pi}{4}g. \qquad \text{(C.77)}$$

Hence, (C.73) reads
$$n_{ij} = \frac{\pi}{12}g\left(g^2\delta_{ij} + 3g_ig_j\right). \qquad \text{(C.78)}$$

With (C.78) and the definition (15.21) of N_{ij} we obtain the second relations (15.23).

Solution of problem 15.2
Prove the second line of (15.66)!

Transforming to time variable t and multiplying (15.63) by $v_T(t)$ we obtain the force
$$\mathcal{F} = v_T(t)\gamma(t)(1-b)L\left[t(\tau)\right]. \qquad \text{(C.79)}$$

Its second moment reads
$$\langle \mathcal{F}(t)\mathcal{F}(t')\rangle = \frac{[v_T(t)\gamma(t)(1-b)]^2 \delta(t-t')}{\gamma(t)(1-b)} \qquad \text{(C.80)}$$

where we use (15.64) and take into account
$$\delta(\tau - \tau') = \frac{1}{\left|\frac{d\tau}{dt}\right|}\delta(t-t') = \delta(t-t')\frac{1}{\gamma(t)(1-b)} \qquad \text{(C.81)}$$

since $d\tau/dt = \gamma(t)(1-b)$. Then from (15.37) we notice that $v_T^2 = \bar{\gamma}/(a\gamma)$ and obtain
$$\langle \mathcal{F}(t)\mathcal{F}(t')\rangle = \frac{\bar{\gamma}}{a\gamma}\gamma(1-b)\delta(t-t') = 2\bar{\gamma}\delta(t-t') \qquad \text{(C.82)}$$

where the relation $a = (1-b)/2$ (15.44) was used.

Solution of problem 17.1
Prove (17.18) for the multiplication of a vector and a dyad!

The multiplication of the tensor $\vec{a}\vec{b}$ by the vector $\vec{c}$ yields a vector whose ith component reads

$$\left(\vec{a}\vec{b}\cdot\vec{c}\right)_i = \left(\vec{a}\vec{b}\right)_{ik} c_k = a_i b_k c_k = a_i \left(\vec{b}\cdot\vec{c}\right), \qquad (\text{C.83})$$

where the summation convention has been used. Similarly

$$\left(\vec{c}\cdot\vec{a}\vec{b}\right)_i = c_k \left(\vec{a}\vec{b}\right)_{ki} = c_k a_k b_i = \left(\vec{c}\cdot\vec{a}\right) b_i. \qquad (\text{C.84})$$

Solution of problem 17.2
Derive the hydrodynamic equation (17.32) from the Boltzmann equation (17.8) by multiplying it with $mv_1^2/2$ and integrating over $\vec{v}_1$!

For the first term in the LHS of (17.8) after this transformation we obtain (using the local velocity $\vec{V} = \vec{v} - \vec{u}$ and omitting the subscript '1')

$$\frac{\partial}{\partial t} \int d\vec{v}\, \frac{mv^2}{2} f(\vec{v})$$

$$= \frac{\partial}{\partial t}\left[\int d\vec{v}\,\frac{mV^2}{2} f(\vec{v}) + \frac{mu^2}{2}\int d\vec{v}\, f(\vec{v}) + m\vec{u}\cdot\int d\vec{v}\,\vec{V} f(\vec{v})\right]$$

$$= \frac{\partial}{\partial t}\left(\frac{3}{2}nT\right) + \frac{\partial}{\partial t}\left(\frac{mu^2}{2}n\right) = \frac{3T}{2}\frac{\partial n}{\partial t} + \frac{3n}{2}\frac{\partial T}{\partial t} + \frac{mu^2}{2}\frac{\partial n}{\partial t} + nm\vec{u}\cdot\frac{\partial \vec{u}}{\partial t}$$

$$= -\left(\frac{3T}{2} + \frac{mu^2}{2}\right)\vec{\nabla}\cdot(n\vec{u}) + \frac{3n}{2}\frac{\partial T}{\partial t} - \frac{nm}{2}\left(\vec{u}\cdot\vec{\nabla}\right)u^2 - \vec{u}\cdot\vec{\nabla}\cdot\hat{P},$$
$$(\text{C.85})$$

where we use (17.14) for $\partial n/\partial t$ and (17.25) for $\partial\vec{u}/\partial t$ and take into account that

$$\vec{u}\cdot\left(\vec{u}\cdot\vec{\nabla}\right)\vec{u} = \frac{1}{2}\left(\vec{u}\cdot\vec{\nabla}\right)u^2. \qquad (\text{C.86})$$

Correspondingly, for the second term on the LHS we can write

$$\int d\vec{v}\,\frac{mv^2}{2}\left(\vec{v}\cdot\vec{\nabla}\right) f(\vec{v}) = \frac{m}{2}\vec{\nabla}\cdot\int\left(\vec{V}+\vec{u}\right)\left(V^2 + 2\vec{u}\cdot\vec{V} + u^2\right) f(\vec{v}) d\vec{v}$$

$$= \vec{\nabla}\cdot\int\frac{mV^2}{2}\vec{V} f(\vec{v}) d\vec{v} + \vec{\nabla}\cdot\left(\int m\vec{V}\vec{V} f(\vec{v}) d\vec{v}\right)\cdot\vec{u}$$

$$+ \vec{\nabla}\cdot\left(\int \vec{V} f(\vec{v}) d\vec{v}\right)\frac{mu^2}{2} + \vec{\nabla}\cdot\vec{u}\int\frac{mV^2}{2} f(\vec{v}) d\vec{v}$$

$$+ \vec{\nabla}\cdot\vec{u}\left(\int \vec{V} f(\vec{v}) d\vec{v}\right)\cdot m\vec{u} + \vec{\nabla}\cdot\vec{u}\left(\int f(\vec{v}) d\vec{v}\right)\frac{mu^2}{2}.$$
$$(\text{C.87})$$

Using the definitions for the heat flux (17.33), the pressure tensor (17.22), density (17.4) and temperature (17.6) and (17.23) we obtain for the LHS

$$\int d\vec{v}\, \frac{mv^2}{2} \left(\vec{v}\cdot\vec{\nabla}\right) f(\vec{v})$$
$$= \vec{\nabla}\cdot\vec{q} + \vec{\nabla}\cdot\hat{P}\cdot\vec{u} + \left(\vec{\nabla}\cdot\vec{u}\right)\left(\frac{3}{2}nT\right) + \left(\vec{\nabla}\cdot\vec{u}\right)\left(n\frac{mu^2}{2}\right). \quad \text{(C.88)}$$

Finally, taking into account that

$$\left(\vec{\nabla}\cdot\vec{u}\right)\left(\frac{3}{2}nT\right) = \frac{3}{2}T\vec{\nabla}\cdot(n\vec{u}) + \frac{3}{2}n\vec{u}\cdot\vec{\nabla}T \quad \text{(C.89)}$$

and

$$\left(\vec{\nabla}\cdot\vec{u}\right)\left(n\frac{mu^2}{2}\right) = \frac{mn}{2}\left(\vec{u}\cdot\vec{\nabla}\right)u^2 + \frac{mu^2}{2}\vec{\nabla}\cdot(n\vec{u}) \quad \text{(C.90)}$$

and

$$\vec{\nabla}\cdot\hat{P}\cdot\vec{u} = \nabla_i P_{ij} u_j = P_{ij}\nabla_i u_j + u_j \nabla_i P_{ij} = \hat{P}:\vec{\nabla}\vec{u} + \vec{u}\cdot\vec{\nabla}\cdot\hat{P}, \quad \text{(C.91)}$$

and summing (C.85) and (C.88) we obtain for the LHS of (17.8)

$$\frac{3}{2}n\frac{\partial T}{\partial t} + \frac{3}{2}n\left(\vec{u}\cdot\vec{\nabla}T\right) + \vec{\nabla}\cdot\vec{q} + \hat{P}:\vec{\nabla}\vec{u}. \quad \text{(C.92)}$$

Evaluation of the RHS as it was already shown in (17.30, 17.31) gives $-(3/2)nT\zeta$. Division of all terms of the obtained equation by $(3/2)n$ yields the hydrodynamic equation (17.32).

Solution of problem 18.1
Derive equations (18.3) and (18.4)!

We assume for simplicity that the kinetic coefficients are constants and consider the ith component of the vector $\vec{\nabla}\cdot\hat{P}$:

$$(\vec{\nabla}\cdot\hat{P})_i = \nabla_j P_{ij} = \nabla_j p \delta_{ij} - \nabla_j \eta \nabla_i u_j - \nabla_j \eta \nabla_j u_i + \frac{2}{3}\eta\delta_{ij}\nabla_j(\vec{\nabla}\cdot\vec{u}), \quad \text{(C.93)}$$

where we use (17.36) and where summation over repeating indices is implied. With

$$\delta_{ij}\nabla_j = \nabla_i, \qquad \nabla_j\nabla_j = \nabla^2, \qquad \nabla_j u_j = \vec{\nabla}\cdot\vec{u} \quad \text{(C.94)}$$

we obtain

$$\vec{\nabla}\cdot\hat{P} = \vec{\nabla}p - \eta\nabla^2\vec{u} - \frac{1}{3}\eta\vec{\nabla}(\vec{\nabla}\cdot\vec{u}). \quad \text{(C.95)}$$

Using the (formal) coefficient λ at each power of the gradient we obtain the second equation in (18.3). Next we consider

$$\hat{P} : \vec{\nabla} \vec{u} = P_{ij} \nabla_j u_i$$
$$= p\delta_{ij} \nabla_j u_i - \eta \left(\nabla_i u_j \right) \left(\nabla_j u_i \right) - \eta \left(\nabla_j u_i \right) \left(\nabla_j u_i \right) + \frac{2}{3} \eta \delta_{ij} \nabla_j u_i (\vec{\nabla} \cdot \vec{u}) \quad \text{(C.96)}$$
$$= p(\vec{\nabla} \cdot \vec{u}) - \eta \left(\nabla_i u_j \right) \left(\nabla_j u_i \right) - \eta \left(\nabla_j u_i \right) \left(\nabla_j u_i \right) + \frac{2}{3} \eta (\vec{\nabla} \cdot \vec{u})^2 ,$$

where we use the definition of the tensor product (17.34) and (C.94). With

$$\vec{\nabla} \vec{q} = -\kappa \nabla^2 T - \mu \nabla^2 n , \quad \text{(C.97)}$$

which follows from (17.36), we obtain

$$\frac{2}{3n} \left(\hat{P} : \vec{\nabla} \vec{u} + \vec{\nabla} \vec{q} \right) = \frac{2}{3n} p (\vec{\nabla} \cdot \vec{u}) - \frac{2}{3n} \left(\kappa \nabla^2 T + \mu \nabla^2 n \right)$$
$$- \frac{2\eta}{3n} \left[\left(\nabla_i u_j \right) \left(\nabla_j u_i \right) - \left(\nabla_j u_i \right) \left(\nabla_j u_i \right) - \frac{2}{3} (\vec{\nabla} \cdot \vec{u})^2 \right] . \quad \text{(C.98)}$$

Using in the above equations λ^k at each kth power of ∇, we arrive at the third equation in (18.3) together with (18.4).

Sometimes the expression for $\hat{P} : \vec{\nabla} \vec{u}$ is written in another form: we use the relation

$$\nabla_i u_j \nabla_j u_i = \left(\nabla_i u_j \right) \left(\nabla_j u_i \right) + u_j \nabla_i \nabla_j u_i = \left(\nabla_i u_j \right) \left(\nabla_j u_i \right) + u_j \nabla_j \nabla_i u_i$$
$$= \left(\nabla_i u_j \right) \left(\nabla_j u_i \right) + (\vec{u} \cdot \vec{\nabla})(\vec{\nabla} \cdot \vec{u}) \quad \text{(C.99)}$$

and the definition of the tensor product (17.34), which allows us to write

$$\left(\nabla_i u_j \right) \left(\nabla_j u_i \right) = \vec{\nabla} \vec{u} : \vec{\nabla} \vec{u} - (\vec{u} \cdot \vec{\nabla})(\vec{\nabla} \cdot \vec{u}) . \quad \text{(C.100)}$$

Similarly, using

$$\nabla_j u_i \nabla_j u_i = \left(\nabla_j u_i \right) \left(\nabla_j u_i \right) + u_i \nabla_j \nabla_j u_i = \frac{1}{2} \nabla_j \nabla_j u_i u_i \quad \text{(C.101)}$$

we obtain

$$\left(\nabla_j u_i \right) \left(\nabla_j u_i \right) = \frac{1}{2} \nabla^2 u^2 - (\vec{u} \cdot \nabla^2 \vec{u}) \quad \text{(C.102)}$$

and finally

$$\hat{P} : \vec{\nabla} \vec{u} = p \left(\vec{\nabla} \cdot \vec{u} \right) - \eta \Big[\vec{\nabla} \vec{u} : \vec{\nabla} \vec{u} - \left(\vec{u} \cdot \vec{\nabla} \right) \left(\vec{\nabla} \cdot \vec{u} \right)$$
$$+ \frac{1}{2} \nabla^2 u^2 - (\vec{u} \cdot \nabla^2 \vec{u}) + \frac{2}{3} \left(\vec{\nabla} \cdot \vec{u} \right)^2 \Big] . \quad \text{(C.103)}$$

This form is frequently used for $\hat{P} : \vec{\nabla} \vec{u}$.

Solution of problem 18.2
Prove (18.14)!

According to the definition of the collision integral $I(a,b)$ (8.1), its dependence on the functions a and b enters only via the factor

$$\frac{1}{\varepsilon^2} a(\vec{v}_1'') b(\vec{v}_2'') - a(\vec{v}_1) b(\vec{v}_2). \tag{C.104}$$

For $a = b = h_1 + \lambda h_2 + \cdots$ this factor reads

$$\frac{1}{\varepsilon^2} \left(h_1(\vec{v}_1'') + \lambda h_2(\vec{v}_1'') + \cdots \right) \left(h_1(\vec{v}_2'') + \lambda h_2(\vec{v}_2'') + \cdots \right)$$
$$- \left(h_1(\vec{v}_1) + \lambda h_2(\vec{v}_1) + \cdots \right) \left(h_1(\vec{v}_2) + \lambda h_2(\vec{v}_2) + \cdots \right)$$
$$= \left[\frac{1}{\varepsilon^2} h_1(\vec{v}_1'') h_1(\vec{v}_2'') - h_1(\vec{v}_1) h_1(\vec{v}_2) \right] + \lambda \left[\frac{1}{\varepsilon^2} h_2(\vec{v}_1'') h_1(\vec{v}_2'') - h_2(\vec{v}_1) h_1(\vec{v}_2) \right]$$
$$+ \lambda \left[\frac{1}{\varepsilon^2} h_1(\vec{v}_1'') h_2(\vec{v}_2'') - h_1(\vec{v}_1) h_2(\vec{v}_2) \right]$$
$$+ \lambda^2 \left[\frac{1}{\varepsilon^2} h_2(\vec{v}_1'') h_2(\vec{v}_2'') - h_2(\vec{v}_1) h_2(\vec{v}_2) \right] + \cdots . \tag{C.105}$$

Substituting this factor into the collision integral we obtain

$$I\left[(h_1 + \lambda h_2 + \cdots), (h_1 + \lambda h_2 + \cdots) \right]$$
$$= I(h_1, h_1) + \lambda I(h_1, h_2) + \lambda I(h_2, h_1) + \lambda^2 I(h_2, h_2) + \cdots, \tag{C.106}$$

which proves (18.14).

Solution of problem 18.3
Find the derivatives of $f^{(0)}$ with respect to n, $\vec{u}$ and T, that is, derive (18.18)!

The first two equations in the system (18.18) follow from the functional form (18.10) of the distribution function in zeroth order $f^{(0)}$. To prove the third equation we write (see also (15.35))

$$\frac{\partial f^{(0)}}{\partial T} = -\frac{\partial v_T}{\partial T} \frac{n}{v_T^4} \left(3 + \vec{c} \cdot \frac{\partial}{\partial \vec{c}} \right) \tilde{f}^{(0)}. \tag{C.107}$$

With

$$\frac{\partial v_T}{\partial T} = \frac{1}{m v_T}, \quad \vec{c} \cdot \frac{\partial}{\partial \vec{c}} = \vec{V} \cdot \frac{\partial}{\partial \vec{V}}, \quad \frac{\partial}{\partial \vec{V}} \cdot \vec{V} = 3 \tag{C.108}$$

we then obtain

$$\frac{\partial f^{(0)}}{\partial T} = -\frac{1}{m v_T^2} \frac{\partial}{\partial \vec{V}} \cdot \vec{V} \frac{n}{v_T^3} \tilde{f}^{(0)} = -\frac{1}{2T} \frac{\partial}{\partial \vec{V}} \cdot \vec{V} f^{(0)}, \tag{C.109}$$

which is the third equation in (18.18).

Solution of problem 18.4
Derive the coefficients $\vec{A}$, $\vec{B}$ and C_{ij} in (18.26)!

To obtain coefficients A and B, we substitute (18.18) into (18.24) and use the property (18.27). Do obtain C_{ij} we start from (18.21) and using $\vec{\nabla} \cdot \vec{u} = \delta_{ij}\nabla_j u_i$ we write all terms in the RHS of (18.21) which contain $\nabla_j u_i$ in the form

$$f^{(0)}\delta_{ij}\nabla_j u_i + \frac{2T}{3}\delta_{ij}\frac{\partial f^{(0)}}{\partial T}\nabla_j u_i + \frac{\partial f^{(0)}}{\partial V_i}V_j\nabla_j u_i$$

$$= \left[V_j\frac{\partial f^{(0)}}{\partial V_i} + f^{(0)}\delta_{ij} - \delta_{ij}\frac{2T}{3}\frac{1}{2T}\frac{\partial}{\partial \vec{V}}\cdot\vec{V}f^{(0)}\right]\nabla_j u_i \quad \text{(C.110)}$$

$$= \left[\frac{\partial}{\partial V_i}\left(V_j f^{(0)}\right) - \frac{1}{3}\delta_{ij}\frac{\partial}{\partial \vec{V}}\cdot\vec{V}f^{(0)}\right]\nabla_j u_i,$$

where we use (18.18) for $\partial_T f^{(0)}$. The expression in square brackets gives the coefficient C_{ij} in (18.26) whose final form again follows from (18.27).

Solution of problem 18.5
Prove that the coefficient $\zeta^{(1)}$, defined by (18.16), vanishes! Use the first-order distribution function $f^{(1)}$ as given on page 201.

$\zeta^{(1)}$ is proportional to the integral

$$\int d\vec{v}_1 d\vec{v}_2 v_{12}^3 f^{(0)}(\vec{v}_1)f^{(1)}(\vec{v}_2) = \int d\vec{V}_1 d\vec{V}_2 V_{12}^3 f^{(0)}(V_1)f^{(1)}(\vec{V}_2). \quad \text{(C.111)}$$

The integration is to be performed over the absolute values of $\vec{V}_1$ and $\vec{V}_2$ and over the directions of these vectors. Integration over the direction of $\vec{V}_2$ may be performed without restrictions, while performing integration over $\vec{V}_1$ we chose the OZ axis along the vector $\vec{V}_2$. Denoting the polar and azimuthal angles as φ and θ, we can write the integral in the form

$$\int d\vec{V}_1 d\vec{V}_2 V_{12}^3 f^{(0)}(V_1)f^{(1)}(\vec{V}_2) = \int_0^\infty dV_1 V_1^2 f^{(0)}(V_1) \int_0^\infty dV_2 V_2^2 \int_0^\pi \sin\theta d\theta$$

$$\times \int_0^{2\pi} d\varphi \left(V_1^2 + V_2^2 - 2V_1 V_2 \cos\theta\right)^{3/2}\left[\int d\vec{V}_2 f^{(1)}(\vec{V}_2)\right] = 0, \quad \text{(C.112)}$$

where we take into account that according to (C.145) from Exercise 20.9 the integral in square brackets vanishes. This proves $\zeta^{(1)} = 0$.

Solution of problem 18.6
Prove (18.36)!

The first equation in (18.36) reads

$$\int D_{ij}(\vec{V})f^{(0)}(V)d\vec{V} = \int_0^\infty dV V^4 f^{(0)}(V)\int d\hat{\vec{V}}\left(\hat{V}_i\hat{V}_j - \frac{1}{3}\delta_{ij}\right) = 0, \quad \text{(C.113)}$$

where the results of Exercise 20.9 have been used. From (18.29, 20.39) and using the definition of $h_1(V)$ from Exercise 20.9 we obtain

$$\vec{\alpha} = \kappa T f_M(V) h_1(V) \hat{\vec{V}}, \qquad \vec{\beta} = \mu n f_M(V) h_1(V) \hat{\vec{V}}. \qquad (C.114)$$

Correspondingly,

$$\int D_{ij}(\vec{V}) \vec{\alpha} d\vec{V} = \int_0^\infty dV V^4 \kappa T f_M(V) h_1(V) \int d\hat{\vec{V}} \, \hat{\vec{V}} \left(\hat{V}_i \hat{V}_j - \frac{1}{3} \delta_{ij} \right) = 0 \qquad (C.115)$$

since

$$\int d\hat{\vec{V}} \hat{V}_i \hat{V}_j \hat{V}_k = 0 \qquad \text{and} \qquad \int d\hat{\vec{V}} \hat{V}_i = 0, \qquad (C.116)$$

because the integrands in the last two equations are antisymmetric functions.

Solution of problem 18.7
Prove $\int d\vec{V} S_i f^{(0)} = 0$ and $\int d\vec{V} S_i \gamma_{kl} = 0$!

Using the definition (17.35) of $\vec{S}(\vec{V})$ we write

$$\int d\vec{V} S_i f^{(0)} = \int_0^\infty dV V^3 \left(\frac{mV^2}{2} - \frac{5}{2} T \right) f^{(0)}(V) \int d\hat{\vec{V}} \hat{V}_i = 0 \qquad (C.117)$$

where (C.116) from Exercise 18.6 is used. Similarly, using (20.6) for γ_{kl} we obtain

$$\int d\vec{V} S_i \gamma_{kl} = \int_0^\infty dV V^3 \left(\frac{mV^2}{2} - \frac{5}{2} T \right) \frac{\gamma_0 m}{T} V^2 f_M(V)$$
$$\int d\hat{\vec{V}} \hat{V}_i \left(\hat{V}_k \hat{V}_l - \frac{1}{3} \delta_{kl} \right) = 0, \qquad (C.118)$$

where again (C.116) has been applied.

Solution of problem 19.1
Prove (19.8)!

Using the scaling form of the velocity distribution function

$$f_M(V) = \frac{n}{v_T^3} \pi^{-3/2} \exp\left(-\frac{V^2}{v_T^2} \right) \qquad (C.119)$$

we write for the integral

$$\frac{2m^2}{3T} \int d\vec{V}_1 V_1^4 f_M(V_1) = \frac{2m^2}{3T} \frac{n}{v_T^3} v_T^3 v_T^4 4\pi \int_0^\infty dc c^2 c^4 \frac{1}{\pi^{3/2}} \exp(-c^2)$$
$$= \frac{2m^2}{3T} \left(\frac{2T}{m} \right)^2 \frac{4n}{\sqrt{\pi}} \frac{15}{16} \sqrt{\pi} = 10nT, \qquad (C.120)$$

where integration over the angles gives 4π, and where we use $v_T^2 = 2T/m$ and the integral

$$\int_0^\infty x^6 \exp(-x^2)\, dx = \frac{15}{16}\sqrt{\pi}. \tag{C.121}$$

Solution of problem 19.2
In general, the coefficient $\vec{\alpha}$ is expressed as an expansion,

$$\vec{\alpha}(\vec{V}) = \vec{V}\left[\alpha_0 + \alpha_1 S_1^{(3/2)}(c^2) + \alpha_2 S_2^{(3/2)}(c^2) + \cdots\right] f_M(V) \tag{C.122}$$

with unknown scalar coefficients α_k. In this series the coefficient α_0 is always trivial, that is, $\alpha_0 = 0$. Explain why!

Since the RHS of (19.30) does not have terms proportional to $S_0^{(3/2)}(c^2) = 1$, we conclude that $\alpha_0 = 0$.

Solution of problem 19.3
Compute the numerical coefficient $\Omega_\kappa^{\text{el}}$ manually and confirm the result using Maple!

With (19.22), which give the pre-collision velocities $\vec{c}_1$, $\vec{c}_2$ as well as after-collision velocities $\vec{c}_1'$, $\vec{c}_2'$ in terms of the centre of mass $\vec{C}$ and the relative $\vec{c}_{12}$ velocity before the collision, we can write

$$\left(c_2^2 - \frac{5}{2}\right)\left[(\vec{c}_1' \cdot \vec{c}_2)(\vec{c}_1')^2 + (\vec{c}_2' \cdot \vec{c}_2)(\vec{c}_2')^2 - (\vec{c}_1 \cdot \vec{c}_2)c_1^2 - (\vec{c}_2 \cdot \vec{c}_2)c_2^2\right]$$
$$= \left[C^2 + \frac{1}{4}c_{12}^2 - (\vec{C} \cdot \vec{c}_{12}) - \frac{5}{2}\right]\left[4\left(\vec{C} \cdot \vec{e}\right)^2 (\vec{c}_{12} \cdot \vec{e})^2 + \left(\vec{C} \cdot \vec{c}_{12}\right)(\vec{c}_{12} \cdot \vec{e})^2\right.$$
$$\left. + c_{12}^2 \left(\vec{C} \cdot \vec{e}\right)(\vec{c}_{12} \cdot \vec{e}) - 2\left(\vec{C} \cdot \vec{e}\right)(\vec{c}_{12} \cdot \vec{e})^3 - 4\left(\vec{C} \cdot \vec{c}_{12}\right)\left(\vec{C} \cdot \vec{e}\right)(\vec{c}_{12} \cdot \vec{e})\right]. \tag{C.123}$$

Substituting (C.123) into (19.42, 19.43) and using the definition of the Basic Integrals (8.31–8.33) together with the properties of the Gamma–function $\Gamma(x+1) = x\Gamma(x)$ and $\Gamma(1/2) = \sqrt{\pi}$ we obtain

$$\Omega_\kappa^{(\text{el})} = \left(\frac{5}{2}I_1 - I_2 + I_3 - \frac{1}{4}I_4\right), \tag{C.124}$$

where

$$I_1 = 4J_{001110} - 4J_{000220} - J_{001020} - J_{020110} + 2J_{000130}$$
$$= \sqrt{2\pi}\left(4 - 4 - 0 - 0 + 2\cdot 0\right) = 0$$
$$I_2 = 4J_{201110} - 4J_{200220} - J_{201020} - J_{220110} + 2J_{200130}$$
$$= \sqrt{2\pi}\left(4\cdot\frac{5}{4} - 4\cdot\frac{5}{4} - 0 - 0 + 2\cdot 0\right) = 0 \quad\quad\quad\text{(C.125)}$$
$$I_3 = 4J_{002110} - 4J_{001220} - J_{002020} - J_{021110} + 2J_{001130}$$
$$= \sqrt{2\pi}\left(4\cdot 0 - 4\cdot 0 - 6 - +2\cdot 4\right) = -4\sqrt{2\pi}$$
$$I_4 = 4J_{021110} - 4J_{020220} - J_{021020} - J_{040110} + 2J_{020130}$$
$$= \sqrt{2\pi}\left(4\cdot 6 - 4\cdot 6 - 0 - 0 + 2\cdot 0\right) = 0\,.$$

Thus,
$$\Omega_\kappa^{(\text{el})} = I_3 = -4\sqrt{2\pi}\,. \quad\quad\quad\text{(C.126)}$$

The same result is obtained by Maple:

```
1  DefDimension(3):
2  unDefJ();
3  Expr:=(c2pDOTc2p-(DIM+2)/2)*(c1aDOTc2p*c1aDOTc1a+c2aDOTc2p*c2aDOTc2a
4      -c2pDOTc2p*c2pDOTc2p-c1pDOTc2p*c1pDOTc1p):
5  OmegaKAPPAel:=getJexpr(0,Expr,0);
6  epsilon:=1;
7  OmegaKAPPAel;
8  DefJ();
9  OmegaKAPPAel;
```

Solution of problem 20.1
Prove (20.3)!

From the definition of the dimensionless velocity, $c = V/v_T$ with $v_T^2 = 2T/m$ follows
$$\frac{1}{V}\frac{\partial c}{\partial V} = \frac{1}{v_T V}\frac{\partial V}{\partial V} = \frac{1}{v_T^2 c} = \frac{m}{T}\frac{1}{2c}\,. \quad\quad\quad\text{(C.127)}$$

Similarly the definition of the Maxwell distribution (19.33) yields
$$\frac{1}{V}\frac{\partial f_M}{\partial V} = \frac{1}{V}\frac{\partial c}{\partial V}\frac{\partial f_M}{\partial c} = \frac{m}{T}\frac{1}{2c}(-2c)\frac{n}{v_T^3}\pi^{-3/2}\exp\left(-c^2\right) = -\frac{m}{T}f_M\,. \quad\quad\quad\text{(C.128)}$$

Solution of problem 20.2
Prove that the second term in the RHS of (20.7) equals $10nT$!

We start with (18.26) for the coefficient C_{ij}:

$$\int d\vec{V} D_{ij}(\vec{V}) C_{ij}(\vec{V})$$
$$= \int d\vec{V} D_{ij}(\vec{V}) \frac{\partial}{\partial V_i} \left(V_j f^{(0)}\right) - \frac{1}{3} \int d\vec{V} \delta_{ij} D_{ij}(\vec{V}) \frac{\partial}{\partial \vec{V}} \cdot \left(\vec{V} f^{(0)}\right). \quad \text{(C.129)}$$

The second term on the RHS vanishes since D_{ij} is the traceless tensor, that is, due to the summation convention

$$\delta_{ij} D_{ij} = D_{ii} = \left(V_x^2 - \frac{V^2}{3}\right) + \left(V_y^2 - \frac{V^2}{3}\right) + \left(V_z^2 - \frac{V^2}{3}\right) = 0. \quad \text{(C.130)}$$

The first term may be integrated by parts. Then, with

$$\frac{\partial}{\partial V_i} D_{ij}(\vec{V}) = \frac{\partial}{\partial V_i} m \left(V_i V_j - \frac{1}{3} \delta_{ij} V_i V_i\right) = mV_j \left(1 + \delta_{ij} - \frac{2}{3} \delta_{ij}\right), \quad \text{(C.131)}$$

according to

$$\frac{\partial}{\partial V_i} V_i V_j = (1 + \delta_{ij}) V_j \quad \text{(C.132)}$$

we arrive at

$$-\int d\vec{V} D_{ij}(\vec{V}) C_{ij}(\vec{V}) = \int d\vec{V} m f^{(0)} V_j V_j \left(1 + \frac{1}{3} \delta_{ij}\right)$$
$$= \frac{10}{3} \int d\vec{V} m V^2 f^{(0)}(\vec{V}) = 10nT, \quad \text{(C.133)}$$

where the definition of temperature has been employed, as well as the relation

$$V_i V_j \left(1 + \frac{1}{3} \delta_{ij}\right) \equiv \sum_{i,j} V_i V_j \left(1 + \frac{1}{3} \delta_{ij}\right) = \sum_j V_j^2 \left(\sum_i 1 + \frac{1}{3} \sum_i \delta_{ij}\right) = \frac{10}{3} V^2$$
$$\quad \text{(C.134)}$$

which follows form the summation convention.

Solution of problem 20.3
The relation $\gamma_0 = -\eta/(nT)$ has been obtained for gases of elastic particles. For which preconditions is it also valid for granular gases with $\varepsilon = \text{const.}$?

The relation $\gamma_0 = -\eta/(nT)$ follows from (19.16), which uses the approximation (19.5) for the coefficient γ_{ij}. Therefore, this relation is valid also for dissipative gases with constant coefficient of restitution as well as for gases of viscoelastic particles. Naturally, this relation is not correct if an expansion for γ_{ij} is used that containes more than one term.

Solution of problem 20.4
The viscosity coefficient (20.18) has been obtained in linear approximation with respect to a_2. How does one derive the viscosity coefficient in second order? Is it enough just to use (8.45) for μ_2, which is correct up to $\mathcal{O}(a_2^2)$?

This may be done if the field gradients are very small, that is, if terms $\propto a_2 \nabla_j u_i$ may be neglected with respect to terms $\propto a_2^2$.

Solution of problem 20.5
Prove (20.20) for $\vec{A}$ by means of (18.26) and (20.4)!

Using the definition (18.26) of $\vec{A}$ and the relation (20.4) for the velocity derivative of the distribution function we write

$$\vec{A} = \vec{V}\left\{-\frac{T}{m}\frac{m}{T}(c^2-1)\left[1 + a_2 S_2(c^2) + a_2 S_1^{(3/2)}(c^2)\right]\right.$$
$$\left. + \frac{3}{2}\left[1 + a_2 S_2(c^2)\right]\right\} f_M(V)$$
$$= \vec{V}\left\{[1 + a_2 S_2(c^2)]\left(1 - c^2 + \frac{3}{2}\right) + (1 - c^2)a_2\left(\frac{5}{2} - c^2\right)\right\} f_M(V)$$
$$= \vec{V}\left(\frac{5}{2} - c^2\right)\{1 + a_2 S_2(c^2) + a_2(1 - c^2)\} f_M(V)$$
$$= \vec{V} S_1^{(3/2)}(c^2)\left[1 + a_2\left(S_2^{(3/2)}(c^2) - 3/2\right)\right] f_M(V),$$
(C.135)

where we use

$$S_1^{(3/2)}(c^2) = \frac{5}{2} - c^2$$
(C.136)
$$S_2(c^2) + 1 - c^2 = \frac{c^4}{2} - \frac{7c^2}{2} + \frac{23}{8} = S_2^{(3/2)}(c^2) - \frac{3}{2}$$

which follow from the definition of the Sonine polynomials (7.14).

Solution of problem 20.6
Prove (20.26)!

With (19.1) for $\vec{A}$ and (19.28) for $\vec{S}$ one can write

$$-\frac{1}{T}\int d\vec{V}_1 \vec{S}(\vec{V}_1) \vec{A}(\vec{V}_1)$$
$$= \int d\vec{V}_1 \vec{V}_1^2 \left[S_1^{(3/2)}(c_1^2)\right]^2 f_M(V_1)\left[1 + a_2\left(S_2^{(3/2)}(c^2) - \frac{3}{2}\right)\right]. \quad (C.137)$$

Changing to dimensionless variables, using the definitions of the Sonine polynomials (7.14) and performing angular integration (which yields 4π) we obtain for the RHS of (C.137)

$$4\pi \left(\frac{n}{v_T^3}\right) v_T^3 v_T^2 \left\{ \int_0^\infty c^4 \phi(c) \left(\frac{5}{2} - c^2\right)^2 dc \right.$$

$$\left. + a_2 \int_0^\infty c^4 \phi(c) \left(\frac{5}{2} - c^2\right)^2 \left[\frac{1}{2}c^4 - \frac{7}{2}c^2 + \frac{23}{8}\right] dc \right\}$$

$$= \frac{15}{4} n v_T^2 + \frac{15}{2} n v_T^2 a_2 = \frac{15}{2} \frac{nT}{m} (1 + 2a_2), \quad \text{(C.138)}$$

where the Maxwell distribution $\phi(c) \equiv \exp(-c^2)/\pi^{3/2}$, the definition of v_T and the integral

$$\int_0^\infty \exp(-x^2) x^{2k} dx = \frac{(2k-1)!!}{2^{k+1}} \sqrt{\pi} \quad \text{(C.139)}$$

have been used.

Solution of problem 20.7
Derive the complete dependence $\kappa(a_2)$, neglecting the terms of order $\mathcal{O}(a_2 \nabla T)$!

The complete dependence of $\kappa(a_2)$ on a_2 without the terms $\mathcal{O}(a_2 \nabla T)$ may be obtained from (20.32) if one substitutes into this equation (20.29) for $\Omega_\kappa^{(\varepsilon)}$ and (8.45) for μ_2, which correspond to the complete dependence of these quantities on a_2.

Solution of problem 20.8
Calculate the coefficient μ, following the same procedure as for the coefficient of thermal conductivity!

The calculation of κ and μ are completely analogous to the derivation of the coefficient of thermal conductivity, i.e. we multiply (18.34) by $-\vec{S}(\vec{V}_1)/T$ and integrate over $\vec{V}_1$. The approximation (20.35) implies

$$\vec{\beta} \simeq \beta_1 \frac{1}{T} \vec{S}(\vec{V}) f_M(V) = \left(\frac{\beta_1}{\alpha_1}\right) \vec{\alpha}, \quad \text{(C.140)}$$

thus we can use the results, which have been obtained for the coefficient κ. The only difference is in the term

$$-\frac{1}{T} \int d\vec{V}_1 \vec{S}(\vec{V}_1) \cdot \vec{B}(\vec{V}_1) = a_2 \int d\vec{V}_1 V_1^2 \left[S_1^{(3/2)}(c_1^2)\right]^2 f_M(V_1)$$

$$= a_2 \frac{15}{4} n v_T^2 = a_2 \frac{15}{2} \frac{nT}{m}, \quad \text{(C.141)}$$

and in the temperature dependence of μ, which scales as $\mu \sim T^{3/2}$.

Solution of problem 20.9
Check the normalization of the distribution function $f(\vec{V})$ given by (20.39)!

For the first part of the distribution function we can write:

$$\int f^{(0)}(V)d\vec{V} = \frac{n}{v_T^3}v_T^3 \int \phi(c)\left[1+a_2 S_2(c^2)\right]d\vec{c}$$
$$= n\int \phi(c)d\vec{c} + na_2 \int \phi(c) S_2(c^2) S_0(c^2) d\vec{c} = n, \qquad \text{(C.142)}$$

where the normalization of $\phi(c)$ and the properties of the Sonine polynomials (7.6) have been used. The second part of the distribution function can be written in the form

$$f^{(1)}(\vec{V}) = f_M(V)\left[h_1(V)\kappa\vec{\nabla}T\cdot\hat{\vec{V}} + h_1(V)\mu\vec{\nabla}n\cdot\hat{\vec{V}} + h_2(V)\nabla_j u_i D_{ij}(\hat{\vec{V}})\right]$$

$$h_1(V) = -\frac{2m\kappa V}{5nT^3}\left(\frac{mV^2}{2} - \frac{5}{2}T\right)$$

$$h_2(V) = -\frac{\eta V^2}{nT^2},$$
$$\qquad \text{(C.143)}$$

where the form of the functions $h_{1/2}(V)$ follows from (20.39) and the definitions (17.27) and (17.35) of $D_{ij}(\vec{V})$ and $\vec{S}$ and the unit vector $\hat{\vec{V}} = \vec{V}/V$ has been introduced. Then we obtain

$$\int f^{(1)}(\vec{V})d\vec{V} = \int_0^\infty dV V^2 \int d\hat{\vec{V}}\, f^{(1)}(\vec{V}), \qquad \text{(C.144)}$$

where

$$\int d\hat{\vec{V}}\, f^{(1)}(\vec{V}) = f_M(V)h_1(V)\left(\kappa\vec{\nabla}T + \mu\vec{\nabla}n\right)\cdot\int d\hat{\vec{V}}\,\hat{\vec{V}}$$
$$+ f_M(V)h_2(V)\nabla_j u_i \int d\hat{\vec{V}}\left(\hat{V}_i\hat{V}_j - \frac{1}{3}\delta_{ij}\right) = 0. \qquad \text{(C.145)}$$

The last equation follows from

$$\int d\hat{\vec{V}}\,\hat{\vec{V}} = 0$$
$$\int d\hat{\vec{V}}\left(\hat{V}_i\hat{V}_j - \frac{1}{3}\delta_{ij}\right) = 0. \qquad \text{(C.146)}$$

To prove (C.146) we consider these identities in components, for example

$$\int d\hat{\vec{V}}\,\hat{V}_x = \int_0^\pi \sin\theta d\theta \int_0^{2\pi} \sin\theta\cos\varphi d\varphi = 0. \qquad \text{(C.147)}$$

Similarly

$$\int d\hat{\vec{V}}\, \hat{V}_x \hat{V}_x = \int_0^\pi \sin\theta d\theta \int_0^{2\pi} \sin^2\theta \cos^2\varphi\, d\varphi = \frac{4\pi}{3}$$

$$\int d\hat{\vec{V}}\, \hat{V}_x \hat{V}_y = \int_0^\pi \sin\theta d\theta \int_0^{2\pi} \sin^2\theta \cos\varphi \sin\varphi\, d\varphi = 0\,.$$
(C.148)

Hence

$$\int d\hat{\vec{V}}\, \hat{V}_i \hat{V}_j = \frac{4\pi}{3}\delta_{ij}\,, \qquad \int d\hat{\vec{V}}\, \frac{1}{3}\delta_{ij} = \frac{4\pi}{3}\delta_{ij}\,,$$
(C.149)

and we arrive at (C.146). From (C.142) and (C.145) it follows that the function $f(V)$ for inhomogeneous granular gases has the same normalization as for a homogeneous gas.

Solution of problem 21.1
Calculate the coefficient μ for a granular gas of viscoelastic particles! Follow the same procedure as for the derivation of the coefficient of thermal conductivity!

According to (18.45) the expression for μ reads

$$\mu = -\frac{1}{3n}\int d\vec{V}\, \vec{S}(\vec{V}) \cdot \vec{\beta}(\vec{V})\,,$$
(C.150)

where $\vec{\beta}$ is the solution of the equation

$$-\zeta^{(0)} T \frac{\partial \vec{\beta}}{\partial T} - \zeta^{(0)} \vec{\alpha} + J^{(1)}\left(f^{(0)}, \vec{\beta}\right) = \vec{B}$$
(C.151)

with

$$\vec{B} = a_2 \vec{V} S_1^{(3/2)}(c^2) f_M(V)\,.$$
(C.152)

Therefore, we seek $\vec{\beta}$ in a form similar to (20.22):

$$\vec{\beta} = \beta_1 \vec{V} S_1^{(3/2)}(c^2) f_M(V) = -\frac{\beta_1}{T}\vec{S}\left(\vec{V}\right) f_M(V)\,.$$
(C.153)

Following the same scheme as in the preceding exercise we multiply (C.151) by $\vec{S}\left(\vec{V}_1\right)\big/T$ and integrate over $\vec{V}_1$. Then with the definition (18.45) of κ and μ we obtain

$$3\zeta^{(0)} n \frac{\partial \mu}{\partial T} + 3\zeta^{(0)} \kappa + \frac{4}{5}n^2\sigma^2 \frac{v_T}{T}\Omega_\kappa^{(\text{vis})}\mu + \frac{15}{4}a_2 n v_T^2 = 0\,,$$
(C.154)

where we evaluate the integral

$$\frac{1}{T}\int d\vec{V}_1 \vec{S}\left(\vec{V}_1\right) \cdot \vec{B}\left(\vec{V}_1\right) = -a_2 \int d\vec{V}_1 \vec{V}_1^2 \left(S_1^{(3/2)}(c_1^2)\right)^2 f_M(V_1)$$
$$= -\frac{15}{4}a_2 n v_T^2\,,$$
(C.155)

which equals (up to the coefficient a_2) the result (19.34). The integral

$$\frac{1}{T}\int d\vec{V}_1 \vec{S}\left(\vec{V}_1\right) J^{(1)}\left(f^{(0)},\vec{\beta}\right) = \beta_1 n^2 \sigma^2 v_T^3 \Omega_\kappa^{(\text{vis})} = \frac{4}{5}\mu n^2 \sigma^2 \frac{v_T}{T}\Omega_\kappa^{(\text{vis})}, \quad \text{(C.156)}$$

is similar to (21.23). We also employ (20.36)

$$\mu = \frac{5}{2}\frac{T^2}{m}\beta_1, \quad \text{(C.157)}$$

which relates μ and β_1. Using for μ the expansion

$$\mu = \frac{\kappa_0 T}{n}\left(\delta'\tilde{\mu}_1 + \delta'^2 \tilde{\mu}_2 + \cdots\right) \quad \text{(C.158)}$$

and substituting it into (C.154) together with

$$T\frac{\partial \mu}{\partial T} = \frac{\kappa_0 T}{n}\left[\delta'\tilde{\mu}_1\left(1+\frac{1}{2}+\frac{1}{10}\right) + \delta'^2 \tilde{\mu}_2\left(1+\frac{1}{2}+\frac{1}{5}\right) + \cdots\right] \quad \text{(C.159)}$$

we find the coefficients $\tilde{\mu}_1$ and $\tilde{\mu}_2$.

The corresponding Maple program reads

```
 1  DefDimension(3);
 2  DefJ();
 3  ExprKAPPA:=(c2pDOTc2p-(DIM+2)/2)*(c1aDOTc2p*c1aDOTc1a+c2aDOTc2p*c2aDOTc2a
 4                -c2pDOTc2p*c2pDOTc2p-c1pDOTc2p*c1pDOTc1p)*(1+a2*S(2,c1p)):
 5  OmegaKAPPA:=getJexpr(1,ExprKAPPA,2):
 6  OmegaKAPPA:=convert(OmegaKAPPA,polynom):
 7  Expr:=Delta2*(-1/2*(1+a2*(S(2,c1p)+S(2,c2p))+a2^2*S(2,c1p)*S(2,c2p))):
 8  mu2:=getJexpr(1,Expr,2):
 9  Expr:=Delta4*(-1/2*(1+a2*(S(2,c1p)+S(2,c2p))+a2^2*S(2,c1p)*S(2,c2p))):
10  mu4:=getJexpr(1,Expr,2):
11  tmpA2:={solve(mu4=mu2/DIM*4*DIM/4*(DIM+2)*(1+a2),a2)};
12  a2:=tmpA2[1];
13  a2:=convert(a2,polynom):
14  mu2:=convert(mu2,polynom):
15  mu2:=convert(taylor(mu2,dprime,3),polynom);
16  xi0:=2*sigma^(DIM-1)/DIM*n*mu2*sqrt(2*T/m):
17  xi01:=coeff(xi0,dprime,1)/(2/DIM*sqrt(2*T/m)*sigma^(DIM-1)*n):
18  xi02:=-coeff(xi0,dprime,2)/(2/DIM*sqrt(2*T/m)*sigma^(DIM-1)*n):
19  Tdxi0dT:=2/DIM*sqrt(2*T/m)*sigma^(DIM-1)*n*(3/5*xi01*dprime
20           -7/10*xi02*dprime^2):
21  a21:=coeff(a2,dprime,1):
22  a22:=coeff(a2,dprime,2):
23  Tda2dT:=1/10*a21*dprime+1/5*a22*dprime^2:
24  OmegaKAPPA:=convert(taylor(OmegaKAPPA,dprime,3),polynom):
25  EqKAPPA:=DIM*(kappa*Tdxi0dT+xi0*TdkappadT+xi0*kappa)
26           +4/(DIM+2)*kappa*n*sigma^(DIM-1)*sqrt(2*T/m)*OmegaKAPPA
27           +(DIM+2)*DIM/4*n*2*T/m*(1+2*a2+Tda2dT):
28  kappa:=kappa0*(1+dprime*kappa1+dprime^2*kappa2):
29  TdkappadT:=kappa0*(1/2+3/5*dprime*kappa1+7/10*dprime^2*kappa2):
30  EqKAPPA:=taylor(EqKAPPA,dprime,3):
```

```
31  kappa0:=solve(coeff(EqKAPPA,dprime,0),kappa0):
32  kappa1:=solve(coeff(EqKAPPA,dprime,1),kappa1):
33  kappa2:=solve(coeff(EqKAPPA,dprime,2),kappa2):
34  EqMU:=DIM*xi0*TdMUdT+DIM*xi0*T/n*kappa
35       +4/(DIM+2)*n*sigma^(DIM-1)*OmegaKAPPA*MU*sqrt(2*T/m)
36       +(DIM+2)*DIM/2*T^2/m*a2:
37  MU:=kappa0*T/n*(dprime*TildaMU1+dprime^2*TildaMU2):
38  TdMUdT:=kappa0*T/n*(dprime*TildaMU1*(1+1/2+1/10)
39       +dprime^2*TildaMU2*(1+1/2+1/5)):
40  EqMU:=taylor(EqMU,dprime,3):
41  TildaMU1:=solve(coeff(EqMU,dprime,1),TildaMU1);
42  TildaMU2:=solve(coeff(EqMU,dprime,2),TildaMU2);
43  omega1:=9.28569:
44  omega0:=6.48562:
45  evalf(TildaMU1);
46  evalf(TildaMU2);
47  plot(MU/(kappa0*T/n),dprime=0..0.2,color=black,linestyle=[1]);
```

Note that on line $\boxed{15}$, the root for $a_2 \propto \delta'$ is to be chosen.

Solution of problem 22.1
Prove that in (22.30) all terms that contain factors $\dot{a}_k \vec{v} \cdot \partial \vec{G}/\partial a_k$ vanish upon integration over $\vec{v}$ for $k \geq 2$!

For the definition of the Sonine expansion

$$f(v) = f_M(v)\left[1 + \sum_{k=2}^{\infty} a_k S_k(c^2)\right], \qquad (\text{C.160})$$

follows from (22.21)

$$\frac{\partial \vec{G}}{\partial a_k} = b_0 \vec{v} f_M(v) S_k(c^2). \qquad (\text{C.161})$$

Hence we can write

$$\dot{a}_k \int d\vec{v}\, \vec{v} \cdot \frac{d\vec{G}}{da_k} = \dot{a}_k b_0 \int d\vec{v}\, \vec{v} \cdot \vec{v} f_M S_k(c^2) = \dot{a}_k b_0 n v_T^2 \int d\vec{c}\, \phi(c) c^2 S_k(c^2)$$
$$= \dot{a}_k b_0 n v_T^2 \nu_{k2} = 0, \qquad (\text{C.162})$$

since, according to (7.17) $\nu_{k2} = 0$ for $k > 1$.

Solution of problem 24.1
Derive (24.3)–(24.5)!

The derivation of the hydrodynamic equations for the space-independent coefficients has been performed in the Exercise 18.1. Therefore, these may be obtained by taking $\lambda = 1$ in the equations (18.3) and (18.4).

Solution of problem 25.1

Find expressions for the coefficients $\eta^*(\varepsilon)$, $\kappa^*(\varepsilon)$, $\mu^*(\varepsilon)$ and $\zeta^*(\varepsilon)$ using the relations for the kinetic coefficients and the cooling rate from Chapter 16!

We substitute the relations (20.17) for μ_2, (20.12) for $\Omega_\eta^{(\varepsilon)}$ and (19.26) for η_0 into (20.16) and obtain

$$\eta^*(\varepsilon) = \left[\frac{(3-\varepsilon)(1+\varepsilon)}{4}\left(1-\frac{a_2}{32}\right) - \frac{5}{24}(1-\varepsilon)^2\left(1+\frac{3a_2}{16}\right)\right]^{-1}. \qquad (C.163)$$

Similarly, from (20.32, 20.29, 19.46) we obtain

$$\kappa^*(\varepsilon) = \frac{4(1+2a_2)}{(1+\varepsilon)\left(\frac{49-33\varepsilon}{8} + \frac{19-3\varepsilon}{256}a_2\right) - 5(1-\varepsilon^2)\left(1+\frac{3a_2}{16}\right)}. \qquad (C.164)$$

The quantity $\zeta^*(\varepsilon)$ may be found from (18.11, 20.17, 19.26):

$$\zeta^*(\varepsilon) = \frac{5}{12}\frac{\mu_2}{\sqrt{2\pi}} = \frac{5}{12}(1-\varepsilon^2)\left(1+\frac{3a_2}{16}\right). \qquad (C.165)$$

Finally, the value of $\mu^*(\varepsilon)$ may be obtained from the results of Exercise 20.8:

$$\mu^*(\varepsilon) = \frac{2(3\zeta^*\kappa^* + 2a_2)}{\left(\frac{49-33\varepsilon}{8} + \frac{19-3\varepsilon}{256}\right) - 9\zeta^*}. \qquad (C.166)$$

The above results are accurate up to $\mathcal{O}(a_2^2)$, since we keep only linear terms with respect to a_2 in ζ^*. Therefore, they coincide with the direct evaluation of these quantities from the general expressions (23.4, 23.6) for the case $d=3$, which are also accurate up to terms $\mathcal{O}(a_2^2)$.

Solution of problem 25.2

Derive the function $k_\perp^*(\varepsilon)$ and show that it depends only on the coefficient of restitution! For $\varepsilon \lesssim 1$ show that $k_\perp^*(\varepsilon) \propto \sqrt{1-\varepsilon^2}$!

From (25.24) and the results of Exercise 25.1 for $\eta^*(\varepsilon)$ and $\zeta^*(\varepsilon)$ it follows that $k_\perp^*(\varepsilon)$ indeed depends only on ε. Moreover, for $1-\varepsilon^2 \ll 1$ we can neglect a_2 as compared to unity and approximate $(3-\varepsilon)(1+\varepsilon)/4 \approx 1$. This gives $\eta^*(\varepsilon) \approx 1$ and

$$k_\perp^*(\varepsilon) = \frac{1}{2}\sqrt{\frac{\zeta^*(\varepsilon)}{\eta^*(\varepsilon)}} \approx \frac{1}{2}\sqrt{\frac{5}{12}(1-\varepsilon^2)} \qquad (C.167)$$

Solution of problem 25.3
Find the function $\vec{w}_{\vec{k}\perp}(t)$, using the relation between the laboratory time t and the time τ measured in the accumulated number of collisions! Find the function $\vec{u}_{\vec{k}\perp}(t)$ and show that the unscaled velocity field decays always with time!

From (25.10), definition of $\tau_c(t)$ (14.18) and evolution of temperature in the homogeneous cooling state (8.64) follows

$$\tau = \frac{8}{5}\int_0^t dt' \tau_c^{-1}(t') = \frac{8}{5}\int_0^t dt' \tau_c^{-1}(0) \sqrt{\frac{T_h(t')}{T_0}}$$

$$= \frac{8}{5\tau_c(0)}\int_0^t \left(1+\frac{t'}{\tau_0}\right)^{-1} dt' = \frac{8}{5}\frac{\tau_0}{\tau_c(0)} \log\left(1+\frac{t}{\tau_0}\right). \quad \text{(C.168)}$$

Therefore, using (25.22, 25.23) we obtain

$$\vec{w}_{\vec{k}\perp}(t) = \vec{w}_{\vec{k}\perp}(0) \exp\left[\left(\frac{1}{4}\zeta^* - \eta^* k^2\right)\tau(t)\right]$$

$$= \vec{w}_{\vec{k}\perp}(0) \left(1+\frac{t}{\tau_0}\right)^{\lambda_\perp(\vec{k}) 8\tau_0/5\tau_c(0)}. \quad \text{(C.169)}$$

Taking into account

$$\frac{\tau_0}{\tau_c(0)} = \frac{6\sqrt{2\pi}}{\mu_2} \quad \text{(C.170)}$$

according to (8.65) and using (25.23) for $\lambda_\perp(\vec{k})$ together with (C.165) for ζ^*, we obtain the exponent in (C.169):

$$\lambda_\perp(\vec{k})\frac{8\tau_0}{5\tau_c(0)} = \left(\frac{1}{4}\frac{5}{12}\frac{\mu_2}{\sqrt{2\pi}} - \eta^* k^2\right)\frac{8}{5}\frac{6\sqrt{2\pi}}{\mu_2}$$

$$= 1 - \frac{48\sqrt{2\pi}}{5\mu_2}\eta^* k^2 = 1 - \frac{48\eta^* k^2}{5(1-\varepsilon^2)(1+3a_2/16)}, \quad \text{(C.171)}$$

where the last equation is written in linear approximation with respect to a_2 for μ_2 from (8.37). With $\vec{u}_{\vec{k}\perp}(t) = \sqrt{2T_h(t)/m}\vec{w}_{\vec{k}\perp}(t)$ and using the temperature dependence $T_h(t)$ we obtain

$$\vec{u}_{\vec{k}\perp}(t) = \sqrt{\frac{T_h(t)}{T_0}}\vec{u}_{\vec{k}\perp}(0)\left(1+\frac{t}{\tau_0}\right)^{\lambda_\perp(\vec{k})8\tau_0/5\tau_c(0)}$$

$$= \vec{u}_{\vec{k}\perp}(0)\left(1+\frac{t}{\tau_0}\right)^{-48\eta^* k^2/[5(1-\varepsilon^2)(1+3a_2/16)]}. \quad \text{(C.172)}$$

The last expression shows that the exponent of the power law dependence of $\vec{u}_{\vec{k}\perp}(t)$ is always negative, that is, this quantity always decays.

Solution of problem 25.4
Derive the dependence of the marginal wave vector on the coefficient of restitution: $k_H^*(\varepsilon) \propto \sqrt{1-\varepsilon^2}$ for $1-\varepsilon^2 \ll 1$!

Similarly as in Exercise 25.2 we notice that for $1-\varepsilon^2 \ll 1$, $\kappa^* \approx 1$, $\mu^* \ll 1$ and $a_2 \ll 1$. Therefore, from (25.29) we obtain the estimate

$$k_H^*(\varepsilon) \approx \sqrt{\frac{1}{10}\frac{5}{12}(1-\varepsilon^2)} \propto \sqrt{1-\varepsilon^2}. \tag{C.173}$$

Solution of problem 27.1
Find the dependence of the non-reduced shear mode $\vec{u}_{\vec{k}\perp}$ on the non-reduced (laboratory) time t for $1-\varepsilon^2 \ll 1$! Hint: use the dependence of temperature on time in the homogeneous cooling state and the relation between the laboratory time t and the reduced time τ.

It is shown in Exercise 25.3 that $\vec{u}_{\vec{k}\perp}(t)$ depends on time as

$$\vec{u}_{\vec{k}\perp}(t) = \vec{u}_{\vec{k}\perp}(0)\left(1+\frac{t}{\tau_0}\right)^{-\nu_k} \tag{C.174}$$

with the exponent ν_k given by (C.172). For $1-\varepsilon^2 \ll 1$ we can approximate $a_2 \approx 0$ and $\eta^* \approx 1$ (see Exercise 25.2). Hence, the exponent ν_k reads

$$\nu_k = \frac{48\eta^* k^2}{5(1-\varepsilon^2)(1+3a_2/16)} \simeq \frac{48}{5(1-\varepsilon^2)}k^2, \tag{C.175}$$

which coincides with (27.1).

REFERENCES

Abramowitz, M. and Stegun, A. (1965). *Handbook of Mathematical Functions*. Dover Publications, New York.

Allen, M. P. and Tildesley, D. J. (1987). *Computer Simulations of Liquids*. Clarendon Press, Oxford.

Baeza-Yates, R., Marín, M., and Cordero, P. (1994). Analysis of an improved priority queue for discrete event simulation of many moving objects. In *Proceedings of the XIV International Conference of the Chilean Computer Science Society* (ed. C. Isaac and R. Peralta), Chile, pp. 29.

Baker, G. A. (1970). *The Padé Approximant in Theoretical Physics*. Academic Press, New York.

Baxter, G. W. and Olafsen, J. S. (2003). Gaussian statistics in granular gases. *Nature*, **425**, 680.

Ben-Naim, E. and Krapivsky, P. (2000). Multiscaling in inelastic collisions. *Phys. Rev. E*, **61**, R5.

Berne, B. J. (1977). Molecular dynamics of the rough sphere fluids. II. Kinetic models of partially sticky spheres, structured spheres, and rough screwballs. *J. Chem. Phys.*, **66**, 2821.

Berne, B. J. and Harp, G. D. (1970). On the calculation of time correlation functions. *Advan. Chem. Phys.*, **17**, 63.

Blair, D. L. and Kudrolli, A. (2001). Velocity correlations in dense granular gases. *Phys. Rev. E*, **64**, R050301.

Blair, D. L. and Kudrolli, A. (2003). Collision statistics of driven granular materials. *Phys. Rev. E*, **67**, 041301.

Bobylev, A. V., Carrillo, J. A., and Gamba, I. M. (2000). On some properties of kinetic and hydrodynamic equations for inelastic interactions. *J. Stat. Phys.*, **98**, 743.

Boltzmann, L. (1896). *Vorlesungen über Gastheorie*. J. A. Barth, Leipzig.

Borderies, N. B., Goldreich, P., and Tremaine, S. (1984). Unsolved problems in planetary ring dynamics. In *Planetary Rings* (ed. R. Greenberg and A. Brahic), Tucson, AZ, pp. 713. Arizona University Press.

Brahic, A. (2001). Dynamical evolution of viscous disks: Astrophysical applications to the formation of planetary systems and to the confinement of planetary rings and arcs. In *Granular Gases* (ed. T. Pöschel and S. Luding), Volume 564 of *Lecture Notes in Physics*, Berlin, pp. 281. Springer.

Brey, J. J. and Cubero, D. (2001). Hydrodynamic transport coefficients of granular gases. In *Granular Gases* (ed. T. Pöschel and S. Luding), Volume 564 of *Lecture Notes in Physics*, Berlin, pp. 59. Springer.

Brey, J. J., Cubero, D., and Ruiz-Montero, M. J. (1996). Homogeneous cooling state of low-density granular flow. *Phys. Rev. E*, **54**, 3664.

Brey, J. J., Cubero, D., and Ruiz-Montero, M. J. (1999a). High energy tail in the velocity distribution of a granular gas. *Phys. Rev. E*, **59**, 1256.

Brey, J. J., Dufty, J. W., Kim, C. S., and Santos, A. (1998). Hydrodynamics for granular flow at low density. *Phys. Rev. E*, **58**, 4638.

Brey, J. J., Dufty, J. W., and Santos, A. (1999b). Kinetic models for granular flow. *J. Stat. Phys.*, **97**, 281.

Brey, J. J. and Ruiz-Montero, M. J. (1999). Direct Monte Carlo simulation of dilute granular gas. *Computer Physics Communications*, **121/122**, 278.

Brey, J. J., Ruiz-Montero, M. J., and Cubero, D. (1999a). Origin of density clustering in a freely evolving granular gas. *Phys. Rev. E*, **60**, 3150.

Brey, J. J., Ruiz-Montero, M. J., Cubero, D., and Garcia-Rojo, R. (2000). Self-diffusion in freely evolving granular gases. *Physics of Fluids*, **12**, 876.

Brey, J. J., Ruiz-Montero, M. J., and Garcia-Rojo, R. (1999b). Brownian motion in granular gases. *Phys. Rev. E*, **60**, 7174.

Bridges, F., Supulver, K., and Lin, D. N. C. (2001). Energy loss and aggregation processes in low speed collisions of ice particles coated with frost or methanol/water mixtures. In *Granular Gases* (ed. T. Pöschel and S. Luding), Volume 564 of *Lecture Notes in Physics*, Berlin, pp. 153. Springer.

Bridges, F. G., Hatzes, A., and Lin, D. N. C. (1984). Structure, stability and evolution of Saturn's rings. *Nature*, **309**, 333.

Brilliantov, N. V. and Pöschel, T. (1998). Rolling friction of a soft sphere on a hard plane. *Europhys. Lett.*, **42**, 511.

Brilliantov, N. V. and Pöschel, T. (2000a). Deviation from Maxwell distribution in granular gases with constant restitution coefficient. *Phys. Rev. E*, **61**, 2809.

Brilliantov, N. V. and Pöschel, T. (2000b). Diffusion in granular gases of viscoelastic particles. In *Stochastic Processes in Physics, Chemistry, and Biology* (ed. J. A. Freund and T. Pöschel), Volume 557 of *Lecture Notes in Physics*, Berlin, pp. 107. Springer.

Brilliantov, N. V. and Pöschel, T. (2000c). Self-diffusion in granular gases. *Phys. Rev. E*, **61**, 1716.

Brilliantov, N. V. and Pöschel, T. (2000d). Velocity distribution of granular gases of viscoelastic particles. *Phys. Rev. E*, **61**, 5573.

Brilliantov, N. V. and Pöschel, T. (2001). Granular gases with impact-velocity dependent restitution coefficient. In *Granular Gases* (ed. T. Pöschel and S. Luding), Volume 564 of *Lecture Notes in Physics*, Berlin, pp. 100. Springer.

Brilliantov, N. V. and Pöschel, T. (2003). Hydrodynamics and transport coefficients for dilute granular gases. *Phys. Rev. E*, **67**, 061304.

Brilliantov, N. V. and Revokatov, O. P. (1984). Relation between momentum and angular momentum correlation times. Analysis of the uncorrelated successive binary collision approximation. *Chem. Phys. Lett.*, **104**, 444.

Brilliantov, N. V., Salueña, C., Schwager, T., and Pöschel, T. (2004). Transient structures in granular gases. *Phys. Rev. Lett.*, **93**, 134301.

Brilliantov, N. V., Spahn, F., Hertzsch, J.-M., and Pöschel, T. (1996). Model for collisions in granular gases. *Phys. Rev. E*, **53**, 5382.

Brito, R. and Ernst, M. H. (1998a). Extension of Haff's cooling law in granular flows. *Europhys. Lett.*, **43**, 497.
Brito, R. and Ernst, M. H. (1998b). Noise reduction and pattern formation in granular flows. *Int. J. Mod. Phys. C*, **8**, 1339.
Busse, W. F. and Starr, F. C. (1960). Change of viscoelastic sphere to a torus by random impacts. *Am. J. Phys.*, **28**, 19.
Cafiero, R., Luding, S., and Herrmann, H.J. (2002). Rotationally driven gas of inelastic rough spheres. *Europhys. Lett.*, **60**, 854.
Carnahan, N. F. and Starling, K. E. (1969). Equation of state for nonattractive rigid spheres. *J. Chem. Phys.*, **51**, 635.
Chandler, D. (1975). Rough hard sphere theory of the self-diffusion constant for molecular liquids. *J. Chem. Phys.*, **62**, 1358.
Chapman, S. and Cowling, T. G. (1970). *The Matematical Theory of Nonuniform Gases*. Cambridge University Press, New York.
Clement, E. and Labous, L. (2001). Pattern formation in a vibrated granular layer. In *Granular Gases* (ed. T. Pöschel and S. Luding), Volume 564 of *Lecture Notes in Physics*, Berlin, pp. 233. Springer.
Constantin, P., Grossman, E., and Mungan, M. (1995). Inelastic collision of three particles on a line as a two-dimensional billiard. *Physica D*, **83**, 409.
Deltour, P. and Barrat, J.-L. (1997). Quantitative study of a freely cooling granular medium. *J. Physique I*, **7**, 137.
Du, Y., Li, H., and Kadanoff, L. P. (1995). Breakdown of hydrodynamics in a one-dimensional system of inelastic particles. *Phys. Rev. Lett.*, **74**, 1268.
Dufty, J. W. and Brey, J. J. (1999). Comment on 'Rapid granular flows as mesoscopic systems'. *Phys. Rev. Lett.*, **82**, 4566.
Ernst, M. H. and Dorfman, J. R. (1972). Non-analytic dispersion relations in classical fluids. I. The hard-sphere gas. *Physica A*, **61**, 157.
Ernst, M. H., Dorfman, J. R., Hoegy, W. R., and van Leeuwen, J. M. J. (1969). Hard-sphere dynamics and binary-collision operators. *Physica A*, **45**, 127.
Esipov, S. E. and Pöschel, T. (1997). The granular phase diagram. *J. Stat. Phys.*, **86**, 1385.
Évesque, P., Falcon, E., Wunenburger, R., Fauve, S., Lecoutre-Chabot, C., Garrabos, Y., and Beysens, D. (2001). Gas-cluster transition of granular matter under vibration in microgravity. In *International Symposium on Microgravity Research and Applications in Physical Sciences and Biotechnology*, Sorrento (Italy), pp. 829. European Space Agency SP-454.
Falcon, E., Fauve, S., and Laroche, C. (1999). Cluster formation, pressure and density measurements in a granular medium fluidized by vibrations. *Eur. Phys. J. B*, **9**, 183.
Falcon, E., Fauve, S., and Laroche, C. (2001). Experimental study of a granular gas fluidized by vibrations. In *Granular Gases* (ed. T. Pöschel and S. Luding), Volume 564 of *Lecture Notes in Physics*, Berlin, pp. 182. Springer.
Falcon, E., Wunenburger, R., Evesque, P., Fauve, S., Chabot, C., Garrabos, Y., and Beysens, D. (1999). Cluster formation in a granular medium fluidized by

vibrations in low gravity. *Phys. Rev. Lett.*, **83**, 440.
Feitosa, K. and Menon, N. (2002). Breakdown of energy equipartition in a 2d binary vibrated granular gas. *Phys. Rev. Lett.*, **88**, 198301.
Ferziger, J. and Kaper, H. (1972). *The Matematical Theory of Transport Processes in Gases*. North-Holland, Amsterdam.
Gardiner, C. W. (1983). *Handbook of Stochastic Methods*, Volume 13 of *Springer Series in Synergetics*. Springer, Berlin.
Garwin, R. L. (1969). Kinematics of an ultraelastic rough ball. *Am. J. Phys.*, **37**, 88.
Garzo, V. and Dufty, J. W. (1999). Dense fluid transport for inelastic hard spheres. *Phys. Rev. E*, **59**, 5895.
Goldhirsch, I. (2001). Granular gases: Probing the boundaries of hydrodynamics. In *Granular Gases* (ed. T. Pöschel and S. Luding), Volume 564 of *Lecture Notes in Physics*, Berlin, pp. 79. Springer.
Goldhirsch, I. and Zanetti, G. (1993). Clustering instability in dissipative gases. *Phys. Rev. Lett.*, **70**, 1619.
Goldman, D., Shattuck, M. D., Bizon, C., McCormick, W. D., Swift, J. B., and Swinney, H. L. (1998). Absence of inelastic collapse in a realistic three ball model. *Phys. Rev. E*, **57**, 4831.
Goldshtein, A. and Shapiro, M. (1995). Mechanics of collisional motion of granular materials. Part 1: General hydrodynamic equations. *J. Fluid Mech.*, **282**, 75.
Grad, H. (1960). *Theory of Rarified Gases in Rarified Gas Dynamics*. Pergamon Press, New York.
Greenberg, R. and Brahic, A. (ed.) (1984). *Planetary Rings*, Tucson, AZ. Arizona University Press.
Grossman, E. L. (1997). Effects of container geometry on granular convection. *Phys. Rev. E*, **56**, 3290.
Haff, P. K. (1983). Grain flow as a fluid-mechanical phenomenon. *J. Fluid Mech.*, **134**, 401.
Hansen, J. P. and McDonald, I. R. (1986). *Theory of Simple Liquids*. Academic Press Limited, London.
Herrmann, H. J., Hovi, J.-P., and Luding, S. (ed.) (1998). *Physics of Dry Granular Media*, NATO ASI Series, Dordrecht. Kluwer.
Hertz, H. (1882). Über die Berührung fester elastischer Körper. *J. reine u. angewandte Math.*, **92**, 156.
Hodkinson, E. (1835). *Report of the 4th Meeting of the British Association for the Advancement of Science*. London.
Huntley, J. M. (1998). Scaling laws for a two-dimensional vibro-fluidized granular material. *Phys. Rev. E*, **58**, 5168.
Huthmann, M., Orza, J. A., and Brito, R. (2000). Dynamics of deviations from the Gaussian state in a freely cooling homogeneous sys tem of smooth inelastic particles. *Granular Matter*, **2**, 189.
Jenkins, J. T. and Louge, M. Y. (1997). Microgravity segregation in binary

mixtures of inelastic spheres. In *Powders and Grains '97* (ed. R. P. Behringer and J. T. Jenkins), Rotterdam, pp. 539. Balkema.

Jenkins, J. T. and Richman, M. W. (1985). Grad's 13-moment system for a dense gas of inelastic spheres. *Arch. Particle Mech. Anal.*, **87**, 355.

Kadanoff, L. P. (1999). Built upon sand: theoretical ideas inspired by granular flows. *Rev. Mod. Phys.*, **71**, 435.

Kudrolli, A. and Henry, J. (2000). Non-gaussian velocity distributions in excited granular matter in the absence of clustering. *Phys. Rev. E*, **62**, R1489.

Kuwabara, G. and Kono, K. (1987). Restitution coefficient in a collision between two spheres. *Jpn. J. Appl. Phys.*, **26**, 1230.

Landau, L. D. and Lifshitz, E. M. (1965). *Theory of Elasticity*. Oxford University Press, Oxford.

Lebowitz, J. L. (1964). Exact solution of generalized Percus–Yevick equation for a mixture of hard spheres. *Phys. Rev. A*, **133**, 895.

Losert, W., Cooper, D. G. W., Delour, J., Kudrolli, A., and Gollub, J. P. (1999). Velocity statistics in excited granular media. *Chaos*, **9**, 682.

Louge, M. Y. and Adams, M. E. (2002). Anomalous behavior of normal kinematic restitution in the oblique impacts of a hard sphere on an elastoplastic plate. *Phys. Rev. E*, **65**, 021303.

Luding, S., Huthmann, M., McNamara, S., and Zippelius, A. (1998). Homogeneous cooling of rough, dissipative particles: theory and simulations. *Phys. Rev. E*, **58**, 3416.

Luding, S. and McNamara, S. (1998). How to handle the inelastic collapse of a dissipative hard-sphere gas with the TC model. *Granular Matter*, **1**, 113.

Luding, S. and Strauss, O. (2001). The equation of state for almost elastic, smooth, polydisperse granular gases for arbitrary density. In *Granular Gases* (ed. T. Pöschel and S. Luding), Volume 564 of *Lecture Notes in Physics*, Berlin, pp. 389. Springer.

Lun, C. K. K., Savage, S. B., Jeffrey, D. J., and Chepurniy, N. (1984). Kinetic theories for granular flow: inelastic particles in Couette flow and slightly inelastic particles in a general flowfield. *J. Fluid Mech.*, **140**, 223.

Marín, M. (1998, Oct.). On the pending event set and binary tournaments. In *Proceedings of the 10th SCS European Simulation Symposium* (ed. A. Bargiela and E. Kerckhoffs), Nottingham, pp. 110. Society for Computer Simulation Europe Publishing House.

Marín, M., Risso, D., and Cordero, P. (1993). Efficient algorithms for many body hard particle molecular dynamics. *J. Comp. Phys.*, **109**, 306.

McNamara, S. (1993). Hydrodynamic modes of a uniform granular medium. *Phys. Fluids A*, **5**, 3056.

McNamara, S. and Young, W. R. (1992). Inelastic collapse and clumping in a one-dimensional granular medium. *Phys. Fluids A*, **4**, 496.

McNamara, S. and Young, W. R. (1993). Inelastic collapse in two dimensions. *Phys. Rev. E*, **50**, R28.

Morgado, W. A. M. and Oppenheim, I. (1997). Energy dissipation for quasielastic granular particle collisions. *Phys. Rev. E*, **55**, 1940.

Olafsen, J. S. and Urbach, J. S. (1998). Clustering, order, and collapse in a driven granular monolayer. *Phys. Rev. Lett.*, **81**, 4369.

Olafsen, J. S. and Urbach, J. S. (1999). Velocity distribution and density fluctuations in a granular gas. *Phys. Rev. E*, **60**, R2468.

Olafsen, J. S. and Urbach, J. S. (2001). Experimental observations of nonequilibrium distributions and transitions in a 2d granular gas. In *Granular Gases* (ed. T. Pöschel and S. Luding), Volume 564 of *Lecture Notes in Physics*, Berlin, pp. 410. Springer.

Pagonabarraga, I., Trizac, E., van Noije, T. P. C., and Ernst, M. H. (2002). Randomly driven granular fluids: collisional statistics and short scale structure. *Phys. Rev. E*, **65**, 011303.

Piel, A. and Melzer, A. (2002). Dusty plasmas – the state of understanding from an experimentalist's view. *Adv. Space Res.*, **29**, 1255.

Pöschel, T. (1999). *Dynamik granularer Systeme – Theorie, Experimente und numerische Experimente*. Logos, Berlin.

Pöschel, T. and Brilliantov, N. V. (2001). Extremal collision sequences of particles on a line: optimal transmission of kinetic energy. *Phys. Rev. E*, **63**, 021505.

Pöschel, T. and Brilliantov, N. V. (ed.) (2003). *Granular Gas Dynamics*, Volume 624 of *Lecture Notes in Physics*, Berlin. Springer.

Pöschel, T., Brilliantov, N. V., and Schwager, T. (2002). Violation of molecular chaos in dissipative gases. *Int. J. Mod. Phys. C*, **13**, 1263.

Pöschel, T. and Luding, S. (ed.) (2001). *Granular Gases*, Volume 564 of *Lecture Notes in Physics*, Berlin. Springer.

Pöschel, T. and Schwager, T. (2004). *Granular Dynamics: Models and Algorithms*. Springer, Berlin.

Ramírez, R. and Cordero, P. (1999). Kinetic description of a fluidized one-dimensional granular system. *Phys. Rev. E*, **59**, 656.

Ramírez, R., Pöschel, T., Brilliantov, N. V., and Schwager, T. (1999). Coefficient of restitution for colliding viscoelastic spheres. *Phys. Rev. E*, **60**, 4465.

Ramírez, R., Risso, D., Soto, R., and P., Cordero (2000). Hydrodynamic theory for granular gases. *Phys. Rev. E*, **62**, 2521.

Rapaport, D. C. (1980). The event scheduling problem in molecular dynamic simulations. *J. Comp. Phys.*, **34**, 184.

Resibois, P. and de Leener, M. (1977). *Classical Kinetic Theory of Fluids*. Wiley & Sons, New York.

Risken, H. (1996). *The Fokker–Planck Equation: Methods of Solution and Applications*, Volume 18 of *Springer Series in Synergetics*. Springer, Berlin.

Rouyer, F. and Menon, N. (2000). Velocity fluctuations in a homogeneous 2d granular gas in steady state. *Phys. Rev. Lett.*, **85**, 3676.

Salo, H. (1992). Gravitational wakes in Saturn's rings. *Science*, **359**, 619.

Salo, H. (2001). Numerical simulations of the collisional dynamics of planetary rings. In *Granular Gases* (ed. T. Pöschel and S. Luding), Volume 564 of *Lecture Notes in Physics*, Berlin, pp. 330. Springer.

Salo, H., Lukkari, J., and Hanninen, J. (1988). Velocity dependent coefficient of restitution and the evolution of collisional systems. *Earth, Moon, Planets*, **43**, 33.

Scheffler, T. and Wolf, D. E. (2002). Collision rates in charged granular gases. *Granular Matter*, **4**, 103.

Schram, P. P. J. M. (1991). *Kinetic Theory of Gases and Plasmas*. Kluwer Academic Publishers, AA Dordrecht, The Netherlands.

Schwager, T. and Pöschel, T. (1998). Coefficient of restitution of viscous particles and cooling rate of granular gases. *Phys. Rev. E*, **57**, 650.

Sela, N. and Goldhirsch, I. (1998). Hydrodynamic equations for rapid flows of smooth inelastic spheres, to Burnett order. *J. Fluid Mech.*, **361**, 41.

Shida, K. and Kawai, T. (1989). Cluster formation by inelastically colliding particles in one-dimensional space. *Physica A*, **162**, 145.

Soto, R. and Mareschal, M. (2001). Statistical mechanics of fluidized granular media: short range velocity correlations. *Phys. Rev. E*, **63**, 041303.

Spahn, F., Petzschmann, O., Schmidt, J., Sremcevic, M., and Hertzsch, J.-M. (2001). Granular viscosity, planetary rings and inelastic particle collisions. In *Granular Gases* (ed. T. Pöschel and S. Luding), Volume 564 of *Lecture Notes in Physics*, Berlin, pp. 363. Springer.

Stroink, G. (1983). Super ball problem. *The Physics Teacher*, **21**, 466.

Stronge, W. I. (1990). Rigid body collision with friction. *Proc. Roy. Soc. A*, **431**, 169.

Sture, S., Costes, N. C., Batiste, S. N., Lankton, M. R., Alshibli, K. A., Jeremic, B., Swanson, R. A., and Frank, M. (1998). Mechanics of granular materials at low effective stresses. *J. of Aerospace Eng.*, **11**, 67.

Taguchi, Y. (1992). Powder turbulence: Direct onset of turbulent flow. *J. Physique*, **2**, 2103.

Tan, M.-L. and Goldhirsch, I. (1998). Rapid granular flows as mesoscopic systems. *Phys. Rev. Lett.*, **81**, 3022.

Tanaka, T., Ishida, T., and Tsuji, Y. (1991). Direct numerical simulatin of granular plug flow in a horizontal pipe: The case of cohesionless particles (in Japanese). For an English presentation of this work see (Taguchi, 1992). *Trans. Jap. Soc. Mech. Eng.*, **57**, 456.

Thornton, C. and Ning, Z. (1998). A theoretical model for the stick/bounce behaviour of adhesive, elastic–plastic spheres. *Powder Technol.*, **99**, 154.

Thornton, C., Ning, Z., Wu, C., and Nasrullah, M. (2001). Contact mechanics and coefficients of restitution. In *Granular Gases* (ed. T. Pöschel and S. Luding), Volume 564 of *Lecture Notes in Physics*, Berlin, pp. 184. Springer.

van Kampen, N. (1992). *Stochastic Processes in Physics and Chemistry*. Elsevier, Amsterdam.

van Noije, T. P. C. and Ernst, M. H. (1998). Velocity distributions in homogeneous granular fluids: The free and the heated case. *Granular Matter*, **1**,

57.

van Noije, T. P. C. and Ernst, M. H. (2000). Cahn–Hillard theory for unstable granular fluids. *Phys. Rev. E*, **61**, 1765.

van Noije, T. P. C., Ernst, M. H., and Brito, R. (1998). Ring kinetic theory for an idealized granular gas. *Physica A*, **251**, 266.

van Noije, T. P. C., Ernst, M. H., Brito, R., and Orza, J. A. G. (1997). Mesoscopic theory of granular fluids. *Phys. Rev. Lett.*, **79**, 411.

Vranjes, J., Tanaka, M. Y., and Pandey, B. P. (2002). Electrostatic interaction in dusty plasma. *Phys. Rev. E*, **66**, 037401.

Weidenschilling, S. J. (1995). Can gravitation instability form planetesimals? *Icarus*, **116**, 433.

Wildman, R. D., Huntley, J. M, and Hansen, J.-P. (1999). Self-diffusion of grains in a two-dimensional vibrofluidized bed. *Phys. Rev. E*, **60**, 7066.

Wildman, R. D., Huntly, J. M., and Hansen, J.-P. (2001). Experimental studies of vibro-fluidised granular beds. In *Granular Gases* (ed. T. Pöschel and S. Luding), Volume 564 of *Lecture Notes in Physics*, Berlin, pp. 215. Springer.

Zhou, T. and Kadanoff, L. P. (1996). Inelastic collapse of three particles. *Phys. Rev. E*, **54**, 623.

Symbol index

A, dissipative material constant, 22
$\vec{A}(\vec{V})$, 181
a_{21}, a_{22}, 204
$a_2^{\rm NE}$, a_2 in a linear approximation, 74
$\mathcal{A}_i$, expansion coefficients in μ_2, 95
$\tilde{\alpha}$, 182
α_1, 194
a_p, Sonine expansion coefficient, 63

B, 69
$\vec{B}(\vec{V})$, 181
$\mathcal{B}_i$, expansion coefficients in μ_4, 95
$\hat{b}^{\vec{e}}$, 131
$\hat{b}^{\vec{e}}_{ij}$, 133
β_1, 200
$\tilde{\beta}$, 182
$B(t)$, 83

$C_{ij}(\vec{V})$, 181
$\vec{C}$, scaled centre of mass velocity, 71
$\vec{c}$, scaled velocity, 62
$\vec{c}_{12}$, scaled relative velocity, 71
C_1, C_2, C_3, C_4, coefficients in the expansion of ε, 26
$\chi, \tilde{\chi}$, 58, 94
$\langle c^p \rangle$, moment of the velocity distribution function, 63

D, coefficient of self-diffusion, 125
D_{ij}, local velocity tensor, 172
d, dimension, 115
D_0, Enskog self-diffusion coefficient, 141
Δ, mass ratio of gas particle and Brownian particle, 150
$\Delta \psi (\vec{v})$, change of $\psi(\vec{v})$ in a direct collision, 61
$\delta, \delta'(t)$, dissipative parameters, 94
$D(t)$, diffusivity, 126

$\vec{e}$, inter-centre unit vector, 18
ε, coefficient of restitution, 17
ε_b, coefficient of restitution for Brownian particles, 150
ε^t, coefficient of tangential restitution, 31
η, packing fraction, 55, 59
η, shear viscosity, 174
η_0, Enskog viscosity, 206
$\tilde{\eta}_1, \tilde{\eta}_2$, 206
$\eta^*(\varepsilon)$, 230

$\tilde{f}(\vec{c})$, scaled velocity distribution function, 62
$f(\vec{r}, \vec{v}, t)$, velocity distribution function, 54
$f^{(0)}$, distribution function in zeroth order, 178
$f^{(1)}$, first-order distribution function, 179
f_b, velocity distribution function of Brownian particles, 150
$F_{\rm diss}$, dissipative part of interaction force, 22
$F_{\rm el}$, elastic part of the interaction force, 21
$\tilde{I}(\tilde{f}, \tilde{f})$, reduced collision integral, 67

$\vec{g}$, normal component of the relative velocity, 18
$g_2(\sigma)$, contact value of the pair correlation function, 59
$\gamma, \bar{\gamma}$, 154
γ_{ij}, 182
γ_0, 194

$\mathcal{H}$, Hamilton function, 127

$I(f, f)$, collision integral, 58
$J_{k,l,m,n,p,\alpha}$, basic integral, 72
$\vec{J}_s(\vec{r}, t)$, 211

κ, thermal conductivity, 174
κ_0, Enskog thermal conductivity, 208
$\tilde{\kappa}_1, \tilde{\kappa}_2$, 209
$\kappa^*(\varepsilon)$, 230
$k_\perp^*(\varepsilon)$, marginal wave vector for stability of shear mode, 234
$k_H^*(\varepsilon)$, marginal wave vector for stability of heat mode, 238
$k_H^*(\tau)$, 242
$K_v(t_1, t_2)$, velocity autocorrelation function, 125

$\mathcal{L}$, Liouville operator, 127
$\mathcal{L}^0$, Liouville operator free-streaming component, 128
$\mathcal{L}'$, Liouville operator interaction component, 128
Li(x), logarithmic integral, 101
l_*, 229

SYMBOL INDEX

m, particle mass, 51
m_b, mass of a Brownian particle, 150
m^{eff}, effective mass, 18
$\hat{\mathbf{M}}(\vec{k})$, hydrodynamic matrix, 236
μ, transport coefficient, 174
$\tilde{\mu}_1, \tilde{\mu}_2$, 210
μ_p, moments of the dimensionless collision integral, 68
$\mu^*(\varepsilon)$, 230

N, total number of particles, 54
n, number density, 51, 54
$n(\vec{r}, t)$, local number density, 168
$n_h(t)$, 227
ν, Poisson ratio, 21
ν_{kp}, moments of Sonine polynomials, 64

ω_0, ω_1, 95
ω_2, 204
$\Omega_\eta [\varphi_1, \varphi_2]$, 189
Ω_η^ε, 196
Ω_η^{el}, 188
Ω_η^{vis}, 205
$\Omega_\kappa [\varphi_1, \varphi_2]$, 193
$\Omega_\kappa^\varepsilon$, 199
$\Omega_\kappa^{\text{el}}$, 192
$\Omega_\kappa^{\text{vis}}$, 207

$\hat{P}(\vec{r}, t)$, pressure tensor, 171
p, hydrostatic pressure, 172
$\varphi_i(a_2)$, 97
$\phi(\vec{C})$, distribution function for the scaled centre of mass velocity, 71
$\phi(c)$, scaled Maxwell distribution function, 63
$\phi(\vec{c}_{12})$, distribution function for the scaled relative velocity, 71

$\vec{q}(\vec{r}, t)$, heat flux, 173
q_0, q_1, 98

r_0, r_2, 98
$\vec{r}_{\text{cm}}$, centre of mass position, 132
R^{eff}, effective radius, 21
ρ, elastic material constant, 21
$\rho(\vec{r}, t)$, scaled fluctuation of density, 227
$\rho_{\vec{k}}$, 232
$\left\langle [\vec{r}(t)]^2 \right\rangle$, mean square displacement, 124

$\vec{S}$, 173
σ, particle diameter, 55
$S_p(c^2)$, Sonine polynomial, 63

T, granular temperature, 51

$T(\vec{r}, t)$, local temperature, 168
τ, reduced time, 83
τ_0, temperature relaxation time, 84
τ_c, mean collision time, 83
τ_H, 243
$\tau_{k\perp}^{\max}$, 241
τ_*, 229
τ_v, velocity correlation time, 125, 137
$\tau_{v,\text{ad}}$, 147
$T_b(t)$, temperature of Brownian particles, 155
$T_h(t)$, 227
$\Theta(x)$, Heaviside step-function, 58
$\theta(\vec{r}, t)$, scaled fluctuation of temperature, 227
$\theta_{\vec{k}}$, 232
$\hat{T}_{ij}$, binary collision operator, 132

$\vec{u}(\vec{r}, t)$, local flow velocity, 168
$u(t)$, reduced temperature, 84

$\vec{V}$, local velocity, 168
$\vec{v}_{12}$, relative particle velocity, 18, 56
$\vec{v}_{\text{cm}}$, centre of mass velocity, 132
v_i, particle velocity, 51
v_T, thermal velocity, 62
v_{Tb}, thermal velocity of Brownian particles, 150

$\vec{w}(\vec{r}, t)$, scaled flow velocity, 227
$\vec{w}_{\vec{k}\|}$, 233
$\vec{w}_{\vec{k}\perp}$, 233
$\vec{w}_{\vec{k}\perp}^{\max}$, 241

ξ, particle compression, 21

Y, Young modulus, 21

ζ, cooling coefficient, 69
$\zeta^{(0)}, \zeta^{(1)}$, cooling coefficient, 178
$\zeta^*(\varepsilon)$, 230

Index

adiabatic approximation, 203
adiabatic cooling, 62
age of granular gases, 104

Basic Integral, 72, 252
binary collision operator, 132
  properties, 133, 137
Boltzmann equation, 54–61, 168
  decomposition, 67
  for scaled velocities, 68, 82
  uniqueness of the solution, 85
  velocity correlations, 59
Boltzmann–Enskog equation, 59
Boltzmann–Lorentz equation, 150
boundary conditions, 13, 236, 245
Brownian motion, 123, 149–164
Brownian particles, 149
  collision rule, 150
  diffusion, 158
  diffusion coefficient, 160
  fluctuation–dissipation relation, 154
  Fokker–Planck equation, 151
  Kramers–Moyal expansion, 153
  Langevin equation, 158
  pair distribution function, 151
  temperature of the ensemble, 157
  velocity distribution, 155
Burnett–order hydrodynamics, 174, 176

Chapman–Enskog expansion, 178
Chapman–Enskog method, 175–185
cluster formation, 144, 223–251
coefficient of restitution
  for translational motion, 17
  for viscoelastic spheres, 26
  normal, 17, 21–31
  Padé approximation, 30
  tangential, 31–35
collision cannon, 40–47
  elastic particles, 40
  particles with $\varepsilon = $ const., 42
Collision Canon
  viscoelastic particles, 44
collision cylinder, 52
collision integral, 58
  loss term for the tail of the distribution, 109
  moments for $\varepsilon = $ const., 80
  moments for viscoelastic particles, 95
  scaled velocities, 67

collision law, 5, 18, 32
  identical spheres, 55
  in one dimension, 18
  in terms of $\vec{C}$ and $\vec{c}_{12}$, 72
  in three dimensions, 19
  inverse collision, 57
  Jacobian, 57
collision of particles, 17–48
  dimension analysis, 23
collision operator, 131
computational formula manipulation, 76
constitutive relations, 174
cooling coefficient, 69, 173
cooling rate, 84
Coulomb friction law, 33

dashpot force model, 25
δ-correlated noise, 158
diffusion, 123, 211
diffusivity, 126, 162
direct collision, 55
direct Monte Carlo simulations, 75
dissipative forces, 21
dissipative parameter, δ, δ', 94
distribution function, see velocity distribution function
driven systems, 13
DSMC, 75
dusty plasma, 13
dyad, 170

effective mass, 18
effective radius, 21
Einstein relation, 158
Enskog factor, 59
Enskog thermal conductivity, 193
Enskog viscosity, 190
Euler equations, 179
event-driven molecular dynamics, 7, 269

finite volume effects, 59
fluctuation–dissipation relation, 126, 154
fluctuation–dissipation theorem, 158
Fokker–Planck equation, 151, 158
forces between colliding spheres, 21

Gaussian white noise, 159
Grad's method, 202
granular temperature, see temperature
grazing collision, 111

Green's theorem, 124

Haff's law, 52
hard-particle approximation, 6
hard-sphere interaction potential, 128
heat flux, 173
heat mode, 237
Heaviside function, 58
Hertz contact force, 4, 21
homogeneous cooling state, 51
  instability, 225
Hooke's law, 4
hydrodynamic description, 9
hydrodynamic equations, 167
  linearized equations, 229
hydrodynamic fields, 168
hydrodynamic modes, 233
hydrostatic pressure, 172

inelastic collapse, 36–40, 144, 281
  viscoelastic particles, 40
inelastic Maxwell model, 70
inherent time scales, 144
inverse collision, 55

Kinetic Integral, 71, 252
  Maple library, 78
Kramers–Moyal expansion, 153

Laguerre polynomials, 64
Langevin equation, 158
linear stability analysis
  shear modes, 234
Liouville operator, 127

macro- and microscales for
  inhomogeneous gases, 167
Markov process, 136
Maxwell distribution function, 63
  moments, 65
Maxwellian molecules, 70
mean collision time, 83
mean square displacement
  for $\varepsilon = $ const., 141
  viscoelastic particles, 143
mode enslaving mechanism, 246
modes of a field, 232
molecular chaos hypothesis, 59, 134
molecular dynamics, 269–282
moment of inertia, 32

Navier–Stokes hydrodynamics, 174
Newton's equation of motion, 5, 19, 32
noise, 158
noise reduction process, 235
number density, 51

Ornstein–Uhlenbeck process, 159

packing fraction, 55
pair correlation function, 59
pairwise collision assumption, 5
particle collisions, 17–48
plastic deformation, 22
Poisson brackets, 127
Poisson ratio, 21
pressure tensor, 171
  collisional component, 171
  kinetic component, 171
pseudo-Liouville operator, 127, 131

quasi-static cooling, 91
quasistatic approximation, 23

Rayleigh equation, 159
reduced collision integral, 67
restituting collision, 55
ring collisions, 60
rotational velocity, 32

scaled time, 83
scaled velocity, 62
scaled velocity distribution function, 81
self-diffusion, 123, 211
self-diffusion coefficient, 137
  $\varepsilon = $ const., 215
  Enskog value, 141
  for $\varepsilon = $ const., 141
  viscoelastic particles, 216
shear modes, 234
shear viscosity, 174
smooth spheres, 35
Sonine coefficient
  first, 70
  second, 69–83
Sonine polynomials, 63
  d-dimensional case, 115
  moments, 64
Sonine polynomials expansion, 62–66
  coefficients, 65
  high-order coefficients, 87–89
  time-dependent coefficients, 82
sound modes, 237
Stoßzahlansatz, 59
stress tensor, 172
structure formation
  scenarios of structure formation, 238

temperature, 51, 62
  cooling coefficient, 69
  decay, 69
  decay for gases of viscoelastic particles, 53

decay rate, 52
  reduced, 97
  scaled, 84
temperature decay
  for $\varepsilon =$ const., 84
  for viscoelastic particles, 99
temperature relaxation time, 84
tensor product, 173
thermal conductivity
  for $\varepsilon =$ const., 200
  viscoelastic particles, 209
thermal conductivity, κ, 174
thermal velocity, 62
time correlation function, 136
tracer particles, 124
transport coefficient, μ, 174
transport coefficients
  relation to the velocity distribution function, 183
transport processes, 123
two-particle correlation function, 134

velocity correlation function, 125
  long-time tails, 162
velocity correlation time, 138
velocity distribution function, 48–120
  Brownian particles, 155
  definition, 54
  high-energy tail, 108
  inhomogeneous gas, 194, 201
  moments, 54, 63, 65
  normalization, 54
  scaled distribution function, 68
  Sonine polynomials expansion, 62–66
  two-particle distribution function, 59
viscoelastic interaction, 21
viscosity coefficient
  for $\varepsilon =$ const., 197
  viscoelastic particles, 206
vortex formation, 60
vortices, 234

white noise, 158

Young modulus, 21

The manufacturer's authorised representative in the EU for product safety is
Oxford University Press España S.A. of el Parque Empresarial San Fernando de
Henares, Avenida de Castilla, 2 – 28830 Madrid (www.oup.es/en or product.
safety@oup.com). OUP España S.A. also acts as importer into Spain of products
made by the manufacturer.